Yaning Zhang
Wenke Zhao
Editors

Characteristics and Applications of Microwave Irradiation

nova science publishers
www.novapublishers.com

DOI: https://doi.org/10.52305/MJXO8111

Library of Congress Cataloging-in-Publication Data

ISBN: 979-8-89530-170-8 (Hardcover)
ISBN: 979-8-89530-273-6 (eBook)

Published by Nova Science Publishers, Inc. † New York

Contents

Preface v

Acknowledgments vii

Chapter 1 **Microwave Irradiation Characteristics** 1
Wenke Zhao, Yaning Zhang and Bingxi Li

Chapter 2 **Microwave Irradiation for Medical Treatments**..............29
Kaihan Xie, Wenke Zhao, Yaning Zhang
and Bingxi Li

Chapter 3 **Microwave Irradiation for Food Drying**...........................61
Zheng Wang, Wenke Zhao, Yaning Zhang
and Bingxi Li

Chapter 4 **Microwave Irradiation for Heating Organic Wastes** 89
Longfei Cui, Wenke Zhao, Yaning Zhang
and Bingxi Li

Chapter 5 **Microwave Irradiation for Pyrolysis of Organic Wastes for Oil Production**............................119
Sichen Fan, Wenke Zhao, Yaning Zhang
and Bingxi Li

Chapter 6 **Microwave Irradiation for Pyrolysis of Organic Wastes for Char Production** 163
Tianhao Qiu, Chaoyue Liu, Yaning Zhang
and Wenke Zhao

Chapter 7 **Microwave Irradiation for the Gasification of Organic Wastes**..237
Wenming Fu, Wenke Zhao, Yaning Zhang
and Bingxi Li

Chapter 8 **Microwave Irradiation for Organic Synthesis** 267
Deepika Dwarakanath and Santosh L. Gaonkar

Index 295

Preface

Microwave is a kind of electromagnetic wave between infrared radiation and radio waves with a wavelength of 0.001 ~ 1 m and a corresponding frequency of 300 MHz ~ 300 GHz. When a material is in the microwave field, the polar or charged particles align themselves as soon as possible to the microwave electrical field direction, and heat is produced during this process due to friction. As compared with conventional electrical heating, microwave irradiation has the advantages of non-contact heating, faster response, quicker heating, less reaction time, more rapid product formation, lower reaction activation energy, etc. This book aims to present the up-to-date knowledge of microwave irradiation, and it is divided to 8 chapters.

In chapter 1 (Microwave Irradiation Characteristics), Wenke Zhao, Yaning Zhang and Bingxi Li introduce the fundamentals of microwave irradiation, and these mainly include heating performances, heating mechanisms, thermal effect, non-thermal effect, etc.

In chapter 2 (Microwave Irradiation for Medical Treatments), Kaihan Xie, Wenke Zhao, Yaning Zhang and Bingxi Li introduce microwave irradiation for medical treatments, and these mainly include hyperthermia, ablation, diagnostics and monitoring.

In chapter 3 (Microwave Irradiation for Food Drying), Zheng Wang, Wenke Zhao, Yaning Zhang and Bingxi Li introduce microwave irradiation for food drying, and these mainly include drying mechanisms and the applications.

In chapter 4 (Microwave Irradiation for Heating of Organic Wastes), Longfei Cui, Wenke Zhao, Yaning Zhang and Bingxi Li introduce microwave irradiation for heating of organic wastes, and these mainly include the heating performances of biomass, plastics, microwave absorbents, etc.

In chapter 5 (Microwave Irradiation for Pyrolysis of Organic Wastes for Oil Production), Sichen Fan, Wenke Zhao, Yaning Zhang and Bingxi Li introduce microwave irradiation for pyrolysis of organic wastes for oil production, and these mainly include the pyrolysis performances of biomass and plastics.

In chapter 6 (Microwave Irradiation for Pyrolysis of Organic Wastes for Char Production), Tianhao Qiu, Chaoyue Liu, Yaning Zhang and Wenke Zhao introduce microwave irradiation for pyrolysis of organic wastes for char production, and these mainly include the properties and applications of biochar, the production of value-added and porous biochar, and the activation of biochar.

In chapter 7 (Microwave Irradiation for Gasification of Organic Wastes), Wenming Fu, Wenke Zhao, Yaning Zhang and Bingxi Li introduce microwave irradiation for gasification of organic wastes for syngas production, and these mainly include syngas properties, gasification performances of biomass and plastics, and simulation of gasification.

In chapter 8 (Microwave Irradiation for Organic Synthesis), Deepika Dwarakanath and Santosh L. Gaonkar introduce microwave assisted organic synthesis, and they mainly concentrate on the reactions.

Hopefully, the contents in this book can supply a good introduction of the up-to-date knowledge of microwave irradiation.

Acknowledgments

The editors (Professor Yaning Zhang and Assistant Professor Wenke Zhao) acknowledge all the authors who contributed to this book. Each piece of work was really a challenging and time-consuming task.

Special thanks are expressed to Haley Wright for providing untiring help for the preparation and publication of this book.

The first editor, professor Yaning Zhang, acknowledges the guidance and support from professor Bingxi Li (who works at the School of Energy Science and Engineering, Harbin Institute of Technology, Harbin, China), professor Abdel Ghaly (who works at the Department of Process Engineering and Applied Science, Faculty of Engineering, Dalhousie University, Canada), professor Roger Ruan (who works at the Department of Bioproducts and Biosystems Engineering, University of Minnesota Twin Cities, USA), etc.

Professor Yaning Zhang acknowledges the financial supports provided by the National Natural Science Foundation of China (52476005, 52076049), Heilongjiang Province "Double First-class" Discipline Collaborative Innovation Achievement Project (LJGXCG2023-080), Heilongjiang Provincial Key R&D Program (2023ZX02C05), and Heilongjiang Provincial Key R&D Program "Unveiling the Leader" Project (2023ZXJ02C04).

Professor Yaning Zhang also acknowledges the love and support from family members: Hongyan Zhang, Jixuan Zhang, Jiye Zhang, etc.

Chapter 1

Microwave Irradiation Characteristics

Wenke Zhao
Yaning Zhang*
and Bingxi Li
School of Energy Science and Engineering, Harbin Institute of Technology, Harbin, China

Abstract

Microwave is a type of electromagnetic wave with a wavelength ranging from 0.001 ~ 1 m and frequency ranging from 300 MHz ~ 300 GHz. This chapter provides the essential information on microwave irradiation, focusing on its heating characteristics and underlying mechanisms. It also explores materials that absorb microwaves, examines both thermal and non-thermal effects, and discusses two effects. Key components of the microwave irradiation system and its applications in various industries, such as food processing, construction, and chemical manufacturing, are highlighted. Finally, the challenges and prospects are briefly discussed, and conclusions are summarized.

Keywords: microwave irradiation, microwave heating mechanism, thermal effect, non-thermal effect, microwave applications

* Corresponding Author's Email: ynzhang@hit.edu.cn.

In: Characteristics and Applications of Microwave Irradiation
Editors: Yaning Zhang and Wenke Zhao
ISBN: 979-8-89530-170-8

1. Introduction of Microwave Irradiation

1.1. Microwave

The microwave is a type of electromagnetic wave that occupies a position between infrared radiation and radio waves according to its wavelength ranging from of 0.001 ~ 1 m and frequency between 300 MHz and 300 GHz, as illustrated in Figure 1 (Fia and Amorim, 2023). Microwaves have the characteristics of easy bunching, highly directional and linear propagation. Microwaves are often classified as ultra-high frequency electromagnetic waves due to their higher frequencies compared to average radio waves. Microwaves are extensively applied to radar, microwave ovens, plasma generators, and wireless networks (Ferrari et al. 2005; Krishna and Padmasine, 2023).

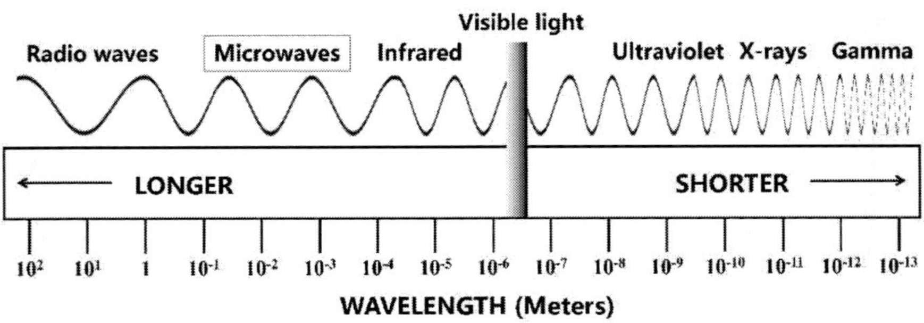

Figure 1. Electromagnetic spectrum (Rhithuparna et al. 2024).

1.2. Microwave Dielectric Heating

When an object is exposed to a microwave environment, polar or charged particles within the object rapidly align with the microwave electrical field, producing fractional heat (Galema, 1997). Microwaves were first applied to dielectric heating in the 1970s as a novel heating technology. Microwave heating offers several advantages over conventional heating, such as shorter reaction times, faster product formation, and lower reaction activation energy (Motevali et al. 2014).

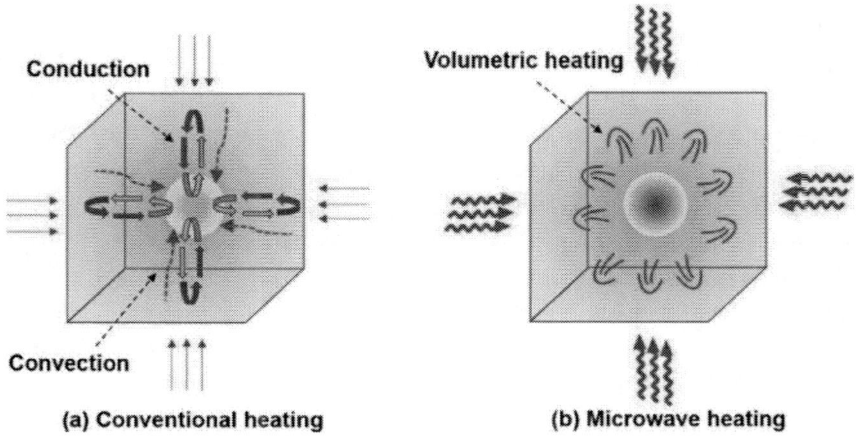

Figure 2. Comparison between conventional and microwave heating patterns (Sun et al. 2021).

As shown in Figure 2, conventional heating delivers heat to the surface of an object through conduction, convection, and radiation, while microwave heating generates heat internally through the alignment of polar particles or molecules. Conventional heating patterns often cause the skin effect, where water on the object's surface rapidly evaporates. This method is efficient for surface heating, but results in non-uniform heating and inefficient energy use because of high thermal gradients and heat conduction or convection (Zlotorzynski, 1995). Microwave heating provides uniform volumetric heating as polar particles or molecules generate heat from the core to the surface of the object (El-Feky et al. 2020). A more detailed explanation of the microwave heating mechanism is provided in subsection 1.3.

1.3. The Mechanism of Microwave Heating

Microwave heating primarily induces thermal energy through three mechanisms: (a) dipolar polarization, (b) ionic conduction, and (c) interactive polarization. The dipolar polarization mechanism involves the formation of rotational torque on polarized molecules in the microwave field, causing the dipoles to rearrange and polarize, as shown in Figure 3 (a). In this process, the dipolar reorient at high speed generates friction, which in turn produces thermal energy (Arpia et al. 2021). Ionic conduction occurs when ions in microwave-absorbing materials attempt to migrate in response to the microwave field but cannot match its high frequency. This lag results in ion vibration, leading to the production of thermal energy, as depicted in Figure 3 (b) (Kumar and Karim, 2019). Interactive polarization, also known as the Maxwell-Wagner-Sillars effect, combines elements of both dipolar polarization and ionic conduction. In interactive polarization, the movement of charged particles creates positive and negative space charges at the microwave material interface, which subsequently causes variations in the electric field distribution. This mechanism is particularly relevant in non-conductive material containing conductive particles, such as mental oxides in a polar solvent (Chemat-Djenni et al. 2007).

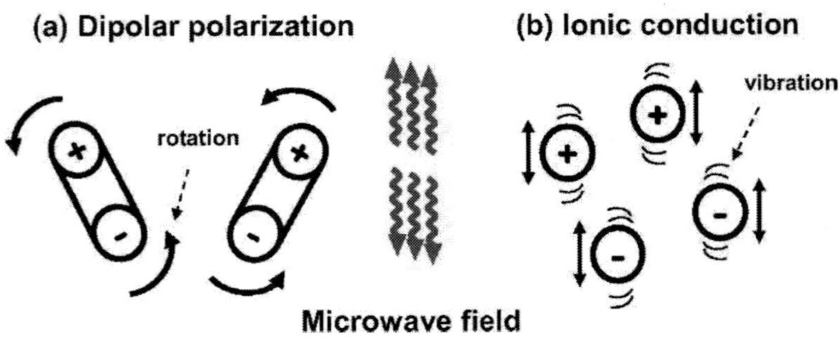

Figure 3. Schematic diagram of microwave heating mechanisms (Sun et al. 2021).

2. Effects of Microwave Irradiation

Microwave irradiation can directly access the reactive material and rapidly increase its temperature, thereby enhancing and accelerating the chemical reaction. Microwave irradiation heating technology has more efficient heat transfer and shorter reaction times than conventional heating technology because (a) conventional heating delivers heat to the material using a furnace or oil bath through conduction and convection, whereas (b) microwave irradiation heating directly heats materials, conserving energy and time, as shown in Figure 4. Consequently, microwave irradiation heating technology can accelerate the reaction rate and simplify the separation process (Mahato et al. 2022). The quantum energy of microwave irradiation, as described by Planck's law, is given by Eq. (1):

$$E = h\nu \quad (1)$$

where h is the Planck's constant and v denotes the wave frequency (Hz). The microwave irradiation is in the range of 0.3 ~ 300 GHz and its energy value fluctuates from 1.24×10^{-6} to 1.24×10^{-3} eV.

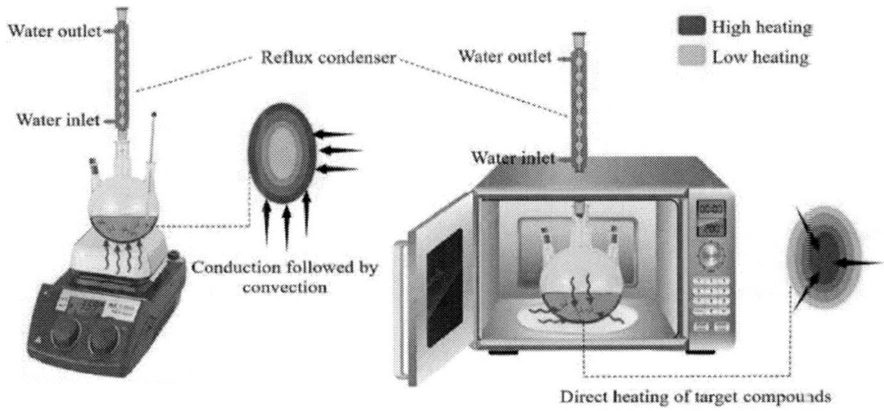

Figure 4. Schematic diagram of microwave irradiation and conventional heating technologies (Rhithuparna et al. 2024).

2.1. Microwave Absorption by Different Materials

Materials can be classified as (a) insulating material, (b) reflective material, and (c) dielectric material based on their capacity to absorb microwave irradiation (Arpia et al. 2021). For insulating materials, microwave irradiation can penetrate with no loss; these materials include silica, quartz, ceramics, and Teflon (Long et al. 2002). For reflective materials, microwave irradiation is reflected, generating no heat; these materials are primarily metals. For dielectric materials, microwave irradiation can be absorbed, and the absorbing capacity is represented by the relative dielectric constant ε'_r (Eq. (2)), while the dielectric loss coefficient ε''_r is expressed as Eq. (3). Meanwhile, the convert ratio of microwave irradiation to thermal energy can be described by Eq. (4), known as dissipation factor tan δ (Nayak et al. 2019).

$$\varepsilon'_r = \varepsilon' / \varepsilon_0 \tag{2}$$

$$\varepsilon''_r = \varepsilon'' / \varepsilon_0 \tag{3}$$

$$\tan\delta = \frac{\varepsilon''_r}{\varepsilon'_r} = \frac{\varepsilon''}{\varepsilon'} \tag{4}$$

where ε_0 denotes the vacuum absolute permittivity, ε' refers to the material dielectric constant, and ε'' is the microwave irradiation loss when penetrating the material.

Materials with a high tan δ can rapidly absorb more microwave irradiation to produce thermal energy due to the presence of a greater number of polar molecules. The tan δ values of some common materials are shown in Table 1 (Sun et al. 2021; Du et al. 2023). These values are measured under controlled conditions at a microwave frequency of 2.45 GHz and an ambient temperature of 20°C. The microwave absorbing capacity of a material is classified into three grades: (a) high absorbing capacity, (b) medium absorbing capacity, and (c) low absorbing capacity based on the tan $\delta > 0.5$, $0.1 < \tan\delta < 0.5$, and tan $\delta < 0.1$, respectively. Water, one of the most common microwave-absorbing materials, has a tan δ of 0.12, placing it in the medium absorbing capacity category.

Table 1. tan δ of some common materials (Sun et al. 2021; Du et al. 2023)

Material	tan δ
Water	0.12
Ethylene glycol	1.35
Dichloromethane	0.042
Sand	0.072
Ethanol	0.94
Chloroform	0.091
Hexane	0.020
Sodium silicate (Na_2SiO_3)	0.19
Methanol	0.66
Acetonitrile	0.062
Toluene	0.04
Fly ash	0.072
Tetrahydrofuran	0.047
Acetic acid	0.17
Acetone	0.054
Sodium hydroxide (NaOH)	0.082

2.2. Thermal Effect of Microwave Irradiation

The thermal and non-thermal effects are two significant impacts of microwave irradiation on material reactions, such as heating, pyrolysis, and gasification. The thermal effect occurs when microwave irradiation interacts with polar molecules within a material, causing them to become a thermal source and raise the material's temperature. Although the thermal effect does not directly break or form chemical bonds, it can initiate chemical reactions by increasing the material's temperature. The mechanism underlying the thermal effect involves dipole motions and interactions with the material, leading to an increase in material temperature and reaction rate under high-frequency microwaves and rapidly changing electric fields (Zhang et al. 2003).

Table 2 illustrates the features of the thermal effect of microwave irradiation (Rhithuparna et al. 2024). The thermal effect of microwave irradiation can rapidly increase a material's temperature (Li et al. 2019), resulting in a uniform temperature distribution and low thermal gradients (Siddique et al. 2022) because of the movement of polar molecules and irons within the material (Arshanitsa et al. 2016). Additionally, the thermal effect is characterized by several unique features, including non-contact heating, penetrative radiation, volumetric heating, and reverse thermal effect (the interior of the material heats faster than the surface) (Patil et al. 2013; Varol et al. 2023).

Table 2. Thermal effect features of microwave irradiation (Rhithuparna et al. 2024)

Feature	Reference
Rapid temperature increase	Li et al. 2019
Low thermal gradient	Siddique et al. 2022
Polar molecules and ions change directions in the microwave field	Arshanitsa et al. 2016
Non-contact heating, penetrative radiation, and volumetric heating	Patil et al. 2013
Inverted heat transfer or reverse thermal effect	Varol et al. 2023

2.3. Non-Thermal Effect of Microwave Irradiation

The non-thermal effect of microwave irradiation is independent of temperature variation and arises directly from molecule interactions during the reaction process (Tian et al. 2020). According to the Arrhenius equation (Eq. (5)), the reaction rate is related to A (pre-exponential factor), ΔE_a (activation energy), and T (temperature). The pre-exponential factor, A, indicates the molecular vibration frequency, which can increase when molecules are accelerated and activated under microwave irradiation (Binner et al. 1995). Microwave irradiation generates an electric driving force that enhances molecule collision frequency and accelerates the chemical reaction due to the microwave electromagnetic wave (Kalhori et al. 2002). The rotation or vibration of molecules under microwave influence can also stimulate energy level transitions, pushing molecules into excited states. This shift can alter ΔE_a and subsequently affect the reaction rate (Wu et al. 2019).

$$K = A \times \exp\left(\frac{-E_a}{RT}\right) \tag{5}$$

where K is the reaction rate, A denotes the pre-exponential factor based on the molecular vibration frequency, E_a refers to the activation energy, R is the universal gas constant, and T is the reaction temperature (K).

According to Eq. (5), the reaction rate can be improved in two ways: raising the pre-exponential factor and reducing the activation energy. Activation energy may also be expressed in terms of Gibbs free energy, as shown in Eq. (6), which consists of entropy, temperature, and enthalpy. The microwave irradiation process leads to molecular interactions and dipole movements, resulting in high ΔS and low ΔH. In other words, microwave irradiation causes the reaction materials to be in a superheating state because of the unique energy dissipating pattern of microwaves. Consequently, the activation energy is reduced, and the reaction rate is improved (Guo et al. 2020).

$$\Delta G = \Delta H - T\Delta S \tag{6}$$

where G is the Gibbs free energy, H refers to the enthalpy, T denotes the temperature, and S is the entropy.

The energy variations of microwave and conventional systems in a reaction case as shown in Figure 5 (Rhithuparna et al. 2024). In this case, E_a and A for the microwave system are 108.6 kJ/mol and 7.3×10^{13}/min, while E_a and A for the conventional system are 24.2 kJ/mol and 5.47/min, respectively. The A value for the microwave system is significantly higher than that of the conventional system, indicating that the microwave reaction is more effective.

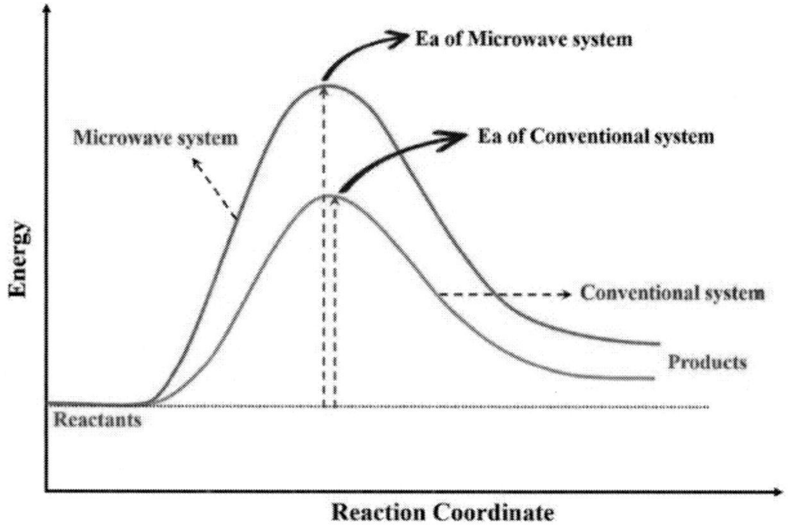

Figure 5. Energy variations of microwave and conventional systems (Rhithuparna et al. 2024).

Table 3 shows the non-thermal effect features of microwave irradiation (Rhithuparna et al. 2024). The most significant feature of the non-thermal effect is that there is no variation in the material temperature (Li et al. 2019), yet the chemical reaction is accelerated (Arshanitsa et al. 2016). Additionally, the non-thermal effect cannot be replicated through conventional heating methods (for example, an electrical heating furnace) (Ge et al. 2022). The non-thermal effect can alter the states of polar molecules and ions states, reduce their activation energy, and thereby enhance the reaction rate (Alarcon et al. 2020).

Table 3. Non-thermal effect features of microwave irradiation (Rhithuparna et al. 2024)

Feature	References
No temperature increase	Li et al. 2019
Chemical reaction acceleration	Arshanitsa et al. 2016
Cannot be obtained by conventional heating	Ge et al. 2022
Reducing the activation energy	Alarcon et al. 2020
Polar molecules and ions response in the microwave field	Wang et al. 2023

2.4. Discussions Between Thermal and Non-Thermal Effects

The thermal and non-thermal effects typically occur simultaneously, making it extremely difficult to accurately distinguish. Many studies present their findings on the thermal and non-thermal effects, leading to three main arguments: (a) Microwave irradiation cannot destroy chemical bonds, (b) The non-thermal effect is difficult to obtain experimentally, and (c) The non-thermal effect is limited to specific conditions. The details are as follows. (a) Microwave irradiation cannot destroy chemical bonds. The energy of a hydrogen bond is far higher than the quantum energy provided by microwave irradiation (Herrero et al. 2008), meaning that microwave irradiation has little impact on bond destruction during the reaction process (Kappe, 2008). However, some scholars believe that while the non-thermal effect cannot destroy chemical bonds, it can cause them to vibrate or rotate, leading to bond weakening (Kalhori et al. 2002). (b) The non-thermal effect is difficult to obtain experimentally. When considering the non-thermal effect, the thermal effect must be removed. However, this is difficult due to the random nature of thermal sources arose from heterogeneous electric field distribution and varying microwave irradiation. Thus, maintaining a certain temperature in the material is significantly difficult (Shazman et al. 2007). Researchers have explored various methods to observe the non-thermal effect, such as comparing microwave with conventional heating methods (Andrey et al. 2021), using microwave constant-temperature irradiation with a cooling reflux system (Wu et al. 2019), and employing silicon carbide (SiC) to stabilize temperature during microwave heating process (Obermayer et al. 2009). (c) The non-thermal effect is limited to specific conditions. Some researchers believe that coherent, polarized, and resonant microwave irradiation is crucial for generating the non-thermal effect in absorbing materials (Tian et al. 2014). This perspective challenges the assertion that the non-thermal effect is observed under inadequately recorded conditions (Stuerga and Gaillard, 1996). Bichot et al. (2020) suggested that the non-thermal effect cannot be detected using specific devices to counteract the thermal effect on absorbing materials, as the heating performance, temperature gradient, and hot spots differ significantly between microwave and conventional heating methods.

Furthermore, the sequence in which the thermal and non-thermal effects appear during microwave heating process remains a subject of ongoing debate. Some researchers argue that the thermal and non-thermal effects emerge simultaneously, with the non-thermal effect having only a minor impact on the chemical reaction assisted by microwave irradiation (Nüchter et al. 2003).

However, as research evolves, a growing body of evidence suggests that the non-thermal effect appears first and influences the interactions among polar molecules to generate thermal energy. The thermal effect then becomes more prominent as this thermal energy accumulates to a certain extent (Zhai et al. 2018).

3. Effects of Microwave Irradiation

The microwave irradiation heating method offers several advantages over conventional heating methods, including (a) non-contact heating, (b) homogeneous temperature distribution, (c) rapid heating rate, and (d) a volumetric heating pattern. Consequently, microwave irradiation has been applied in the food, construction, and chemical industries.

3.1. Key Components of Microwave Irradiation

Microwave irradiation systems are crucial for microwave applications, and typically comprise a magnetron, waveguide, high-voltage transformer, and cooling fan. The most commonly used system is the microwave oven, which is frequently modified to serve as key reactor equipment in research and industrial applications.

Magnetrons are the most common microwave generators and consist of vacuum tubes designed to consistently produce microwave irradiation within a specific frequency band (Zhou et al. 2023). Electrons are first emitted from a hot cathode, then whirled into the resonant cavities of the anode to produce microwave irradiation, which is then coupled out of the vacuum via a probe (connected to a waveguide) to deliver the microwave energy to the oven (Anilkumar et al. 2022). Figure 6 shows the working principle of a magnetron in a microwave reactor. When a high voltage is applied to the cathode of the magnetron, the cathode becomes hot and emits electrons. As the electrons begin to move, they are repelled to the nearest cavity of the anode, causing the surface of the cavity to acquire a negative charge. Eventually, all cavities are filled, forming an alternating arrangement of positive and negative charges. The distribution of positive and negative charges oscillates in the inductor and capacitor, eventually generating microwave irradiation through the metal loop and antenna.

The magnetron's operational frequency is influenced by several factors, including the power supply, the configuration of anode cavities, the magnetic field from permanent magnets, load impedance, and magnetron's age. Fluctuations in microwave frequencies pose significant challenges in predicting and regulating the distribution of cold and hot spots (heating patterns) within reaction mixtures, thus constraining the limiting of microwave heating (Zhou et al. 2023). The power output of the magnetron determines the intensity of the microwave; higher power outputs result in more intense microwave irradiation and consequently faster heating. Effective real-time control of microwave output power requires meticulous attention to the design of magnetron hardware. Ye et al. (2020) reported a real-time temperature control system utilizing microwave irradiation to maintain a constant reaction flow without interruptions (Ye et al. 2020).

A waveguide is a hollow metallic conduit used to guide microwaves, and it transmits microwave signals via continuous reflections from the inner walls of the tube. The waveguide is typically constructed from materials like copper, aluminum, or brass, and its inner surfaces are often coated with gold or silver. Several common waveguides, including rectangular, circular, coaxial, elliptical, single-ridged, and double-ridged types, are shown in Figure 7 (Barapatre et al. 2020). The most common type of waveguide used in microwave systems is the rectangular waveguide, known for its compatibility and ease of manufacture (Varshney and Sharma, 2020). Circular waveguides supporting a TE11 (transverse electric) mode are also favored due to their simpler installation. In TE mode, all electric fields are perpendicular to the direction of propagation with no longitudinal electric field, making circular waveguides straightforward to install and operate (Zhu et al. 2021). Coaxial waveguides are used in microwave reactors because they do not exhibit a cut-off frequency in the transverse electric-magnetic mode, allowing them to operate over a broad range of frequencies. Each mode of propagation in waveguides has its own cut-off frequency, determining the limit beyond which waves cannot propagate through the waveguide. Therefore, selecting the appropriate waveguide is critical in the design of microwave systems. For instance, Xu et al. (2023) used a microwave reactor based on the leaky waveguide to produce biodiesel using the methanol and oleic acid (Xu et al. 2023).

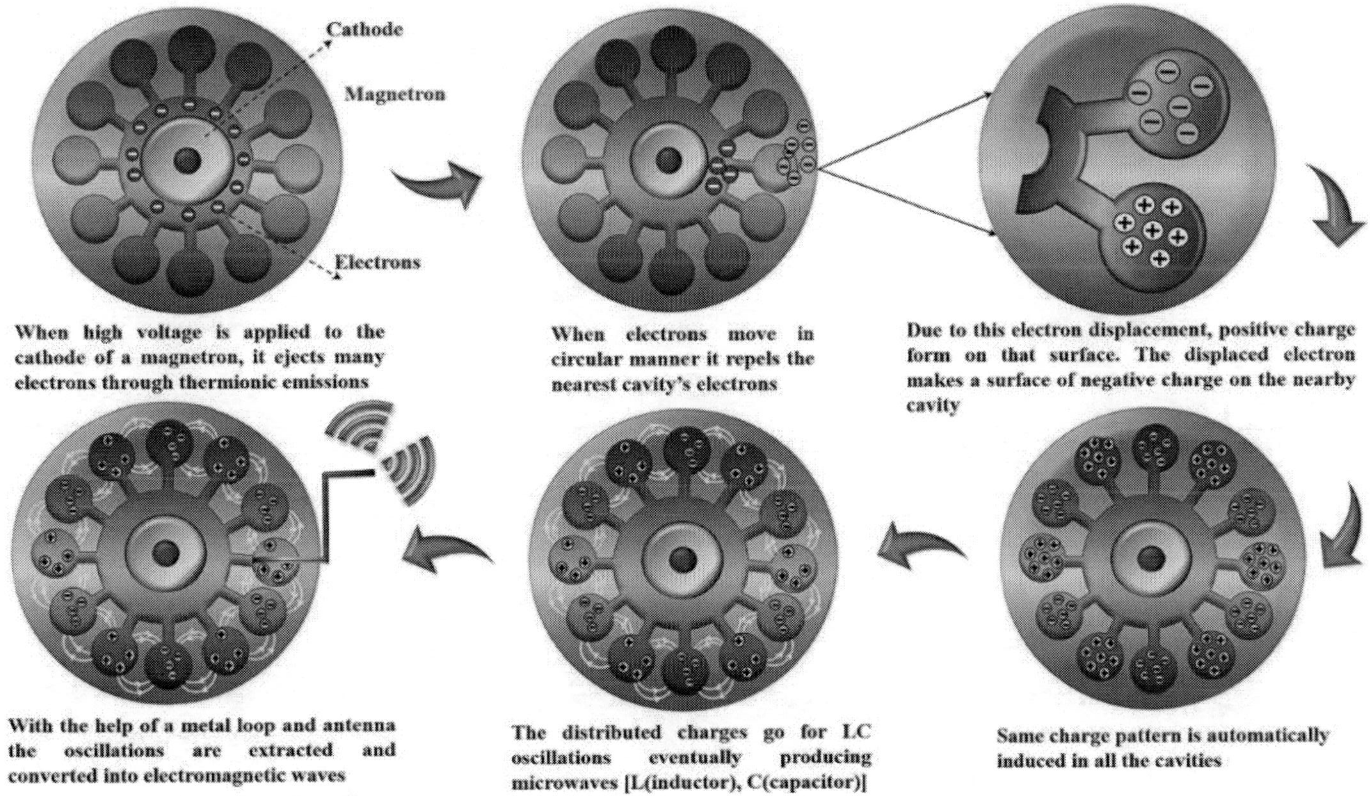

Figure 6. Working principle of the magnetron (Rhithuparna et al. 2024).

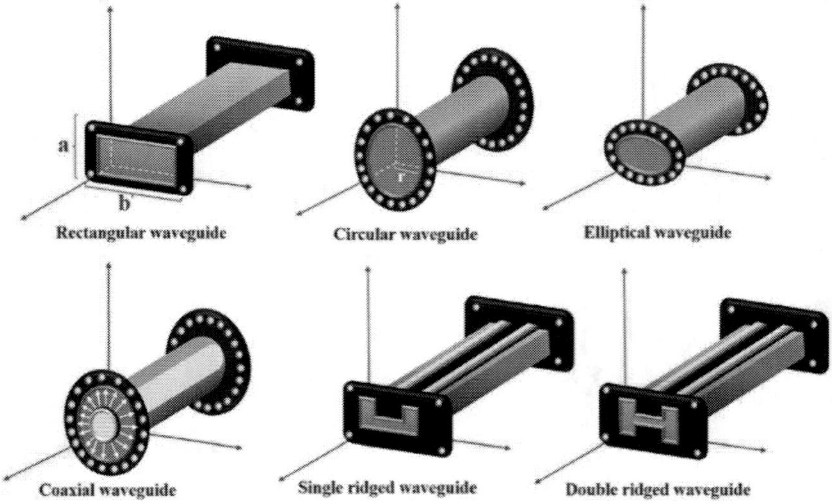

Figure 7. Different types of waveguides in microwave systems (Rhithuparna et al. 2024).

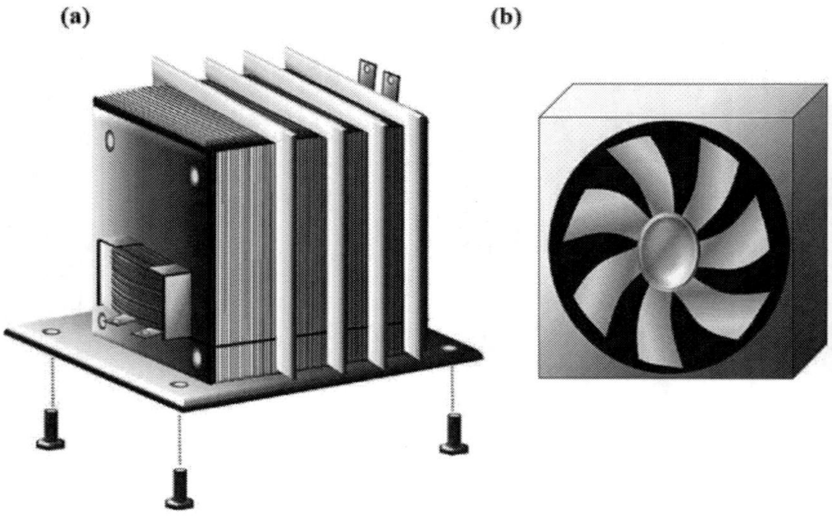

Figure 8. (a) High-voltage transformer and (b) cooling fan (Rhithuparna et al. 2024).

Two other important auxiliaries of the microwave irradiation system are the high-voltage transformer and the cooling fan, as shown in Figure 8. The high-voltage transformer provides high voltage to ensure the effective operation of the magnetron (Reja et al. 2023). The high-voltage transformer comprises multiple iron plates arranged to form a core, with a coil wound around it. Heat radiation fins, crafted from a material with high thermal conductivity, are inserted between the iron plates, extending beyond the core's outer periphery. These fins feature flange portions bent to create airflow apertures, facilitating direct contact between cooling air and the core. This high-voltage output from the transformer enables the magnetron to generate microwaves at a predetermined frequency (Cooper, 2020).

The cooling fan plays a vital role in maintaining system temperature and preventing overheating. As microwave irradiation generates heat in parts that do not require heating, the cooling fan helps to manage this heat and ensures the safe operation of the microwave reaction system (Khan, 2020). In older microwave models, cooling was managed by a stirrer fan rather than a dedicated cooling fan.

3.2. Food Application of Microwave Irradiation

When food is exposed to microwave radiation, it absorbs energy, producing heat that evaporates internal moisture. This process aligns the temperature gradient, pressure gradient, and moisture concentration gradient, enabling effective moisture removal without crust formation. At the beginning of microwave drying, high-moisture agricultural products release substantial water vapor into the resonant cavity. The water vapor is rapidly removed through the exhaust fan to ensure efficient drying of agricultural products (Lv et al. 2016). However, certain fresh foods require steam treatment to deactivate enzymes. In such applications, induced draft fans operate intermittently to supply sufficient steam containing small water droplets during the initial drying phase, as illustrated in Figure 9.

During the initial phase of enzyme inactivation via microwaves, a significant quantity of white steam appears on the surface of the food material, and this steam is a blend of water vapor and minute water droplets (see Figure 9 (a)). These water droplets absorb microwave energy, convert microwave into heat, and eventually evaporate. As shown in Figure 9 (b) and (c), the microwave heating process is uneven due to the heterogeneous dielectric properties of food materials and non-uniform distribution of microwave field (Huang et al. 2016). Rapid heat transfer occurs between the high-temperature water vapor and the adjacent gas to mitigate the uneven microwave heating process because water is a good absorbent. Meanwhile, as the temperature of the moist air around the materials rises, the food materials heat up in a relatively humid environment and obtain a homogeneous temperature distribution as shown in Figure 9 (d). Therefore, microwave radiation can enhance the heating uniformity. Generally, microwave-based techniques show great heat transfer and heating efficiency compared to conventional drying methods. Microwave-assisted enzyme inactivation is widely used in high-moisture agricultural products like fruits, vegetables, and tea.

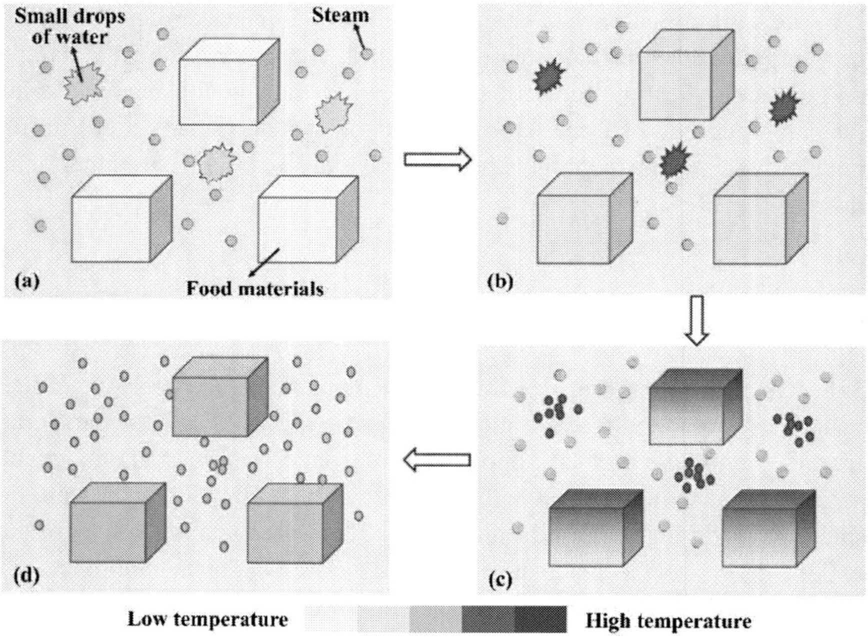

Figure 9. Variations of water drops and steam in the microwave heating process (Wu et al. 2022).

Figure 10 shows the preheating and hot-air drying process of fresh day-lily buds (Wu et al. 2022). Figure 10 (a) illustrates the transportation of fresh day-lily buds from the warehouse to the enzymatic inactivation section. Enzyme inactivation involves two methods: high-temperature steam and microwave heating. Figure 10 (b) depicts the use of high-temperature steam for this process. The introduction of high-temperature steam from external sources swiftly halts the biological activity of fresh day-lily buds. Figure 10 (c) illustrates the enzyme inactivation process using microwave heating. By evaporating a substantial amount of water vapor under the influence of microwaves, as shown in Figure 9, efficient enzyme inactivation of fresh day-lily buds is achieved. Figure 10 (d) and (e) show the dehydrated lily product obtained through subsequent hot air drying. Notably, there were no significant differences in color and morphology between the two enzyme inactivation methods. However, the microwave heating significantly reduces drying time (20-30 minutes) and energy consumption (Wu et al. 2022).

Figure 10. Pretreatment and drying process of the day-lily buds (Wu et al. 2022).

Water vapor, generated from high-moisture materials in the microwave heating process, swiftly migrates to the surface of the material. The material is processed within an environment characterized by high temperature and high humidity gas (Tian et al. 2021). Currently, hot steam produced by microwave irradiation is widely used in various applications such as food processing, drying biomaterials, and conducting heat treatments (Bouxin et al. 2019). As an illustration, Xu et al. (2021) introduced a novel microwave drying system featuring relative humidity control. This system comprised a temperature control unit and an inlet air unit, maintaining constant temperature while monitoring and adjusting relative humidity. The drying schematic diagram of the system is depicted in Figure 11. In this setup, air is generated from the air compressor of a mass flow controller, then heated by both the microwave oven and a heater coil to produce hot air. The new unit allows control of both the temperature of the material, and the relative humidity and temperature of the hot air. A fiber optic sensor positioned near the inlet air within a Teflon container measures the inlet air temperature. The relative humidity of the hot air is gauged using an electronic hygrometer before it enters the microwave oven. Slices of finger citron are heated and dried using this apparatus. When the measured relative humidity exceeds the set value, the microwave oven automatically switches off. The research indicated that hot air with a relative humidity of 40% was considered optimal (Xu et al. 2021).

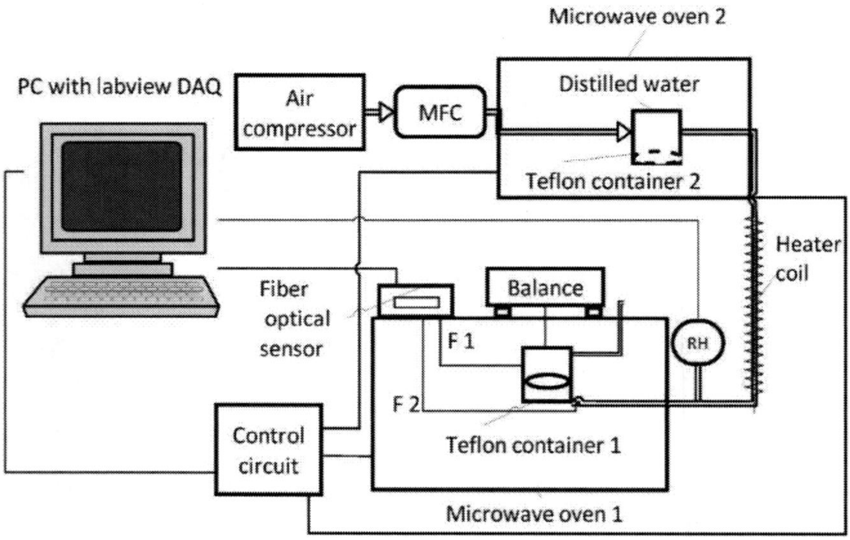

Figure 11. Schematic diagram of a microwave heating system (Xu et al. 2021).

3.3. Geopolymer Application of Microwave Irradiation

The processes involved in the effects of microwave irradiation on geopolymerization and the resulting mechanical properties of geopolymer products are intricate. Geopolymers are typically produced from raw materials such as active aluminosilicate precursors like fly ash, metakaolin, and slag, in combination with alkaline activators such as NaOH and Na_2SiO_3. In the alkali activation reaction, a highly alkaline solution is necessary to dissolve sufficient SiO_2 and Al_2O_3 from the aluminosilicate feedstock, and water serves as a lubricating agent to facilitate the alkaline activation process to ensure efficient hydrolysis of Al^{3+} and Si^{4+}. The geopolymerization process, involving dissolution, speciation equilibrium, gelation, reorganization, polymerization, and hardening, leads to the formation of geopolymers (Vu and Gowripalan, 2018).

Microwave irradiation plays a crucial role in promoting the formation of geopolymers during the geopolymerization process (see Figure 12). Initially, the dissolution of aluminosilicate source minerals occurs through alkali digestion and hydrolysis, resulting in the formation of aluminate and silicate monomers (Favier et al. 2015). During this phase, polar water molecules absorb microwaves and rapidly adjust their dipolar orientation within a fluctuating electric field, causing friction and internal heating of the geopolymer (Arpia et al. 2021). This heating effect accelerates the dissolution of geopolymer raw materials. Subsequently, the Al^{3+} and Si^{4+} ions are transformed into oligomers, and a gel with a more extensive network structure is formed through condensation.

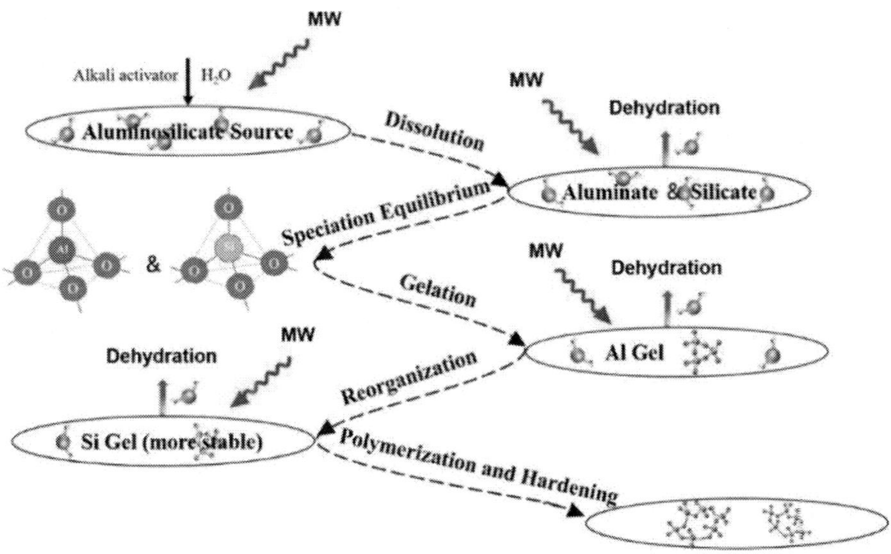

Figure 12. Mechanisms of microwave irradiation to geopolymer formation (Sun et al. 2021).

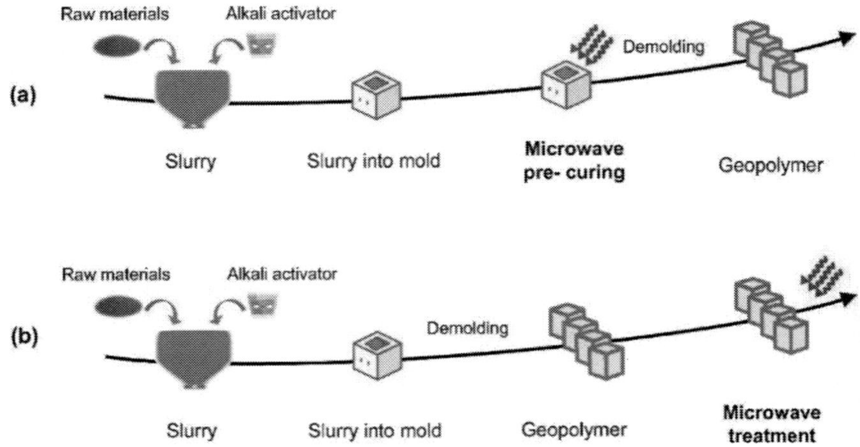

Figure 13. The timing sequence of microwave treatment on the geopolymer production process: (a) Microwave pre-curing; (b) Microwave treatment (Sun et al. 2021).

The expanding gel continues to merge, creating an amorphous aluminosilicate structure through polycondensation (Vu and Gowripalan, 2018). Microwave processing of geopolymer blocks expedites water evaporation in the dehydration reaction of geopolymerization. Following the initial dissolution phase, microwaves facilitate the reorganization of the expanding gel within a short timeframe. This process establishes a three-dimensional aluminosilicate network structure, enhancing the densification of geopolymers and improving their mechanical properties.

The timing of microwave treatment significantly influences the mechanical properties of geopolymers. Microwave irradiation applied to geopolymer preparation can be categorized into two phases (see Figure 13): Pre-curing, emerging as geopolymer precursors are freshly cast into molds or in a plastic state, and microwave curing occurs after demolding. Microwave pre-curing has been shown to reduce mortar porosity and enhance compressive strength (Kong et al. 2018).

Additionally, applying microwave can eliminate free water while the slurry is still in its plastic state. This reduces the permeability and water-cement ratio of the mixture, potentially mitigating strength reduction during the heating process (Wu et al. 1987). However, some researchers have observed that geopolymers pre-treated with microwaves sometimes exhibited lower compressive strength. This discrepancy may stem from localized overheating of microwaves in high-conductivity zones, leading to water evaporation, increased pore formation, and consequently weakened compressive strength (Hájková, 2018).

3.4. Biodiesel Application of Microwave Irradiation

Rapid microwave heating offers several benefits in biodiesel production, including improved quality, higher yield, increased efficiency, reduced reaction time, effective mixing, and lower energy consumption. Figure 14 illustrates a potential setup for microwave-irradiated biodiesel production, which can potentially be scaled up for commercial use. This portable system involves the pre-treatment of feedstock and alcohol in a flask surrounded by a rectangular coil. The flow rate of the reaction mixture can be regulated using a peristaltic pump. Transesterification can occur while the mixture is exposed to microwave radiation. The final product can be obtained from the outlet of this supernatant reactor, followed by rotary evaporation to separate the product from the by-product. This biodiesel meets fuel-grade standards and is suitable for use as vehicle fuel (Rhithuparna et al. 2024).

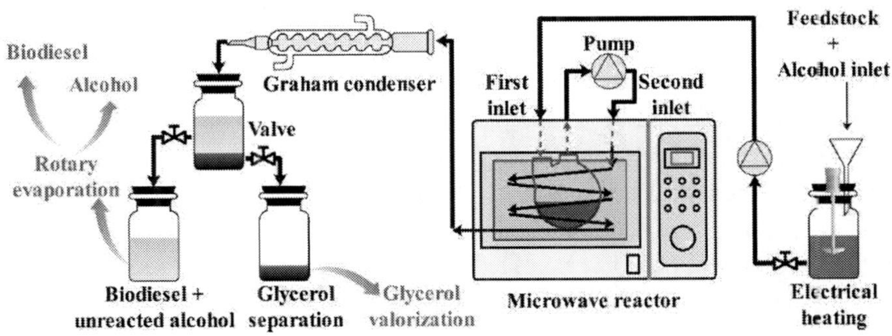

Figure 14. A typical microwave biodiesel production process (Rhithuparna et al. 2024).

4. Challenges and Prospects

Although microwave irradiation heating methods are more effective than conventional heating methods, its applications in the medical, food, and chemical industries are still limited. One of the biggest challenges is that the energy conversion efficiency from electricity to microwave energy (50% ~ 65%) (Moseley and Woodman, 2009) is considerably lower than the conventional electric heat conversion (90%) (Gude et al. 2012). Therefore, it is essential to design microwave irradiation devices to improve microwave generation efficiency.

Most materials are non-uniform in component and structure distribution, resulting in thermal damage during microwave irradiation processing. Additionally, microwave irradiation inside materials may not be uniform due to unevenly distributed waveguides or obstacles reflecting microwaves. Thermal damage is primarily caused by abnormal temperature gradients induced by uneven material properties or non-uniform microwave irradiations (Mishra and Sharma, 2016). Understanding microwave and material behaviors through physical interpretations and numerical analysis is crucial for improving microwave irradiation and optimizing materials processing.

Most studies on microwave irradiation are conducted in laboratory settings (Zhao et al. 2024). Microwave leakage remains a serious concern due to its harmful effects on humans (Mishra and Sharma, 2016). Therefore, caution is necessary in applying microwave irradiation for safety reasons, and further study is needed for industrial applications.

Conclusion

Microwaves are electromagnetic waves with wavelengths of 0.001 ~ 1 m and frequencies of 300 MHz ~ 300 GHz, positioned between infrared radiation and radio waves in the electromagnetic spectrum. Microwave heating provides uniform volumetric heating by interacting with polar particles or molecules inside the target object. Mechanisms of microwave irradiation heating include dipolar polarization, ionic conduction, and interactive polarization.

Microwave irradiation can only be absorbed by dielectric materials and does not affect insulating and reflective materials. It has two main effects on material reactions like heating, pyrolysis, and gasification: thermal and non-thermal effects, which often occur simultaneously, making their accurate distinction challenging. There are three main arguments: (a) Microwave irradiation cannot destroy chemical bonds, (b) The non-thermal effect is difficult to obtain experimentally, and (c) The non-thermal effect is limited to specific conditions.

Microwave irradiation systems, essential for microwave applications, typically include a magnetron, waveguide, high-voltage transformer, and cooling fan. Microwave irradiation has found applications in the food, construction, and chemical industries because of advantages such as non-contact heating, homogeneous temperature distribution, rapid heating rate, and volumetric heating pattern.

Acknowledgments

The following financial supports are acknowledged: Financial support was provided by the National Natural Science Foundation of China (52076049, 52476005), Heilongjiang Province "Double First-class" Discipline Collaborative Innovation Achievement Project (LJGXCG2023-080), Heilongjiang Provincial Key R&D Program (2023ZX02C05), and Heilongjiang Provincial Key R&D Program "Unveiling the Leader" Project (2023ZXJ02C04).

References

Alarcon RT, Gaglieri C, de Souza OA, Rinaldo D, Bannach G (2020). Microwave-assisted syntheses of vegetable oil-based monomer: a cleaner, faster, and more energy efficient route. *J Polym Environ* 28: 1265-1278.

Andrey J and Gut W (2021). Inactivation of polyphenol oxidase by microwave and conventional heating: Investigation of thermal and non-thermal effects of focused microwaves. *Food Chem* 340: 127911.

Anilkumar P, Pamu D, Tiwari T (2022). Historical developments and recent advances in high-power magnetron: A review. *IETE Tech Rev* (Institution Electron Telecommun Eng India) 39: 1309-1323.

Arpia AA, Chen W, SS Lam, P Rousset, de Luna DG (2021). Sustainable biofuel and bioenergy production from biomass waste residues using microwave-assisted heating: A comprehensive review. *Chem Eng J* 403: 126233.

Arshanitsa A, Akishin Y, Zile E, Dizhbite T, Solodovnik V, Telysheva G (2016). Microwave treatment combined with conventional heating of plant biomass pellets in a rotated reactor as a high rate process for solid biofuel manufacture. *Renew Energy* 91: 386-396.

Barapatre U, Panchal S, Rathod JM, Panchal PH, Parikh K (2020). Design and analysis of quad ridged horn antenna for high gain application. *J Phys Conf Ser* 1706: 012099.

Bichot A, Lerosty M, Radoiu M, Méchin V, Bernet N, Delgenès JP, GarcíaBernet D (2020). Decoupling thermal and non-thermal effects of the microwaves for lignocellulosic biomass pretreatment. *Energy Convers Manag* 203: 112220.

Binner JGP, Hassine NA, Cross TE (1995). The possible role of the pre-exponential factor in explaining the increased reaction rates observed during the microwave synthesis of titanium carbide. *J Mater Sci* 30: 5389-5393.

Bouxin FP, Clark JH, Fan J, Budarin V (2019). Combining steam distillation with microwave-assisted pyrolysis to maximise direct production of levoglucosenone from agricultural wastes. *Green Chem* 21: 1282-1291.

Chemat-Djenni Z, Hamada B, Chemat F (2007). Atmospheric pressure microwave assisted heterogeneous catalytic reactions. *Molecules* 12: 1399-1409.

Cooper NS (2020). 4 - Microwave ovens, Editor(s): Ulrich Erle, Peter Pesheck, Matthew Lorence, In *Woodhead Publishing in Materials*, *Development of Packaging and Products for Use in Microwave Ovens* (Second Edition), Woodhead Publishing, 177-199. https://doi.org/10.1016/B978-0-08-102713-4.00004-9.

Du S, Zhang Q, Ye W, Wang Z, Liu Z, Huang J, Xu X, Wu Y, Lin X, He QS. Nanda S, Changotra R, Hu Y, Zhao Z, Yang J (2023). Advancing hydrochar production and application: A critical examination of microwave irradiation and blended feedstocks synergy. *J Clean Prod* 426: 139049.

El-Feky MS, Kohail M, El-Tair AM, Serag MI (2020). Effect of microwave curing as compared with conventional regimes on the performance of alkali activated slag pastes. *Constr Build Mater* 233: 117268.

Favier A, Habert G, Roussel N, D'Espinosede Lacaillerie JB (2015). A multinuclear static NMR study of geopolymerisation. *Cem Concr Res* 75: 104-109.

Ferrari G, Fumagalli L, Sampietro M, Prati E, Fanciulli M (2005). CMOS fully compatible microwave detector based on MOSFET operating in resistive regime. *IEEE Microw Wirel Compon Lett* 15: 445-447.

Fia AZ and Amorim J (2023). Microwave pretreatment of biomass for conversion of lignocellulosic materials into renewable biofuels. *J Energy Inst* 106: 101146.

Galema SA (1997). Microwave chemistry. *Chem Soc Rev* 26: 233-238.

Ge L, Liu X, Feng H, Jiang H, Zhou T, Chu H, Zhang Y, Xu C, Wang Z (2022). The interaction between microwave and coal: A discussion on the state-of-the-art. *Fuel* 314: 123140.

Gude VG, Patil P, Deng S (2012). Microwave energy potential for large scale biodiesel production. World Renew. Energy Forum, WREF 2012, Incl. World Renew. Energy Congr. XII Color. *Renew Energy Soc Annu Conf* 1: 751-759.

Guo L, Lan J, Du Y, Zhang TC, Du D (2020). Microwave-enhanced selective leaching of arsenic from copper smelting flue dusts. *J Hazard Mater* 386: 121964.

Hájková P (2018). Kaolinite claystone-based geopolymer materials. Effect of chemical composition and curing conditions. *Minerals* 8: 17-19.

Herrero MA, Kremsner JM, Kappe CO (2008). Nonthermal microwave effects revisited: On the importance of internal temperature monitoring and agitation in microwave chemistry. *J Org Chem* 73: 36-47.

Huang J, Zhang M, Adhikari B, Yang Z (2016). Effect of microwave air spouted drying arranged in two and three-stages on the drying uniformity and quality of dehydrated carrot cubes. *J Food Eng* 177: 80-89.

Kalhori S, Minaev B, Stone-Elander S, Elander N (2002). Quantum chemical model of an S_N2 reaction in a microwave field. *J Phys Chem A* 106 8516-8524.

Kappe CO (2008). Microwave dielectric heating in synthetic organic chemistry. *Chem Soc Rev* 37: 1127-1139.

Khan T (2020). An intelligent microwave oven with thermal imaging and temperature recommendation using deep learning. *Appl Syst Innov* 3: 1-17.

Kong Y, Wang P, Liu S (2018). Microwave pre-curing of Portland cement-steel slag powder composite for its hydration properties. *Constr Build Mater* 189: 1093–1104.

Krishna NV and Padmasine KG (2023). A review on microwave band pass filters: Materials and design optimization techniques for wireless communication systems. *Mater Sci Semicond Process* 154: 107181.

Kumar C and Karim MA (2019). Microwave-convective drying of food materials: A critical review. *Crit Rev Food Sci Nutr* 59: 379-394.

Li H, Zhao Z, Xiouras C, Stefanidis GD, Li X, Gao X (2019). Fundamentals and applications of microwave heating to chemicals separation processes. *Renew Sustain Energy Rev* 114: 109316.

Long S, Dong J, Yan C (2002). Microwave promoted clinkering of sulfoaluminate cement. *Cem Concr Res* 32: 1653-1656.

Lv W, Li S, Han Q, Zhao Y, Wu H (2016). Study of the drying process of ginger (Zingiber officinale Roscoe) slices in microwave fluidized bed dryer. *Dry Technol* 34: 1690-1699.

Mahato TK, Bhatt RJ, Agnihotri V, Saraf S (2022). Heating using microwave technique: An approach towards green chemistry. *Int J Sci Res Updat* 4: 340-345.

Mishra RR and Sharma AK (2016). Microwave-material interaction phenomena: Heating mechanisms, challenges and opportunities in material processing. *Compos Part A Appl S* 81: 78-97.

Moseley JD and Woodman EK (2009). Energy efficiency of microwave- and conventionally heated reactors compared at meso scale for organic reactions. *Energy Fuels* 23: 5438-5447.

Motevali A, Minaei S, Banakar A, Ghobadian B, Khoshtaghaza MH (2014). Comparison of energy parameters in various dryers. *Energy Convers Manage* 87: 711-725.

Nayak SN, Bhasin CP, Nayak MG (2019). A review on microwave-assisted transesterification processes using various catalytic and non-catalytic systems. *Renew Energy* 143: 1366-1387.

Nüchter M, Müller U, Ondruschka B, Tied A, Lautenschläger W (2003). Microwave-assisted chemical reactions. *Chem Eng Technol* 26: 1207-1216.

Obermayer D, Gutmann B, Kappe CO (2009). Microwave chemistry in silicon carbide reaction vials: Separating thermal from nonthermal effects. *Angew Chemie* 121: 8471-8474.

Patil PD, Reddy H, Muppaneni T, Schaub T, Holguin FO, Cooke P, Lammers P, Nirmalakhandan N, Li Y, Lu X, Deng S (2013). *In situ* ethyl ester production from wet algal biomass under microwave-mediated supercritical ethanol conditions. *Bioresour Technol* 139: 308-315.

Reja AH, Hussein AA, Hussain SA (2023). Design a microwave transmitter using magnetron and two layers waveguides. *Bull Electr Eng Informatics* 12: 2837-2845.

Rhithuparna D, Ghosh N, Rokhum SL, Halder G (2024). Current progress and future outlooks of microwave-irradiated biodiesel production: A holistic review. *Chem Eng J* 482: 149033.

Shazman A, Mizrahi S, Cogan U, Shimoni E (2007). Examining for possible nonthermal effects during heating in a microwave oven. *Food Chem* 103:444-453.

Siddique IJ, Salema AA, Antunes E, Vinu R (2022). Technical challenges in scaling up the microwave technology for biomass processing. *Renew Sustain Energy Rev* 153: 111767.

Stuerga DAC and Gaillard P (1996). Microwave athermal effects in chemistry: A myth's autopsy Part I: Historical background and fundamentals of wave-matter interaction. *J Microw Power Electromagn Energy* 31: 87-100.

Sun Y, Zhang P, Hu J, Liu B, Yang J, Liang S, Xiao K, Hou H (2021). The microwave heating mainly induces energy of polar molecules and ionic conduction. *Constr Build Mater* 294: 123491.

Tian S, Guo L, Gu Y, Ju S, Xu L (2021). Energy transfer and efficiency analysis of microwave flash evaporation with tap water as medium. *Desalination* 511: 115095.

Tian W, Li Z, Wu L (2020). Experimental demonstration of a microwave nonthermal effect in DMSO-NaCl aqueous solution. *Chem Phys* 528: 110523.

Tian WY, Huang KM, Yang LJ, Guo YN, Liu FH (2014). Investigate the microscopic properties and the non-thermal effect of the electrolyte solution under microwave irradiation. *Chem Phys Lett* 607: 15-20.

Varol PM, Çakan A, Kiren B, Ayas N (2023). Microwave-assisted catalytic transesterification of soybean oil using KOH/γ-Al_2O_3. *Biomass Convers Biorefinery* 13: 633-645.

Varshney A and Sharma V (2020). A comparative study of microwave rectangular waveguide-to-microstrip line transition for millimeterwave, wireless communications and radar applications. *Microw Rev* 26: 26-37.

Vu TH and Gowripalan N (2018). Mechanisms of heavy metal immobilisation using geopolymerisation techniques. *A review J Adv Concr Technol* 16: 124-135.

Wang K, Zhu G, Li Y, Chen S, Rashid A, Wang X, Wu X (2023). Non-thermal effects of microwave irradiation alleviates postharvest chilling injury of peach fruit by retarding phenolic accumulation and enhancing membrane stability. *Food Chem* 411: 135448.

Wu J, Liao W, Zhang J, Chen W (2019). Microwave-assisted reaction of gelatin with tannic acid: non-thermal effect on the crosslinking process. *Chem Eng Commun* 206: 1152-1158.

Wu X, Dong J, Tang M (1987). Microwave curing technique in concrete manufacture. *Cem Concr Res* 17: 205-210.

Wu Y, Mu R, Li G, Li M, Lv W (2022). Research progress in fluid and semifluid microwave heating technology in food processing. *Compr Rev Food Sci Food Saf* 21: 3436-3454.

Xu C, Lan J, Ye J, Yang Y, Huang K, Zhu H (2023). Design of continuous-flow microwave reactor based on a leaky waveguide. *Chem Eng J* 452: 139690.

Xu W, Islam M N, Cao X, Tian J, Zhu G (2021). Effect of relative humidity on drying characteristics of microwave assisted hot air drying and qualities of dried finger citron slices. *LWT* 137: 110413.

Ye J, Zhang C, Zhu H (2020). A temperature-control system for continuous-flow microwave heating using a magnetron as microwave source. *IEEE Access* 8: 44391-44399.

Zhai C, Teng N, Pan B, Chen J, Liu F, Zhu J, Na H (2018). Revealing the importance of non-thermal effect to strengthen hydrolysis of cellulose by synchronous cooling assisted microwave driving. *Carbohydr Polym* 197: 414-421.

Zhang X, Hayward DO, Mingos DMP (2003). Effects of microwave dielectric heating on heterogeneous catalysis. *Catal Letters* 88: 33-38.

Zhao Y, Tu D, Wang D, Xu J, Zhuang W, Wu F (2024). Structural and property changes of starch derivatives under microwave field: A review. *Int J Biol Macromol* 256: 128465.

Zhou X, Pedrow PD, Tang Z, Bohnet S, Sablani SS, Tang J (2023). Heating performance of microwave ovens powered by magnetron and solid-state generators. *Innov Food Sci Emerg Technol* 83: 103240.

Zhu Y, Chen X, Yuan W, Chu Z, Wong KY, Lei D, Yu Y (2021). A waveguide metasurface based quasi-far-field transverse-electric superlens. *Opto-Electronic Adv* 4: 1-6.

Zlotorzynski A (1995). The application of microwave radiation to analytical and environmental chemistry. *Crit Rev Anal Chem* 25: 43-76.

Chapter 2

Microwave Irradiation for Medical Treatments

Kaihan Xie
Wenke Zhao
Yaning Zhang*
and Bingxi Li
School of Energy Science and Engineering, Harbin Institute of Technology, Harbin, China

Abstract

Microwaves can be utilized in the medical fields of direct therapy and diagnostic imaging. This chapter provides a comprehensive introduction to the use of microwave irradiation for medical treatment. The clinical applications of microwave thermal effects, including microwave thermotherapy and microwave ablation, are introduced initially. Subsequently, the principles and methods of applying the non-thermal effects of microwaves to imaging, monitoring and diagnosis are described. Additionally, microwave sterilization, the promotion of protein and peptide synthesis, and the promotion of blood coagulation are discussed. Finally, the conclusions and future prospects are briefly elaborated and summarized.

Keywords: microwave irradiation, medical treatment, hyperthermia, ablation, diagnostics, monitoring

* Corresponding Author's Email: ynzhang@hit.edu.cn.

In: Characteristics and Applications of Microwave Irradiation
Editors: Yaning Zhang and Wenke Zhao
ISBN: 979-8-89530-170-8

1. Introduction

As science and technology continue to advance, many countries are researching new treatments and diagnostic methods to improve treatment accuracy and reduce harm to the human body. Among the new technologies, microwave technology has garnered increasing attention due to its high treatment efficiency and imaging accuracy. Microwaves are electromagnetic waves with wavelengths ranging from 1 m to 1 mm in free space and frequencies between 27 MHz and 300 GHz (Tang, 2015). Microwaves with frequencies ranging from 27 MHz to 10 GHz have significant biological importance as they propagate easily in tissues, are readily absorbed by biological tissues, and exhibit varying degrees of reflection based on tissue characteristics. The most commonly used microwave frequency in industrial production and medicine is concentrated at 2.45 GHz (Gong et al. 2024). Currently, microwave irradiation for medical treatment is categorized into three main types: microwave thermotherapy, microwave ablation, and microwave diagnostics and monitoring.

Microwave thermotherapy is a method that utilizes the thermal effects of microwaves in human tissues to heat the target area to an effective therapeutic temperature of 41~45 °C while ensuring that the temperature of the normal tissue area is maintained at 41 °C or below for 30~60 minutes to destroy the target tissue (Nielsen et al. 2001; Vernon et al. 1996; Kapp, 1996). Currently, microwave thermotherapy is widely used in the treatment of breast cancer, liver cancer, and other diseases.

The microwave ablation technique uses a microwave ablation antenna to accurately radiate microwaves to the target tissue, enabling polar molecules such as water and protein within the target tissue to absorb microwave energy. Under the influence of an electromagnetic field, polar molecules oscillate at high speed and generate heat through friction, rapidly warming the tissues and forming an ellipsoidal heat field. When the temperature exceeds 46 °C, the cells will undergo irreversibly coagulate and necrosis, achieving the purpose of ablating cell tissues (Simon et al. 2005; Sapareto and Dewey, 1984). Additionally, the high temperature can reduce blood flow around the tissue, leading to blockage and thrombus formation, which not only reduces heat loss but also cuts off the blood supply to the tissue (Wu, 2016).

Microwave diagnostics and monitoring take advantage of the differences in dielectric constants between the body's tissues and the monitored object. As microwaves propagate through the human body, they reflect, attenuate, and scatter upon encountering interfaces between different tissues. The changes in these parameters can be measured for non-invasive monitoring and diagnosis (Villena et al. 2019; Narang et al. 2018; Choi et al. 2015). In this chapter, the applications of microwaves in the medical field, including microwave thermotherapy, microwave ablation, and microwave diagnostics, and monitoring are introduced. The principles and advantages of these technologies are described in detail. Finally, the conclusions and future outlook are summarized.

2. Microwave Irradiation for Hyperthermia

2.1. Cancer Hyperthermia Technology

According to the World Health Organization, the number of new cases of malignant neoplasms worldwide increased from 12.7 million in 2008 to 14.1 million in 2012 (Jemal et al. 2011; Torre et al. 2015). This number is expected to exceed 20 million by 2025, with 80% of these cases occurring in lower-middle-income countries (Conner et al. 2022). The most mainstream cancer treatments today include radiotherapy, surgery, and chemotherapy (Altintas et al. 2021). Chemotherapy is the most widely used modality, but its drawbacks are also significant. It is not selective in its destruction of cells, affecting both normal and tumor cells. It can lead to side effects such as hair loss and damage to the oral mucosa (Yamada et al. 2023). Moreover, the longer the treatment is administered, the more severe the side effects become.

Combining chemotherapy with microwave thermotherapy can enhance treatment effectiveness. This is because microwave irradiation raises the temperature of the tumor tissue. Blood flow in tissues varies in response to temperature changes within the tissue (Rossmanna and Haemmerich, 2014). This thermoregulatory mechanism in the body increases blood perfusion to the heated area when tissue temperature rises. This blood, at the body's core temperature (approximately 37 °C), counteracts the temperature increase. Blood perfusion thus becomes an effective heat dissipation pathway, removing heat from the tumor tissue while maintaining normal temperature in healthy tissue. Additionally, the high temperature in the target area attracts temperature-sensitive chemotherapy drugs, increasing their efficacy on the target area and reducing damage to healthy tissues (Issels, 2008). Furthermore, thermotherapy can activate the body's immune system, similar to the process of fever (Evans et al. 2015). Increased blood flow to the tumor tissue improves the pathway between the tumor tissue and lymph nodes, attracting immune cells. Both the duration and temperature of thermotherapy affect the immune system (van den Tempel et al. 2016). The temperature of microwave hyperthermia is typically 42-45 °C (Jeong and Son, 2023).

In the literature, researchers utilized the system shown in Figure 1 to administer microwave thermotherapy to ten patients (Gardner et al. 2002). The results indicated that 6 out of 10 patients exhibited significant tumor shrinkage (average 41%) as seen through ultrasound imaging after 5-18 days of hyperthermia. Eighty percent of patients demonstrated a remarkable tumor response, primarily referring to a reduction in tumor volume or significant killing of tumor cells.

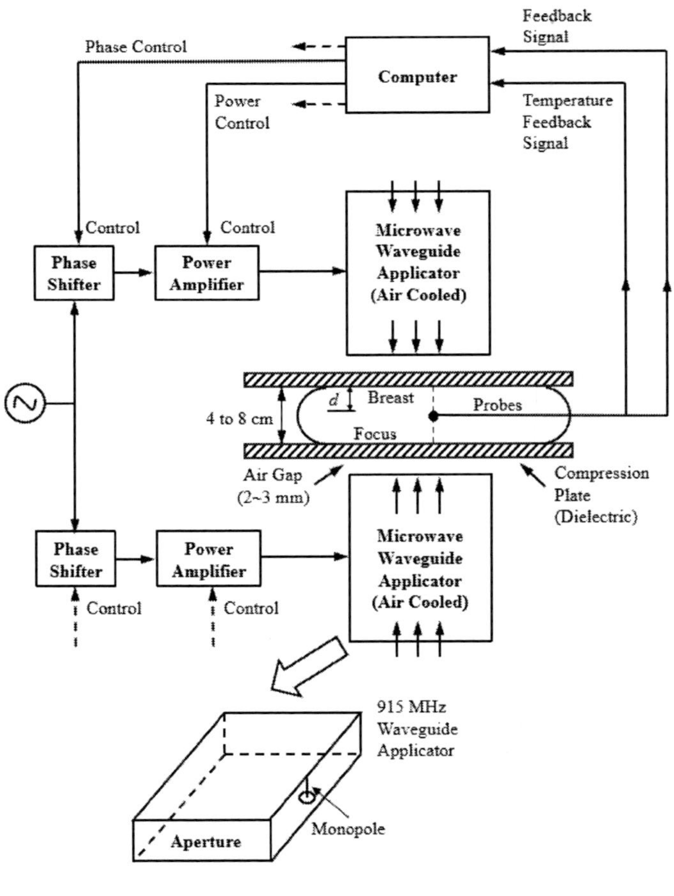

Figure 1. Block diagram of a dual-channel adaptive microwave phased array hyperthermia system for the treatment of deep breast cancer (Gardner et al. 2002).

2.2. Microwave Urethroplasty Technology

An enlarged prostate can lead to a pathological condition that severely blocks the flow of urine and, in the worst-case scenario, can even cause urinary retention. The traditional treatment is transurethral resection of the prostate, but its objective improvement and symptom relief are limited (Stovsky et al. 2006; Rosario et al. 2007). Microwave urethroplasty is an emerging microwave technology used to target prostate disorders.

Microwave urethroplasty can enlarge the urethral lumen where the urethra is narrowed due to prostate enlargement. The procedure involves heating the prostate with microwave power through the urethra while simultaneously dilating the surrounding part of the urethra with an expanding balloon (Sterzer, 1991). The catheter used for the procedure is called the transurethral microwave balloon catheter. Figure 2 illustrates the structural features of this catheter. The catheter is composed of a plastic tube and a lumen for the insertion of a miniature coaxial cable, which ends in a microwave antenna used to inflate a compressed balloon. The compressed balloon is inflated with circulating cooled deionized water (Sterzer et al. 2000). Microwaves are fed into the coaxial cable and transmitted through the microwave antenna to the prostate. The microwave frequency used in the procedure is 915 MHz for heating. Additionally, a fixed balloon is also implanted to prevent the catheter from dislodging during the procedure.

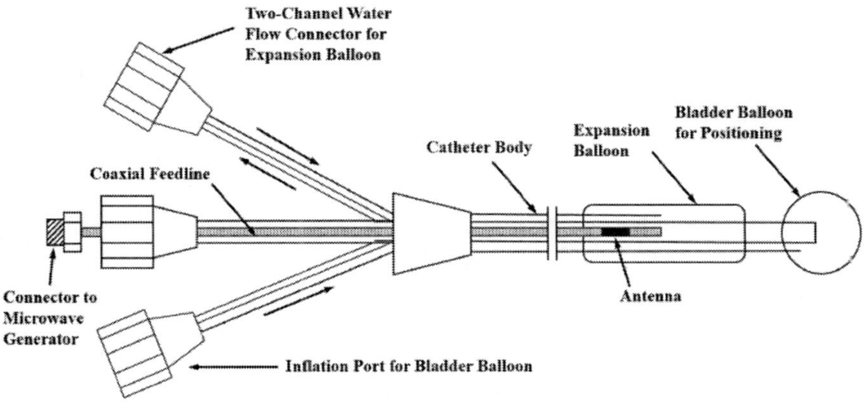

Figure 2. The structure features of the transurethral microwave balloon catheter (Sterzer et al. 2000).

2.3. Peptide Synthesis Technology

Microwave synthesis was first reported in 1986 (Gedye et al. 1986 a; Gedye et al. 1986 b). As research progresses, it has been found that the initial microwave power, reaction temperature, and reaction pressure significantly affect the synthesis process (Collins and Leadbeater, 2007).

Microwaves promote protein synthesis primarily by enhancing peptide synthesis (Santagada et al. 2001; Pramanik et al. 2002). The process of microwave-promoted peptide synthesis, as mentioned in the literature (Aslan et al. 2006; Wiesinger et al. 2004), is described as follows. The mixture was heated to 110 °C using a temperature controller and maintained at this temperature for 20 minutes. Significant pressure was generated during the reactions due to the volatile nature of the coupling agents. However, the efficiency of the solid-phase reaction is greatly reduced by the need to constantly change reaction vessels during the coupling reaction and washing steps. Microwave heating can be extended to the deprotection phase and the coupling reaction period to improve reaction efficiency. Typically, the standard peptide used to test synthetic methods is the ACP peptide. The method of preparing high-purity ACP peptides involves a two-minute deprotection reaction, a three-minute coupling reaction, and a ten-minute cleavage reaction (Collins et al. 2010). Figure 3 illustrates a schematic diagram of the use of microwaves to promote the synthesis of small tripeptides.

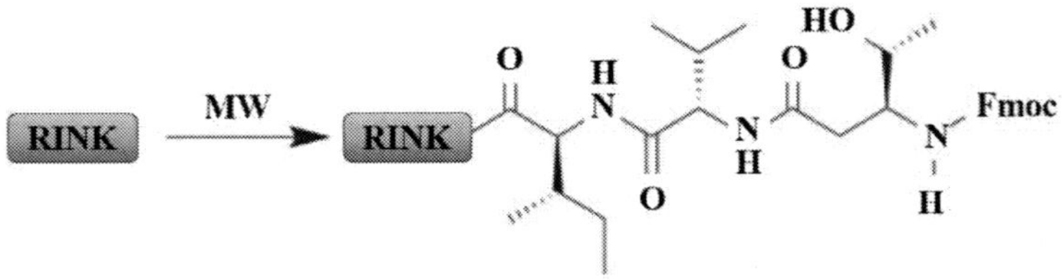

Figure 3. Schematic diagram of using microwaves to promote the synthesis of small tripeptides (Erdélyi and Gogoll, 2003).

2.4. Improvement of Enzymatic Efficiency

Microwave heating can enhance the effectiveness of enzymes (Elmoslemany et al. 2023). The differences between conventional and microwave heating were compared in the process of breaking down cytochrome. The results showed that the decomposition effect of microwave heating for ten minutes is similar to that of conventional heating at 37°C for six hours (Tai et al. 2023). However, the pigment cells remained intact after 20 minutes of microwave heating without the addition of protease. This proves that microwave heating enhances the catabolic capacity of protease rather than directly causing a catabolic reaction.

The secondary structure of large proteins is mainly composed of α-helices and β-sheets together. The α-helix structure is based on a molecular axis around which the peptide backbone is wound. The amino acid side chains extend outward and downward from the backbone. α-helices contain large net dipoles due to the orderly arrangement of hydrogen bonds, creating dipole stacking between the hydrogen bonds of the helices. Table 1 shows the magnitudes of these dipole moments. The presence of large net dipoles increases the digestibility of certain proteins under microwave heating conditions. Figure 4 illustrates the perturbation of the three-dimensional structure of proteins due to the interaction of microwave energy with α-helical dipoles. This interaction can promote the decomposition of closed regions of proteins (Arakawa et al. 1991; Chen et al. 1991; Sandoval et al. 2007).

Table 1. Sizes of dipole moments (Arakawa et al. 1991)

Molecule	Dipole moment (Debye)
Horse serum albumin	380
Water	1.8
Horse carboxyhemoglobin	480
Myoglobin	170
Peptide bond	3.5

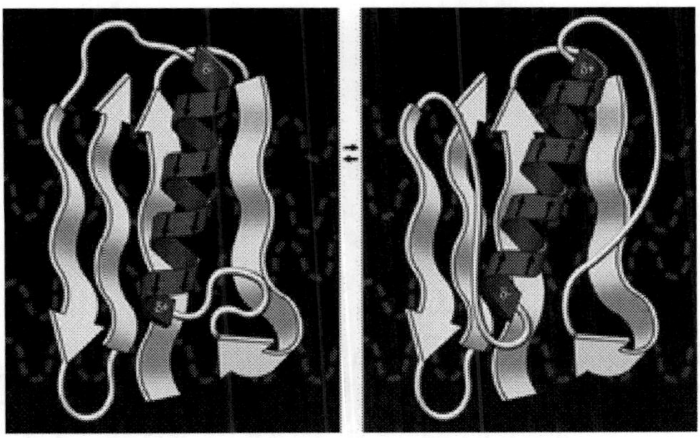

Figure 4. Perturbation of protein structure (Arakawa et al. 1991).

2.5. Microwave Hyperthermia Equipment

Figure 5 illustrates four different types of microwave antennas used in thermal therapy. The "Radiant type" and "Contact type" are categorized as "External antennas", while the "Intracavity type" and "Interstitial type" are referred to as "Internal antennas" (Ito and Saito, 2011; Kawai et al. 2006).

For external heating devices, the highest temperature is generated at the surface of the body unless cooling is applied. Therefore, to prevent the surface of the body from overheating and causing harm, it is necessary to incorporate a surface cooling device during the design process to make the heating system practical, as shown in Figure 6 (Ito and Saito, 2011). Meanwhile, antennas used for external microwave heating are categorized into "guided wave antennas" and "microstrip antennas". Microstrip antennas are more promising and widely used due to their thin and lightweight construction. Figure 7 illustrates two examples of treatment using microstrip antennas (Jacobsen et al. 2000; Stauffer et al. 1995).

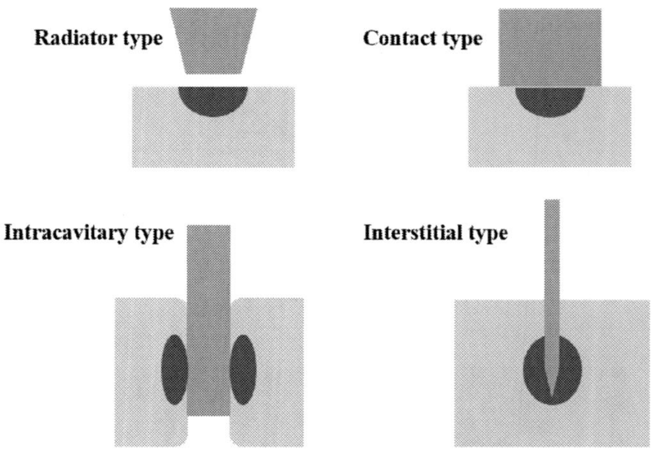

Figure 5. Four types of antennas for microwave heating (Ito and Saito, 2011).

For intracavitary heating devices, if the target location is near the abdominal cavity, such as the esophagus or urethra, a microwave antenna can be used for intracavitary microwave thermotherapy. Figure 8 shows an endoluminal heating protocol for cholangiocarcinoma (Saito et al. 2006). This treatment is performed by extending an endoscope into the duodenum allowing the flexible microwave antenna to enter the intestine without difficulty. The antenna is then guided to the bile ducts through the papillae of Vater located in the duodenum. With this approach, the combined use of microwave antennas and endoscopes enhances the effectiveness and reliability of thermotherapy (Zhang and Chang, 2023).

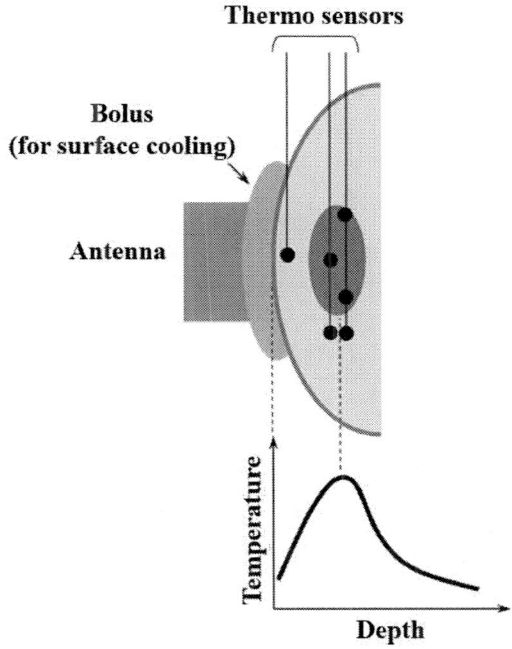

Figure 6. Microwave external heating system (Ito and Saito, 2011).

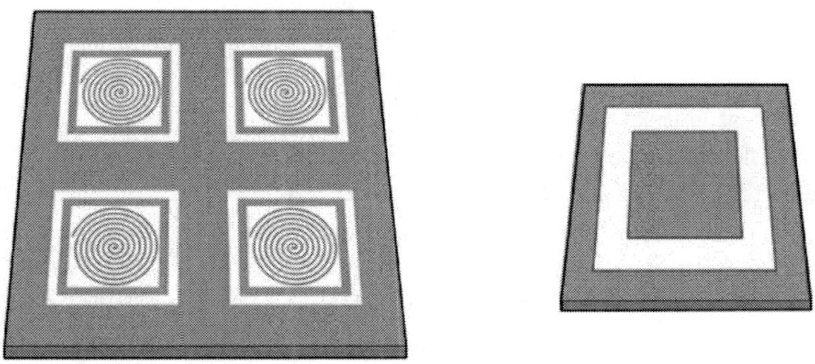

(a) Spiral type. (b) Square type.

Figure 7. Schematic diagram of microstrip antenna (Jacobsen et al. 2000; Stauffer et al. 1995).

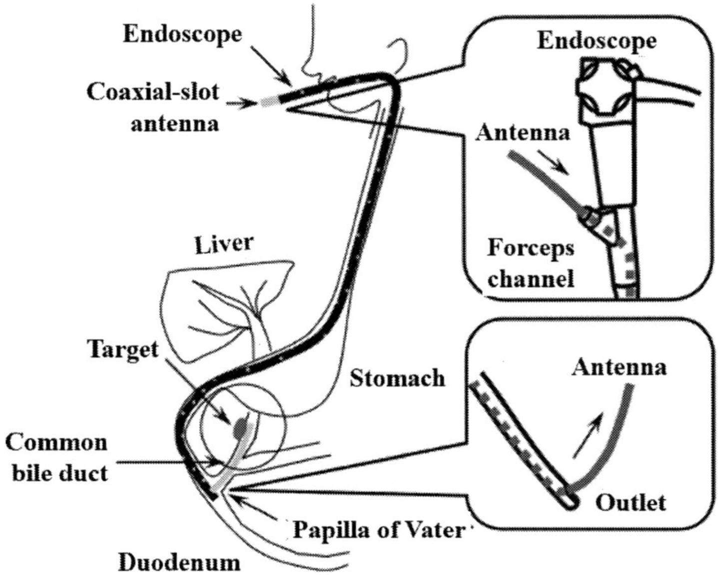

Figure 8. The sample for intracavity heating equipment (Saito et al. 2006).

3. Microwave Irradiation for Ablation

3.1. Tumor Tissue Ablation Technology

Microwave ablation technology is a minimally invasive tumor ablation procedure based on the principle that the deposition of electromagnetic (EM) energy causes cancer cells to coagulate and undergo necrosis due to increased temperature (Karampatzakis et al. 2013; Xi et al. 2023). It has been reported that temperatures between 50°C and 60°C denature cellular proteins, leading to cell death, while temperatures above 60 °C can cause instantaneous cell death (Ibitoye et al. 2015; Hancock et al. 2013).

As shown in Figure 9, the fundamental microwave ablation system consists of three components: a microwave generator, a coaxial cable, and a cooling system (Tammam et al. 2020). The microwave signals are transmitted to the antenna through the coaxial cable and are continuously fed from the generator. The antenna distributes the microwave energy during the tissue ablation process.

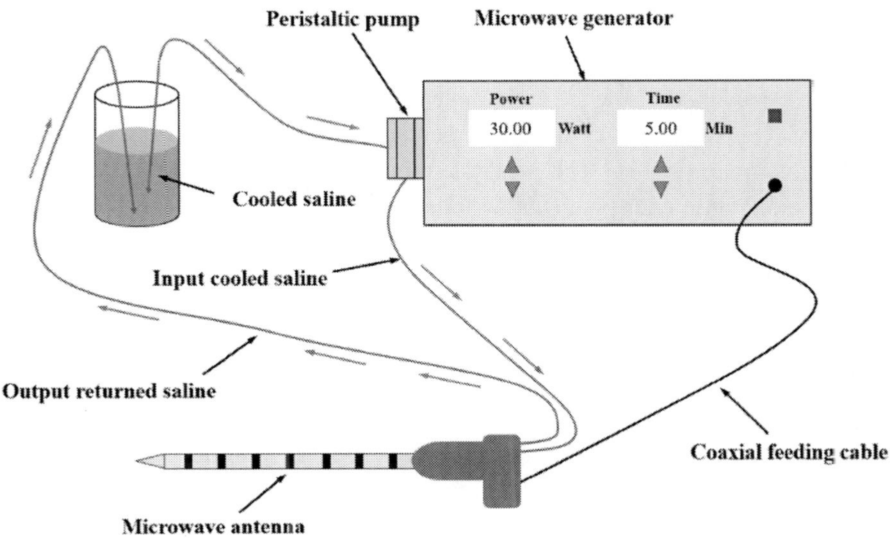

Figure 9. Microwave ablation system (Tammam et al. 2020).

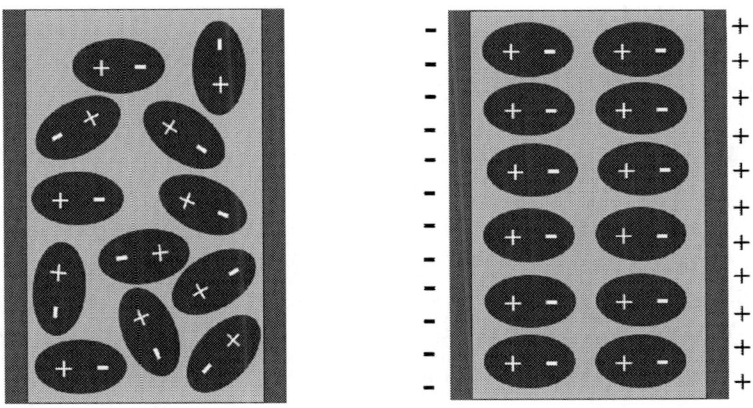

(a) Unpolarized (b) Polarized by the applied electric field

Figure 10. Polar molecules are polarized by electromagnetic fields (Tammam et al. 2020).

The fundamental principle of microwave ablation is the heating of the medium using an applied electric field. In this context, the dielectric material is human tissue. As shown in Figure 10, dielectric heating occurs due to the continuous rotation of polar molecules (e.g., water molecules). These polar molecules continuously rotate in the direction of the applied electric field, but the high frequency of the field causes them to alternate billions of times per second. Friction prevents the polar molecules from keeping pace with the electromagnetic field, resulting in a damping effect on the rotating dipole. The excess energy generated by the damping effect is deposited as heat in the tissues, causing an increase in tissue temperature (Carrafiello et al. 2008; Lara et al. 2015). Microwaves are highly penetrating and can traverse a wide range of biological tissues. The tissue's moisture content affects the rate of electromagnetic energy accumulation. Higher moisture content in the tissue results in a higher rate of energy accumulation (Lubner et al. 2010; Reimann et al. 2019). Tumor tissue has a significantly higher moisture content than normal cells, allowing microwave ablation to specifically target and remove tumor tissue.

For tumor tissue, maintaining a temperature of 46 °C for 60 minutes results in irreversible cell damage. At approximately 52 °C, cell damage occurs within about 5 minutes. Above 52 °C, tumor cells are killed immediately (Wang et al. 2015). However, if the temperature exceeds 105 °C, the tissue will boil, leading to vaporization and carbonization. This reduces the rate of energy transfer, thereby increasing the time required to completely ablate the affected area. Therefore, it is crucial to maintain the temperature within the ablation volume between 50 and100 °C during therapy (Liu et al. 2006).

3.2. Modalities and Keratinized Cells Treatment Technology

Microwave ablation is widely used for the treatment of condyloma acuminatum (Hagon et al. 2023; Kimura et al. 2014; Park and Choi, 2008; Tosti and Piraccini, 2001). Direct thermal ablation is infrequently used due to extensive scarring and high recurrence rates. Currently, cryotherapy, involving the application of liquid nitrogen to diseased skin, is the most widely used physical method. This treatment can lead to tissue destruction. Studies have shown that the efficacy of cryotherapy on common warts is low, with an average clearance rate of 49% across various sites (Sterling et al. 2014; Kwok et al. 2011). Microwave therapy, which involves delivering thermal energy to induce ablation of the wart tissue, is highly effective for common warts (Vaiman et al. 2011). Figure 11 shows a comparison before and after microwave treatment.

Hagon et al. (Hagon et al. 2023) conducted a study on the efficacy of microwave therapy for warts and keratinized cells. Their studies found that microwave therapy has a 75.9% complete cure rate, compared to cryotherapy's 23-33% cure rate (Bruggink et al. 2010). Moreover, microwave therapy specifically induces keratinocytes to synthesize IL-6, a pro-inflammatory mediator crucial for antiviral immunity (Böttcher et al. 2014).

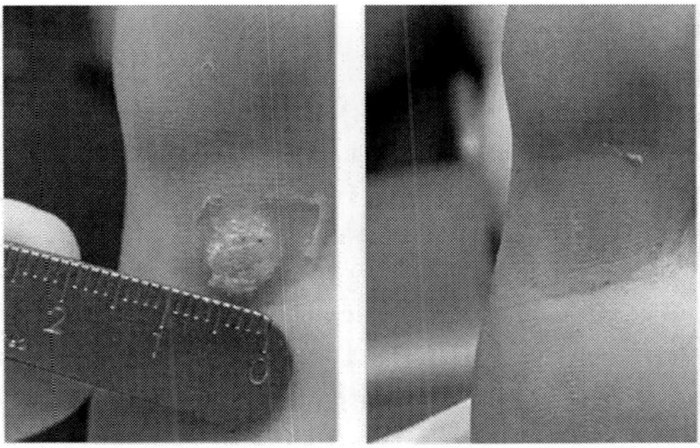

Figure 11. Comparison chart before and after microwave treatment (Hagon et al. 2023).

3.3. Microwave Balloon Angioplasty Technology

With the continuous advancement of medical technology, numerous new treatments for narrowing arteries or atherosclerosis have been developed. Among these methods, in addition to percutaneous atherectomy (de Waha et al. 2016; Nakabayashi et al. 2019) and lithotripsy (Serruys et al. 2017), stenting and angioplasty techniques are commonly used in the clinical treatment of arterial obstruction (Levine, 2012).

Percutaneous transluminal balloon angioplasty (PTBA) dilates arteries by first advancing a balloon catheter into an arterialstenosis, inflating the balloon, and using the generated pressure to dilate the artery (Amstutz et al. 2023; Benetti et al. 1995). Microwave balloon angioplasty (MBA) is a primary enhancement for PTBA (Rosen et al. 1989 a; Rosen et al. 1989 b; Walinsky et al. 1991). The MBA catheter consists of a microwave cable and an antenna. This antenna is wrapped around the catheter's end and located at the balloon's center, similar to the structure in a catheter-based microwave ablation tool (Fallahi and Prakash, 2018). The balloon's expansion occurs simultaneously with the release of microwave energy, allowing the vascular tissue to dilate while being uniformly heated. The combination of pressure and heat has been shown to enhance vasodilatation and reduce resistance to blood flow, maintaining patency (Landau et al. 1994).

3.4. Microwave Coagulation Technology

Microwaves can influence blood clotting. Uncontrolled bleeding is one of the most significant risks during surgery. Traditional compression and suture hemostasis have limited effectiveness (Poulin et al. 1995). Microwave coagulation, a thermogenesis-based method, promotes coagulation and can significantly reduce bleeding during surgery (Ryu et al. 2019; Chong et al. 2018).

Zhang et al. (2020) investigated the adjuvant effect of 915 MHz microwaves on coagulation. They found that microwave-assisted coagulation is a safe and effective technique. The heat production efficiency and heating rate of microwaves exceed those of radiofrequency (Andreano and Brace, 2013). The average operative time using microwave-assisted coagulation was 205 minutes, with minimal blood loss of 30 to 50 ml, no complications, and a hospital stay of 5 to 10 days. Table 2 provides specific information on patients undergoing 915 MHz microwave-assisted coagulation procedures.

Table 2. Specific information on patients undergoing 915 MHz microwave-assisted coagulation procedures (Zhang et al. 2020)

Case	1	2	3	4
Gender	Female	Male	Female	Female
Age (years)	19	20	25	33
Diagnosis	Epidermoid cyst	Sclerosing angiomatoid nodular transformation of the spleen	Haemangioma	Non-Hodgkin's disease
Size of specimen (cm)	6×5	3.7×3.2	4.5×4.5	4×3
Segmentomy	Lower pole	Lower pole	Upper pole	Upper pole
Blood loss (ml)	50	30	50	40
Operative time (min)	218	180	195	225
Length of stay (days)	10	6	5	8
Complications	None	None	None	None

3.5. Neurostimulation Technology

Neuroplasticity is the fundamental mechanism underlying learning and memory, and it facilitates functional recovery after brain nerve damage (Kiourti et al. 2022). Traditional rehabilitation methods do not directly alter the brain but improve brain function through environmental changes (Harid et al. 2023). Noninvasive brain stimulation is a neuromodulation method that stimulates specific brain regions using magnetic fields or electrical currents without requiring surgery. Transcranial magnetic stimulation (TMS) is currently used clinically (Makaroff et al. 2023). However, its disadvantages include the requirement for expensive equipment and the generation of noise. Microwave stimulation can address these issues.

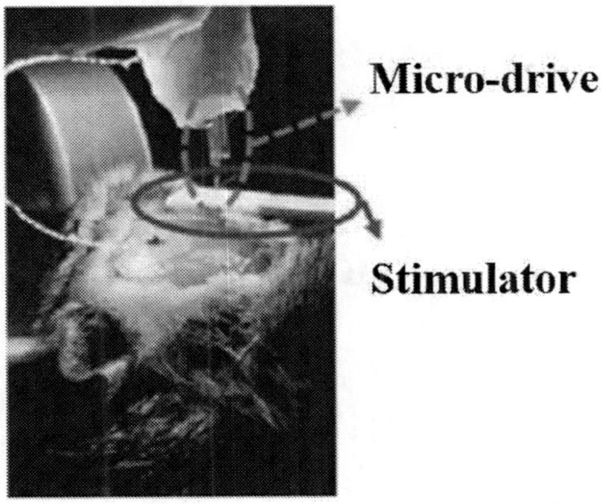

Figure 12. Experimental setup for mouse brain stimulation (Seo et al. 2018).

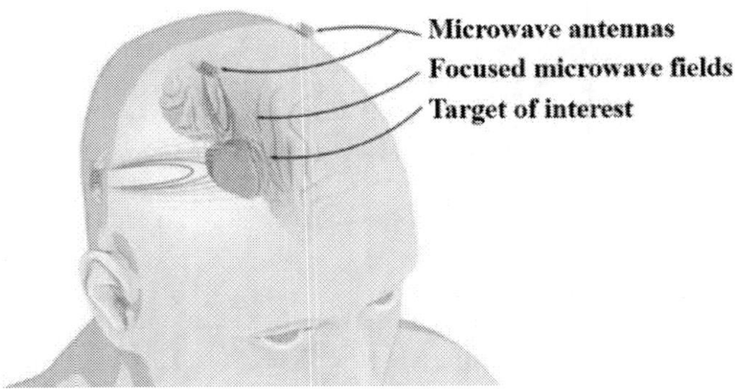

Figure 13. The schematic diagram of non-invasive microwave stimulation (Harid et al. 2023).

Microwave stimulation typically uses radio signals in the 100 MHz to 10 GHz band, based on the principle of microwave ablation. Microwave stimulation modalities are categorized as invasive and non-invasive. As shown in Figure 12, Seo et al. investigated the effects of invasive microwave stimulation in mice (Seo et al. 2018). They found that microwave signals could effectively alter the firing rate of neurons in mice. Figure 13 presents a schematic diagram of non-invasive microwave stimulation (Harid et al. 2023). Non-invasive microwave stimulation can create a hot spot in a specific area to stimulate nerves (Nepa and Buffi, 2017; George and Aston-Jones, 2010).

3.6. Microwave Medical Waste Sterilization Technology

The treatment of medical waste is categorized into incineration and non-incineration methods. Currently, the main methods for treating medical waste include high-temperature incineration, chemical sterilization, microwave sterilization, and high-temperature steam (Al-Khatib et al. 2020).

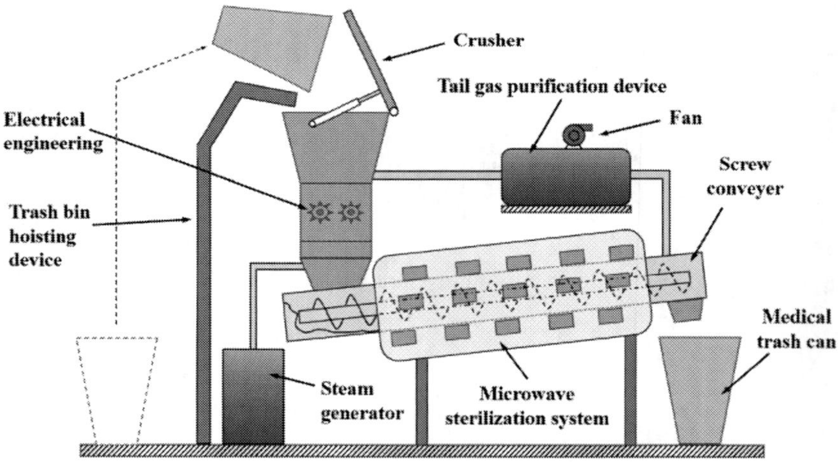

Figure 14. The structure of the microwave sterilization experimental bench (Liu et al. 2022).

Currently, the most widely used method for medical waste treatment worldwide is direct high-temperature incineration. However, this method has many drawbacks. Due to the complex composition of medical waste, high-temperature combustion produces significant quantities of dioxins, furans, and other toxic substances, as well as fine particles, which severely damage human health and cause environmental pollution (Cai and Du, 2020). High-temperature steam sterilization produces toxic volatile organic compounds (VOCs) during disinfection (Giakoumakis et al. 2021). Additionally, the large amount of wastewater generated significantly increases the treatment cost.

Chemical disinfection offers fast and broad-spectrum sterilization with stable performance. However, its drawbacks include the toxicity of chemical reagents to the human body and strict requirements for temperature and pH during the disinfection process (Giakoumakis et al. 2021; Ilyas et al. 2020). Compared to the aforementioned methods, microwave sterilization technology has broader development prospects. It offers higher sterilization efficiency and can safely and efficiently handle large volumes of waste liquids, maximizing waste minimization and harmless treatment (Giakoumakis et al. 2021).

Microwave sterilization technology utilizes the combined thermal and field effects of microwaves to achieve sterilization (Xu et al. 2020; Celandroni et al. 2004). The thermal effect of microwaves causes polar molecules in microorganisms to rotate and rub together, generating sufficient heat to quickly kill the microorganisms. The field effect of microwaves rapidly penetrates microorganisms and destroys their cellular structure. The field effect of microwaves also disrupts the protein structure of microorganisms, resulting in rapid and complete sterilization (Lv et al. 2019).

Liu et al. (2022) constructed a professional microwave sterilization experimental bench to investigate the effects of various factors on the microwave sterilization process of medical waste. Figure 14 shows the structure of the microwave sterilization experimental bench. Experimental results indicate that microwave irradiation time has the greatest influence on sterilization effectiveness. Higher microwave power, longer irradiation time, and higher water content improve the bactericidal rate. Additionally, the detection of the experimental exhaust gas revealed that no harmful substances, such as dioxins, were produced during the sterilization process.

4. Microwave Irradiation for Diagnostics and Monitoring

4.1. Microwave Imaging Technology

Microwave imaging is a novel method for real-time treatment monitoring to improve accuracy (Scapaticci et al. 2018; Wang et al. 2021 a; Wang et al. 2021 b). Microwave imaging analyzes and distinguishes changes in backscattered signals caused by differences in dielectric properties between cells of different tissues (Hussein et al. 2019). Typical microwave imaging systems use multiple antennas surrounding the measurement area to achieve high imaging resolution (Di Benedetto et al. 2008; Bucci et al. 2016). Microwave imaging offers several advantages over conventional modalities, including portability, non-invasiveness, and the non-ionizing nature of the electromagnetic field (Di Benedetto et al. 2008).

Microwave imaging techniques are categorized as passive, hybrid, and active, with active methods being widely used due to their higher sensitivity (Wang, 2023). Typical active microwave imaging techniques include microwave tomography, radar imaging techniques, and machine learning imaging techniques, as shown in Figure 15 (Meaney et al. 2000; Ruiz et al. 2023).

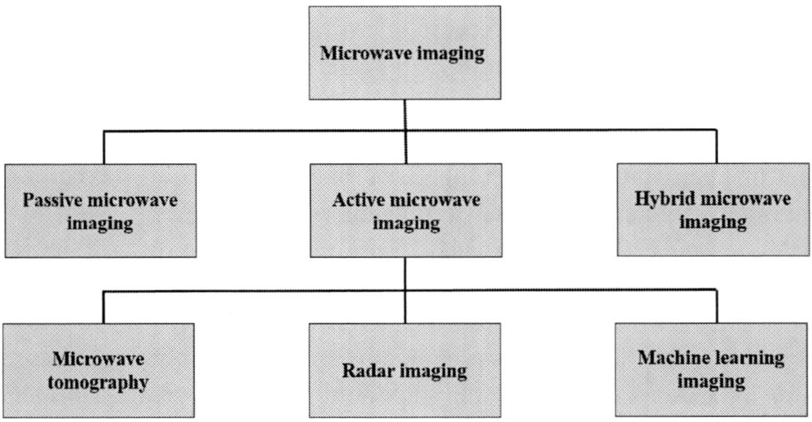

Figure 15. Classification of microwave imaging methods (Meaney et al. 2000; Ruiz et al. 2023).

Microwave tomography creates cross-sectional images of tissues using varying microwave frequencies. It exploits the differing dielectric properties of various tissue types. Therefore, reconstructing the dielectric properties within the human body is critical for improving imaging quality (Simonov and Son, 2021). Gradient-based methods (e.g., conjugate gradient least squares and Landweber) and global techniques (e.g., particle swarm optimization) are used to process collected data and solve inverse problems (Edwards et al. 2023; Estatico et al. 2018; Xiao and Yabe, 1998; Noghanian et al. 2006). Incorporating image reconstruction algorithms enhances the accuracy and reliability of imaging techniques (Ozgun and Kuzuoglu, 2018; Purkait and Datta, 1996; Liu and Zhang, 2011; Guo et al. 2006; Bindu et al. 2006). Microwave tomography is now widely used in clinical practice. For example, a newly developed prototype of a mobile electromagnetic tomography brain scanner from EM Tensor GmbH is operationally safe, commercially viable, and has demonstrated its capability to detect brain pathologies in stroke patients (Henriksson et al. 2022).

Microwave confocal imaging algorithms are among the fastest growing and most widely used radar imaging techniques in medical imaging. The imaging principle involves sending an incident signal to the imaging region through an antenna, where scatterers scatter and absorb the electromagnetic waves to varying degrees (Elahi et al. 2018). Multiple antennas are arranged around the imaging region to receive electromagnetic waves. After receiving the scattered electromagnetic waves, the scattering intensity of each point in the region is calculated using specific methods. Finally, the energy value of each point in the reconstructed image is determined based on the level of scattered intensity (Zamani et al. 2016).

4.2. Microwave Blood Glucose Testing Technology

Diabetes is a chronic disease that necessitates diagnosis and appropriate treatment to prevent complications (Bae et al. 2022; Borgnakke, 2019). The traditional method of measuring blood glucose involves pricking a finger and measuring blood glucose directly using test strips or instruments. Frequent finger pricking causes discomfort and increases the risk of infection (Chen et al. 2017).

Non-invasive glucose monitoring techniques are generally categorized into optical and non-optical methods. Optical methods commonly use photoacoustic spectroscopy, optical coherence tomography, and Raman spectroscopy (Li and Li, 2016; Lü et al. 2016; Pao et al. 2018; Shih et al. 2015; De Pretto et al. 2016). These methods involve passing a beam of light through a specific part of the body, which causes a spectral change that is calibrated and used to measure blood glucose concentration (Yang et al. 2022). Non-optical monitoring techniques are less common, with microwave blood glucose testing being the primary method.

Microwave blood glucose testing utilizes the interaction of electromagnetic waves with biological tissues (Losoya-Leal et al. 2012; Defronzo et al. 1992). Due to the weak variation of the blood dielectric constant with glucose concentration (Karacolak et al. 2013; Venkataraman and Freer, 2011; Yilmaz et al. 2014; Mustafa et al. 2014; Gabriel et al. 1996; Yilmaz et al. 2019), studies often focus on designing sensors with a high-Q factor, whose sensitive response can more easily track the resonance frequency shift caused by changes in glucose concentration (Kandwal et al. 2021; Navaei et al. 2023; Kandwal et al. 2022; Vélez et al. 2021).

Yang et al. (2022) optimized a 3D model of the earlobe using a diffusion-limited aggregation approach to represent the continuous and complex nature of blood flow distribution. Experimental results showed a maximum detection sensitivity of 0.0198 dB per mg/dL in the ultra-wideband range of 8.0 to 10.0 GHz, with the prediction error of blood glucose levels not exceeding 8%. Figure 16 illustrates the experimental system.

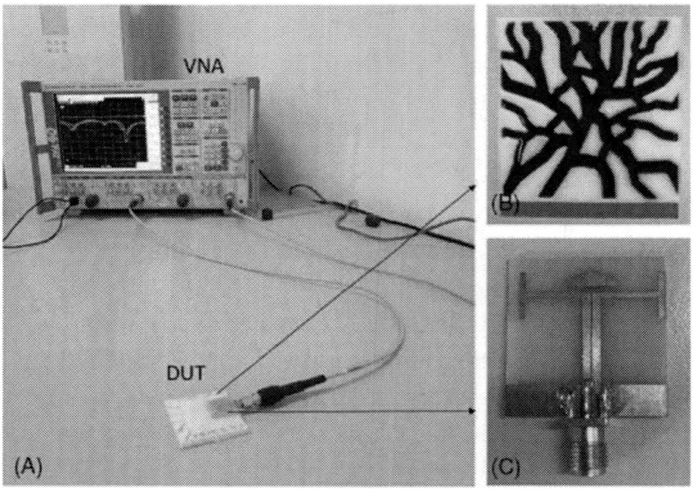

Figure 16. The experimental system of microwave blood glucose (Yang et al. 2022).

4.3. Microwave Rheumatism Detection Technology

Rheumatic disease is a common, progressive clinical condition typically associated with significant pain and dysfunction. Temperature is a key physical property for assessing the severity of rheumatic disease and directly reflects the degree of joint inflammation. Elevated temperature is one of the four classic signs of inflammation (Schiavon et al. 2022; Salisbury et al. 1983; Kastberger and Stachl, 2003; Ring et al. 1981). The traditional diagnostic modality is ultrasound. Although this method is more sensitive, it is hampered by operator dependency and limited availability (Saleem et al. 2011).

Microwave radiometry (MWR) is a non-invasive, rapid detection method. It effectively measures the microwave electromagnetic radiation generated by human tissues within 7 cm of the skin. The measurement mechanism converts microwave signals generated by the human body into temperature signals for statistical purposes (Barrett and Myers, 1975). MWR can detect joint inflammation because the local tissue temperature in inflamed areas is higher than in normal tissues, producing more microwave radiation (Fraser et al. 1987; MacDonald et al. 1994).

Conclusion

Microwaves are a versatile type of electromagnetic wave with numerous applications in medicine. Currently, the thermal and non-thermal effects of microwaves are widely utilized in medicine for thermotherapy, tissue ablation, diagnostics, imaging, sterilization, and promoting peptide and protein synthesis. Additionally, microwaves can be used to heat blood plasma and promote blood coagulation during surgery.

Despite the achievements in the use of microwaves, the understanding of their mechanisms and effects is limited. However, the application of microwave technology in diagnostic and therapeutic medicine will drive and inspire innovative healthcare solutions for a wide range of challenging diseases, bring comfort and convenience to patients, enable effective disease management for both patients and physicians, and improve the overall well-being of our society.

Acknowledgments

The following financial supports are acknowledged: Financial support was provided by the National Natural Science Foundation of China (52076049, 52476005), Heilongjiang Province "Double First-class" Discipline Collaborative Innovation Achievement Project (LJGXCG2023-080), Heilongjiang Provincial Key R&D Program (2023ZX02C05), and Heilongjiang Provincial Key R&D Program "Unveiling the Leader" Project (2023ZXJ02C04).

References

Al-Khatib, I. A., Khalaf, A. S., Al-Sari, M. I. & Anayah, F. (2020). Medical waste management at three hospitals in Jenin district. Palestine. *Environmental Monitoring and Assessment*, *192*(1), 10.

Altintas, G., Akduman, I., Janjic, A. & Yilmaz, T. (2021). A Novel Approach on Microwave Hyperthermia. *Diagnostics*, *11*(3), 493.

Amstutz, C., Behr, J., Krebs, S., Haeberlin, A., Vogel, R., Zurbuchen, A. & Burger, J. (2023). Design of percutaneous transluminal coronary angioplasty balloon catheters. *Biomedical Engineering Online*, *22*(1), 94.

Andreano, A. & Brace, C. L. (2013). A comparison of direct heating during radiofrequency and microwave ablation in ex vivo liver. *Cardiovascular and Interventional Radiology*, *36*(2), 505-511.

Arakawa, T., Kita, Y. & Carpenter, J. F. (1991). Protein–Solvent Interactions in Pharmaceutical Formulations. *Pharmaceutical Research*, *8* (3), 285-291.

Aslan, K., Holley, P. & Geddes, C. D. (2006). Microwave-Accelerated Metal-Enhanced Fluorescence (MAMEF) with silver colloids in 96-well plates: Application to ultra fast and sensitive immunoassays, High Throughput Screening and drug discovery. *Journal of Immunological Methods*, *312*(1-2), 137-147.

Bae, J. H., Han, K. D., Ko, S. H., Yang, Y. S., Choi, J. H., Choi, K. M., Kwon, H. S., Won, K. C. (2022). Diabetes Fact Sheet in Korea 2021. *Diabetes & Metabolism Journal*, *46*(3). 417-426.

Barrett, A. H. & Myers, P. C. (1975). Microwave thermography: a method of detecting subsurface thermal patterns. *Bibliotheca Radiologica*, *6*, 456.

Benetti, F. J., Ballester, C., Sani, G., Doonstra, P. & Grandjean, J. (1995). Video-assisted coronary bypass surgery. *Journal of Cardiac Surgery*, *10*(6), 620-625.

Bindu, G., Lonappan A., Thomas, V., Aanandan, C. K., Mathew, K. T. & Abraham, S. J. (2006). Active microwave imaging for breast cancer detection. *Progress in Electromagnetics Research-Pier*, *58*, 149-169.

Borgnakke, W. S. (2019). IDF diabetes atlas: diabetes and oral health- a two-way relationship of clinical importance. *Diabetes Research and Clinical Practice*, *157*, 107839.

Böttcher, J. P., Schanz, O., Garbers, C., Zaremba, A., Hegenbarth, S., Kurts, C., Beyer, M., Schultze, J. L., Kastenmüller, W., Rose-John, S. & Knolle, P. A. (2014). IL-6 trans-signaling dependent rapid development of cytotoxic $CD8^+$ T cell function. *Cell Reports*, *8*(5), 1318-1327.

Bruggink, S. C., Gussekloo, J., Berger, M. Y., Zaaijer, K., Assendelft, W. J. J., de Waal, M. W. M., Bavinck, J. N. B., Koes, B. W. & Eekhof, J. A. H. (2010). Cryotherapy with liquid nitrogen versus topical salicylic acid application for cutaneous warts in primary care: randomized controlled trial. *Canadian Medical Association Journal*, *182*(15), 1624-1630.

Bucci, O. M., Crocco, L., Scapaticci, R. & Bellizzi, G. (2016). On the design of phased arrays for medical applications. *Proceedings of the IEEE*, *104*(3), 633-648.

Cai, X. W. & Du, C. M. (2020). Thermal plasma treatment of medical waste. *Plasma Chemistry and Plasma Processing*, *41*(1), 1-46.

Carrafiello, G., Laganà, D., Mangini, M., Fontana, F., Dionigi, G., Boni, L. G., Rovera, F., Cuffari, S. & Fugazzola, C. (2008). Microwave tumors ablation: Principles, clinical applications and review of preliminary experiences. *International Journal of Surgery*, *6*, 65-69.

Celandroni, F., Longo, I., Tosoratti, N., Giannessi, F., Ghelardi, E., Salvetti, S., Baggiani, A. & Senesi, S. (2004). Effect of microwave radiation on Bacillus subtilis spores. *Journal of Applied Microbiology*, *97*(6), 1220-1227.

Chen, C., Zhao, X. L., Li, Z. H., Zhu, Z. G., Qian, S. H. & Flewitt, A. J. (2017). Current and emerging technology for continuous glucose monitoring. *Sensors.*, *17*(1), 182.

Chen, S. T., Chiou, S. H. & Wang, K. T. (1991). Enhancement of Chemical Reactions by Microwave Irradiation. *Journal of The Chinese Chemical Society*, *38*(1), 85-91.

Choi, H. J., Naylon, J., Luzio S., Beutler, J., Birchall, J., Martin C. & Porch, A. (2015). Design and in vitro interference test of microwave non-invasive blood glucose monitoring sensor. *IEEE Transactions on Microwave Theory and Techniques*, *63*(10), 3016-3025.

Chong, C. C. N., Lee, K. F., Chu, C. M., Chan, A. W. H., Wong, J., Chan, S. L., Lok, H. T., Fung, A. K. Y., Fong, A. K. W., Cheung, Y. S., Yu, S. C. H., Johnson, P. & Lai, P. B. S. (2018). Microwave ablation provides better survival than liver resection for hepatocellular carcinoma in patients with borderline liver function: application of ALBI score to patient selection. *HPB*, *20*(6), 546-554.

Collins, C. M., Liu, W., Swift, B. J. & Smith, M. B. (2010). Combination of optimized transmit arrays and some receive array reconstruction methods can yield homogeneous images at very high frequencies. *Magnetic Resonance in Medicine*, *54*(6), 1327-1332.

Collins, J. M. & Leadbeater, N. E. (2007). Microwave energy: a versatile tool for the biosciences. *Organic & Biomolecular Chemistry*, *5*(8), 1141-1150.

Conner, T., Arguello, R., Gyory, T. & Hauser, R. (2022). Real-world symptom severity reported by genomically-tested gastric cancer patients in the united states between 2014-2017. *Value in Health*, *25*(7), S547-S547.

De Pretto, L. R., Yoshimura, T. M., Ribeiro, M. S. & de Freitas, A. Z. (2016). Optical coherence tomography for blood glucose monitoring in vitro through spatial and temporal approaches. *Journal of Biomedical Optics*, *21*(8), 086007.

de Waha, S., Allali, A., Büttner, H. J., Toelg, R., Geist, V., Neumann, F. J., Khattab, A. A., Richardt, G. & Abdel-Wahab, M. (2016). Rotational atherectomy before paclitaxel-eluting stent implantation in complex calcified coronary lesions: Two-year clinical outcome of the randomized ROTAXUS trial. *Catheterization and Cardiovascular Interventions*, *87*(4), 691-700.

Defronzo, R. A., Bonadonna, R. C. & Ferrannini, E. (1992). Pathogenesis of NIDDM: a balanced overview. *Diabetes Care*, *15*(3), 318-368.

Di Benedetto, F., Estatico, C., Nagy, J. G. & Pastorino, M. (2008). Numerical linear algebra for nonlinear microwave imaging. *Electronic Transactions on Numerical Analysis*, *33*, 105-125.

Edwards, K., Martin, B., Gilmore, C. & Jeffrey, I. (2023). Deep-learning enabled uncertainty estimation applied to experimental near-field microwave imaging of 2-D dielectric cylinders. *2023 IEEE USNC-URSI Radio Science Meeting, Joint with AP-S Symposium*, 37-38.

Elahi, M. A., O'Loughlin, D., Lavoie, B. R., Glavin, M., Jones, E., Fear, E. C. & O'Halloran, M. (2018). Evaluation of image reconstruction algorithms for confocal microwave imaging: application to patient data. *Sensors*, *18*(6), 1678.

Elmoslemany, A. M., Ghamry, H. I., Awad, A. A., EL-Kholy, R. I., Almami, I. S. M., Alyamani, N. M. & Zedan, A. M. G. (2023). Liver tissues oxidative status, epigenetic and molecular characteristics in rats administered magnetic and microwave treated water. *Scientific Reports*, *13*, 1

Erdélyi, M. & Gogoll, A. (2003). Rapid microwave promoted sonogashira coupling reactions on solid phase. *Journal of Organic Chemistry*, *68*(16), 6431-6434.

Estatico, C., Fedeli, A., Pastorino, M. & Randazzo, A. (2018). Comparison between conjugate gradient and landweber based regularization approaches in L^p banach spaces for microwave tomography. *2018 2nd URSI Atlantic Radio Science Meeting (AT-RASC), Gran Canaria, Spain*, 1-1.

Evans, S. S., Repasky, E. A. & Fisher, D. T. (2015). Fever and the thermal regulation of immunity: The immune system feels the heat. *Nature Reviews Immunology*, *15*(6), 335-349.

Fallahi, H. & Prakash, P. (2018). Antenna designs for microwave tissue ablation. *Critical Reviews in Biomedical Engineering*, *46*(6), 495-521.

Fraser, S., Land, D. & Sturrock, R. D. (1987). Microwave thermography—an index of inflammatory joint disease. *British Journal of Rheumatology*, *26*(1), 379.

Gabriel, S., Lau, R. W. & Gabriel, C. (1996). The dielectric properties of biological tissues: III. Parametric models for the dielectric spectrum of tissues. *Physics in Medicine and Biology*, *41*(11), 2271-2293.

Gardner, R. A., Vargas, H. I., Block, J. B., Vogel, C. L., Fenn, A. J., Kuehl, G. V. & Doval, M. (2002). Focused Microwave Phased Array Thermotherapy for Primary Breast Cancer. *Annals of Surgical Oncology*, *9*(4), 326-332.

Gedye, R., Smith, F., Westaway, K., Baldisera, L., Laberge, L. & Rousell, J. (1986 a). The use of microwave ovens for rapid organic synthesis. *Tetrahedron Letters*, *27*(3), 279-282.

George, M. S. & Aston-Jones, G. (2010). Noninvasive techniques for probing neurocircuitry and treating illness: vagus nerve stimulation (VNS), transcranial magnetic stimulation (TMS) and transcranial directcurrent stimulation (tDCS). *Neuropsychopharmacol*, *35*(1), 301-316.

Giakoumakis, G., Politi, D. & Sidiras, D. (2021). Medical waste treatment technologies for energy, fuels, and materials production: a review. *Energies*, *14*(23), 8065.

Giguere, R. J., Bray, T. L., Duncan, S. M. & Majetich G. (1986 b). Application of commercial microwave ovens to organic synthesis. *Tetrahedron Letters*, *27*(41), 4945-4948.

Gong, Z., Chen, Y. F., Ding, Y. H. & Zhang, H. (2024). Perspective: microwave medical imaging using space-time-frequency a priori knowledge for health monitoring. *IEEE Journal of Electromagnetics RF and Microwaves in Medicine and Biology*, *8*(1), 2-14.

Guo, B., Wang, Y., Li, J., Stoica, P. & Wu, R. (2006). Microwave imaging via adaptive beamforming methods for breast cancer detection. *Journal of Electromagnetic Waves and Applications*, *20*(1), 53-63.

Hagon, W., Hagon, J., Noble, G., Brenton-Rule, A., Stewart, S. & Bristow, I. (2023). Microwave therapy for the treatment of plantar warts. *Journal of Foot and Ankle Research*, *16*(1), 37.

Hancock, C. P., Dharmasiri, N., White, M. & Goodman, A. M. (2013). The design and development of an integrated multi-functional microwave antenna structure for biological applications. *IEEE Transactions on Microwave Theory and Techniques*, *61*(5), 2230-2241.

Harid, V., Kim, H., Li, B. Z. & Lei, T. (2023). A method for non-destructive microwave focusing for deep brain and tissue stimulation. *Plos One*, *18*(2), e0278765.

Henriksson, T., Sahebdiyan, S., Manus, R., Brunner, C., Kellermair, L., Guger, M., Struhal, W., Fuchs, P., Reichl, J., Stelzer, A., Semenoy, S., Pearce, S. & Vosko, M. R. (2022). Human brain imaging by electromagnetic tomography: a mobile brain scanner for clinical settings. *2022 16th European Conference on Antennas and Propagation (EUCAP)*, 1-5.

Hussein, M., Awwad, F., Jithin, D., El Hasasna, H., Athamneh, K. & Iratni, R. (2019). Breast cancer cells exhibits specific dielectric signature in vitro using the open-ended coaxial probe technique from 200MHz to 13.6GHz. *Scientific Reports*, *9*, 4681.

Ibitoye, Z. A., Nwoye, E. O., Aweda, M. A., Oremosu, A. A., Annunobi, C. C. & Akanmu, O. N. (2015). Optimization of dual slot antenna using floating metallic sleeve for microwave ablation. *Medical Engineering & Physics*, *37*(4), 384-391.

Ilyas, S., Srivastava, R. R. & Kim, H. (2020). Disinfection technology and strategies for COVID-19 hospital and bio-medical waste management. *Science of the Total Environment*, *749*, 141652.

Issels, R. D. (2008). Hyperthermia adds to chemotherapy. *European Journal of Cancer*, *44*(17), 2546-2554.

Ito, K. & Saito, K. (2011). Development of Microwave Antennas for Thermal Therapy. *Current Pharmaceutical Design*, *17*(22), 2360-2366.

Jacobsen, S., Stauffer, P. R. & Neuman, D. G. (2000). Dual-mode antenna design for microwave heating and noninvasive thermometry of superficial tissue disease. *IEEE Transactions on Biomedical Engineering*, *47*(11), 1500-1509.

Jemal, A., Bray, F., Center, M. M., Ferlay, J., Ward, E. & Forman, D. (2011). Global cancer statistics. *CA-A Cancer Journal for Clinicians*, *61*(2), 69-90.

Jeong, J. & Son, S. H. (2023). Body boundary measurement using multiple line lasers for a focused microwave thermotherapy system: a proof-of-concept study. *Sensors*, *23*(17), 7438.

Kandwal, A., Li, J. Z., Igbe, T., Liu, Y. H., Das, R., Kanaujia, B. K. & Nie, Z. D. (2022). Young's double slit method-based higher order mode surface plasmon microwave antenna sensor: modeling, measurements, and application. *IEEE Transactions on Instrumentation and Measurement*, *71*, 8006711.

Kandwal, A., Liu, L. W., Igbe, T., Li, J. Z., Liu, Y. H., Das, R., Kanaujia, B. K., Wang, L. & Nie, Z. D. (2021). A novel method of using bifilar spiral resonator for designing thin robust flexible glucose sensors. *IEEE Transactions on Instrumentation and Measurement,* 70, 8004810.

Kapp, D. S. (1996). Efficacy of adjuvant hyperthermia in the treatment of superficial recurrent breast cancer: Confirmation and future directions. *International Journal of Radiation Oncology Biology Physics*, *35*(5), 1117-1121.

Karacolak, T., Moreland, E. C. & Topsakal, E. (2013). Colecole model for glucose-dependent dielectric properties of blood plasma for continuous glucose monitoring. *Microwave and Optical Technology Letters*, *55*(5), 1160-1164.

Karampatzakis, A., Kühn, S., Tsanidis, G., Neufeld, E., Samaras, T. & Kuster, N. (2013). Heating characteristics of antenna arrays used in microwave ablation: A theoretical parametric study. *Computers in Biology and Medicine*, *43*(10), 1321-1327.

Kastberger, G. & Stachl, R. (2003). Infrared imaging technology and biological applications. Behavior Research *Methods Instruments & Computers*, *35*(3), 429-439.

Kawai, H., Ito, K., Takahashi, M., Saito, K., Ueda, T., Saito, M., Ito, H., Osada, H., Koyanagi, Y. & Ogawa, K. (2006). Simple modeling of an abdomen of pregnant women and its application to SAR estimation. *IEICE Transactions on Communications*, *E89B*(12), 3401-3410.

Kimura, U., Takeuchi, K., Kinoshita, A., Takamori, K. & Suga, Y. (2014). Long-pulsed 1064-nm neodymium: yttrium-aluminum-garnet laser treatment for refractory warts on hands and feet. *Journal of Dermatology*, *41*(3), 252-257.

Kiourti, A., Abbosh, A. M., Athanasiou, M., Bjorninen, T., Eid, A., Furse, C., Ito, K., Lazzi, G., Manoufali, M., Pastorino, M., Tentzeris, M. M., Tisdale, K., Topsakal, E., Ukkonen, L., Whittow, W. G., Zhang, H. N. & Nikita, K. S. (2022). Next-generation healthcare: enabling technologies for emerging bioelectromagnetics applications. *IEEE Open Journal of Antennas and Propagation*, *3*, 363-390.

Kwok, C. S., Holland, R. & Gibbs, S. (2011). Efficacy of topical treatments for cutaneous warts: a meta-analysis and pooled analysis of randomized controlled trials. *British Journal of Dermatology*, 165(2), 233-246.

Landau, C., Currier, J. W., Haudenschild, C. C., Minihan, A. C., Heymann, D. & Faxon, D. P. (1994). Microwave balloon angioplasty effectively seals arterial dissections in an atherosclerotic rabbit model. *Journal of the American College of Cardiology*, *23*(7), 1700-1707.

Lara, J. E., Vera, A., Leija, L. & Gutiérrez, M. I. (2015). Modeling of electromagnetic and temperature distributions of an interstitial coaxial-based choked antenna for hepatic tumor microwave ablation. *12th International Conference on Electrical Engineering, Computing Science and Automatic Control (CCE)*, 1-5.

Levine, G. N. (2012). 2011 ACCF/AHA/SCAI guideline for percutaneous coronary intervention. *Circulation*, *125*(8), E412-E412.

Li, X. L. & Li, C. W. (2016). Improved CEEMDAN and PSO-SVR modeling for near-infrared noninvasive glucose detection. *Computational and Mathematical Methods in Medicine*, *2016*, 8301962.

Liu, G. D. & Zhang, Y. R. (2011). Three-dimensional microwave-induced thermo-acoustic imaging for breast cancer detection. *ACTA Physica Sinica*, *60*(7), 074303.

Liu, J., Li, H. R., Liu, Z. B., Meng, X. W., He, Y. & Zhang, Z. T. (2022). Study on the process of medical waste disinfection by microwave technology. *Waste Management*, *150*, 13-19.

Liu, Y. J., Qiao, A. K., Nan, Q. & Yang, X. Y. (2006). Thermal characteristics of microwave ablation in the vicinity of an arterial bifurcation. *International Journal of Hyperthermia*, *22*(6), 491-506.

Losoya-Leal, A., Camacho-León, S., Dieck-Assad, G. & Martínez-Chapa, S. O. (2012). State of the art and new perspectives in non-invasive glucose sensors. *Revista Mexicana de Ingeniería Biomédica*, *33*(1), 41-52.

Lü, X. F., Zhang, T. L., Xiao, F., Li, G. & Wang, Y. (2016). Noninvasive blood glucose analysis based on near-infrared reflectance spectroscopy. *Spectroscopy and Spectral Analysis*, *36*(7), 2312-2317.

Lubner, M. G., Brace, C. L., Hinshaw, J. L. & Lee, F. T. (2010). Microwave tumor ablation: Mechanism of action, clinical results, and devices. *Journal of Vascular and Interventional Radiology*, *21*(8), 192-203.

Lv, Y., Hu, G. Y., Liang, J. Y., Chen, X., Chen, B., Zhao, T. K., Lu, X. Y., Wang, B., Yuan, W. J. & Li, Y. M. (2019). Study on microwave sterilization technology of humidifier in central air conditioning system. *Building and Environment*, *160*, 106220.

MacDonald, A., Land, D. & Sturrock, R. (1994). Microwave thermography as a noninvasive assessment of disease activity in inflammatory arthritis. *Clinical Rheumatology*, *13*(4), 58992.

Makaroff, S. N., Nguyen, H., Meng, Q. L., Lu, H. B., Nummenmaa, A. R. & Deng, Z. D. (2023). Modeling transcranial magnetic stimulation coil with magnetic cores. *Journal of Neural Engineering*, *20*(1), 016028.

Meaney, P. M., Fanning, M. W., Li, D., Poplack, S. P. & Paulsen, K. D. (2000). A clinical prototype for active microwave imaging of the breast. *IEEE Transactions on Microwave Theory and Techniques*, *48*(11), 1841-1853.

Mustafa, S., Abbosh, A. M. & Nguyen, P. T. (2014). Modeling human head tissues using fourth-order Debye model in convolution-based three-dimensional finite-difference time-domain. *IEEE Transactions on Antennas and Propagation*, *62*(3), 1354-1361.

Nakabayashi, K., Sunaga, D., Kaneko, N., Matsui, A., Tanaka, K., Ando, H. & Shimizu, M. (2019). Simple percutaneous coronary interventions using the modification of complex coronary lesion with excimer laser. *Cardiovascular Revascularization Medicine*, *20*(4), 293-302.

Narang, R., Mohammadi, S., Ashani, M. M., Sadabadi, H., Hejazi, H., Zarifi, M. H. & Sanati-Nezhad, A. (2018). Sensitive, real-time and non-intrusive detection of concentration and growth of pathogenic bacteria using microfluidic-microwave ring resonator biosensor. *Scientific Reports*, *8*, 15807.

Navaei, M., Rezaei, P. & Kiani, S. (2023). Measurement of low-loss aqueous solutions permittivity with high detection accuracy by a contact and free label resonance microwave sensor. *International Journal of Communication Systems*, *36*(5), 5417.

Nepa, P. & Buffi, A. (2017). Near-field-focused microwave antennas: near-field shaping and implementation. *IEEE Antennas and Propagation Magazine*, *59*(3), 42-53.

Nielsen, O. S., Horsman, M. & Overgaard, J. (2001). A future for hyperthermia in cancer treatment. *European Journal of Cancer*, *37*(13), 1587-1589.

Noghanian, S., Sabouni, A. & Pistorius, S. (2006). A numerical approach to microwave imaging based on genetic algorithm optimization. *Health Monitoring and Smart Nondestructive Evaluation of Structural and Biological Systems V*, *6177,* 61771B.

Ozgun, O. & Kuzuoglu, M. (2018). Coordinate transformation aided finite element method for contour detection of breast tumors in microwave imaging. *International Journal for Numerical Methods in Biomedical Engineering*, *34*(10), e3124.

Pai, P. P., De, A. & Banerjee, S. (2018). Accuracy enhancement for noninvasive glucose estimation using dual-wavelength photoacoustic measurements and kernel-based calibration. *IEEE Transactions on Instrumentation and Measurement*, *67*(1), 126-136.

Park, H. S. & Choi, W. S. (2008). Pulsed dye laser treatment for viral warts: a study of 120 patients. *Journal of Dermatology, 35*(8), 491-498.

Poulin, E. C., Thibault, C., Descoteaux, J. G. & Cote, G. (1995). Partial laparoscopic splenectomy for trauma: technique and case report. *Surgical Laparoscopy Endoscopy & Percutaneous Techniques*, *5*(4), 306-310.

Pramanik, B. N., Mirza, U. A., Ing, Y. H., Liu, Y. H., Bartner, P. L., Weber, P. C. & Bose, M. K. (2002). Microwave-enhanced enzyme reaction for protein mapping by mass spectrometry: A new approach to protein digestion in minutes. *Protein Science*, *11*(11), 2676-2687.

Purkait, K. & Datta, A. N. (1996). An improved form of iterative reconstruction algorithm for first order and second order microwave image reconstruction. *Indian Journal of Pure & Applied Physics*, *34*(6), 420-424.

Reimann, C. H. N., Bazrafshan, B., Hübner, F., Schmidt, S., Schüler, M., Panahi, B., Kaltenbach, B., Polkowski, C., Vogl, T. J. & Jakoby, R. (2019). Dielectric contrast between normal and tumor ex-vivo human liver tissue. *IEEE Access*, *7*, 164113-164119.

Ring, E. F., Dieppe, P. A. & Bacon, P. A. (1981). Thermographic assessment of inflammation and anti-inflammatory drugs in osteoarthritis. *British Journal of Clinical Practice*, *35*(7-8), 263-264.

Rosario, D. J., Phillips, J. T. & Chapple, C. R. (2007). Durability and cost-effectiveness of transurethral needle ablation of the prostate as an alternative to transurethral resection of the prostate when alpha-adrenergic antagonist therapy fails. *Journal of Urology*, *177*(3), 1047-1051.

Rosen, A., Walinsky, P., Smith, D., Shi, Y., Kosman, Z., Martinez, A., Rosen, H., Sterzer, F., Mawhinney, D., Presser, A., Chou, J. S., Goth, P. & Lowery, G. (1989 a). Microwave balloon angioplasty. *IEEE 19th Eur. Microwave Conference*, 913-917.

Rosen, A., Walinsky, P., Smith, D., Shi, Y., Kosman, Z., Martinez., A. Rosen, H., Sterzer, F., Mawhinney, D., Presser, A., Chou, J. S., Goth, P. & Lowery, G. (1989 b). Percutaneous transluminal microwave catheter angioplasty. *MTT-S International Microwave Symposium Digest*, *1*, 167-170.

Rossmanna, C. & Haemmerich, D. (2014). Review of temperature dependence of thermal properties, dielectric properties, and perfusion of biological tissues at hyperthermic and ablation temperatures. *Critical Reviews in Biomedical Engineering*, *42*(6), 467-492.

Ruiz, A. Y., Cavagnaro, M. & Crocco, L. (2023). An effective framework for deep-learning-enhanced quantitative microwave imaging and its potential for medical applications. *Sensors*, *23*(2), 643.

Ryu, T., Takami, Y., Wada, Y., Hara, T., Sasaki, S. & Saitsu, H. (2019). Hepatic resection versus operative microwave ablation for single hepatocellular carcinoma ≤5 cm: A propensity score-matched analysis. *Surgery*, *166*(3), 254-262.

Saito, K., Hiroe, A., Kikuchi, S., Takahashi, M. & Ito, K. (2006). Estimation of heating performances of a coaxial-slot antenna with endoscope for treatment of bile duct carcinoma. *IEEE Transactions on Microwave Theory and Techniques*, *54*(8), 3443-3449.

Saleem, B., Brown, A. K., Keen, H., Nizam, S., Freeston, J., Wakefield, R., Karim, Z., Quinn, M., Hensor, E., Conaghan, P. G. & Emery, P. (2011). Should imaging be a component of rheumatoid arthritis remission criteria? A comparison between traditional and modified composite remission scores and imaging assessments. *Annals of the Rheumatic Diseases*, *70*(5), 792-798.

Salisbury, R. S., Parr, G., De Silva, M., Hazleman, B. L. & Page-Thomas, D. P. (1983). Heat distribution over normal and abnormal joints: thermal pattern and quantification. *Annals of the Rheumatic Diseases*, *42*(5), 494-499.

Sandoval, W. N., Arellano, F., Arnott, D., Raab, H. Vandlen, R. & Lill, J. R. (2007). Rapid removal of N-linked oligosaccharides using microwave assisted enzyme catalyzed deglycosylation. *International Journal of Mass Spectrometry*, *259*(1-3), 117-123.

Santagada, V., Fiorino, F., Perissutti, E., Severino, B., De Filippis, V., Vivenzio, B. & Caliendo, G. (2001). Microwave-enhanced solution coupling of the α, α-dialkyl amino acid, Aib. *Tetrahedron Letters*, *42*(31), 5171-5173.

Sapareto, S. A. & Dewey, W. C. (1984). Thermal dose determination in cancer therapy. *International Journal of Radiation Oncology Biology Physics*, *10*,(6), 787-800.

Scapaticci, R., Lopresto, V., Pinto, R., Cavagnaro, M. & Crocco, L. (2018). Monitoring thermal ablation via microwave tomography: an ex vivo experimental assessment. *Diagnostics*, *8*(4), 81.

Schiavon, G., Capone, G., Frize, M., Zaffagnini, S., Candrian, C. & Filardo, G. (2022). Infrared thermography for the evaluation of inflammatory and degenerative joint diseases: a systematic review. *Cartilage, 13*(2), 1790S-1801S.

Seo, T., Oh, S., Jung, D., Huh, Y., Cho, J. & Kwon, Y. (2018). Noninvasive brain stimulation using a modulated microwave signal. *Journal of Electromagnetic Engineering and Science*, *18*(1), 70-72.

Serruys, P. W., Katagiri, Y. & Onuma, Y. (2017). Shaking and breaking calcified plaque: Lithoplasty, a breakthrough in interventional armamentarium. *JACC-Cardiovascular Imaging*, *10*(8), 907-911.

Shih, W. C., Bechtel, K. L. & Rebec, M. V. (2015). Noninvasive glucose sensing by transcutaneous Raman spectroscopy. *Journal of Biomedical Optics*, *20*(5), 051036.

Simon, C. J., Dupuy, D. E. & Mayo-Smith, W. W. (2005). Microwave ablation: principles and applications. *Radiographics*, *25*, 69-83.

Simonov, N. & Son, S. H. (2021). Overcoming Insufficient Microwave Scattering Data in Microwave Tomographic Imaging. *IEEE Access*, *9*, 111231-111237.

Stauffer, P. R., Leoncini, M., Manfrini, V., Gentili, G. B., Diederich, C. J. & Bozzo, D. (1995). Dual concentric conductor radiator for microwave hyperthermia with improved field uniformity to periphery of aperture. *IEICE Transactions on Communications*, *E78B*(6), 826-835.

Sterling, J. C., Gibbs, S., Haque Hussain, S. S., Mustapa, M. F. M. & Handfield-Jones, S E. (2014). British Association of Dermatologists guidelines for the management of cutaneous warts 2014. *British Journal of Dermatology*, *171*(4), 696-712.

Sterzer, F. (1991). Localized heating of deep-seated tissues using microwave balloon catheters. *Electro International*, 7-8.

Sterzer, F., Mendecki, J., Mawhinney, D. D., Friedenthal, E. & Melman, A. (2000). Microwave treatments for prostate disease. *IEEE Transactions on Microwave Theory and Techniques*, *48*(11), 1885-1891.

Stovsky, M. D., Griffiths, R. I. & Duff, S. B. (2006). A clinical outcomes and cost analysis comparing photoselective vaporization of the prostate to alternative minimally invasive therapies and transurethral prostate resection for the treatment of benign prostatic hyperplasia. *Journal of Urology*, *176*(4), 1500-1506.

Tai, J. J., Qiao, D., Huang, X., Hu, H., Li, W. Z., Liang, X. L., Zhang, F. M., Lu, Y. B. & Zhang, H. (2023). Structural Property, Immunoreactivity and gastric digestion characteristics of glycated parvalbumin from mandarin fish (siniperca chuaisi) during microwave-assisted maillard reaction. *Foods*, *12*(1), 52.

Tammam, E., Said, A. M., Ibrahim, A. A. & Galal, A. I. A. (2020). About the interstitial microwave cancer ablation: principles, advantages, and challenges. *IEEE Access*, *8*, 49685-49694.

Tang, J. (2015). Unlocking Potentials of Microwaves for Food Safety and Quality. *Journal of Food Science*, *80*, 1776-1793.

Torre, L. A., Bray, F., Siegel, R. L., Ferlay, J., Lortet-Tieulent, J. & Jemal, A. (2015). Global cancer statistics, *CA-A Cancer Journal for Clinicians*, *65*(2), 87-108.

Tosti, A. & Piraccini, B. M. (2001). Warts of the nail unit: surgical and nonsurgical approaches. *Dermatologic Surgery*, *27*(3), 235-239.

Vaiman, M., Shilco, P., Roitblat, Y., Padilla-Raygoza, N., Leit, A., Kavin, A., Schonberger, E., Nehuliaieva, L., Buchris, N. & Shterenshis, M. (2011). Screening Procedure for Hemihypertrophy: Preliminary Results of International Multicenter Prospective Study. *Central Asian Journal of Global Health*, *8*(1), 336.

van den Tempel, N., Horsman, M. R. & Kanaar, R. (2016). Improving efficacy of hyperthermia in oncology by exploiting biological mechanisms. *International Journal of Hyperthermia*, *32*(4), 446-454.

Vélez, P., Muñoz-Enano, J., Ebrahimi, A., Herrojo, C., Paredes, F., Scott, J., Ghorbani, K. & Martin, F. (2021). Single-frequency amplitude-modulation sensor for dielectric characterization of solids and microfluidics. *IEEE Sensors Journal*, *21*(10), 12189-12201.

Venkataraman, J. & Freer, B. (2011). Feasibility of non-invasive blood glucose monitoring: *In-vitro* measurements and phantom models. *2011 IEEE International Symposium on Antennas and Propagation (APSURSI)*, 603-606.

Vernon, C. C., Hand, J. W., Field, S. B., Machin, D., Whaley, J. B., vanderZee, J., vanPutten, W. L. J., vanRhoon, G. C., vanDijk, J. D. P., Gonzalez, D. G., Liu, F. F., Goodman, P. & Sherar, M. (1996). Radiotherapy with or without hyperthermia in the treatment of superficial localized breast cancer: Results from five randomized controlled trials. *International Journal of Radiation Oncology Biology Physics*, *35*(4), 731-744.

Villena, G. W., Mobashsher, A. T. & Abbosh, A. (2019). The progress of glucose monitoring—A review of invasive to minimally and non-invasive techniques, devices, and sensors. *Sensors*, *19*(4), 800.

Walinsky, P., Rosen, A., Martinez-Hernandez, A., Smith, D. L., Nardone, D. D. & Brevette, B. (1991). Microwave balloon angioplasty. *Journal of Invasive Cardiology*, *3*(3), 152-156.

Wang, L. L. (2023). Microwave Imaging and Sensing Techniques for Breast Cancer Detection. *Micromachines*, *14*(7), 1462.

Wang, M. C., Crocco, L. & Cavagnaro, M. (2021 a). On the design of a microwave imaging system to monitor thermal ablation of liver tumors. *IEEE Journal of Electromagnetics RF and Microwaves in Medicine and Biology*, *5*(3), 231-237.

Wang, M. C., Scapaticci, R., Cavagnaro, M. & Crocco, L. (2021 b). Towards a microwave imaging system for continuous monitoring of liver tumor ablation: Design and in-silico validation of an experimental setup. *Diagnostics*, *11*(5), 866.

Wang, T., Zhao, G. & Qiu, B. (2015). Theoretical evaluation of the treatment effectiveness of a novel coaxial multi-slot antenna for conformal microwave ablation of tumors. *International Journal of Heat and Mass Transfer*, *90*, 81-91.

Wiesinger, F., Van De Moortele, P. F., Adriany, G., De Zanche, N., Ugurbil, K. & Pruessmann, K. P. (2004). Parallel imaging performance as a function of field strength—an experimental investigation using electrodynamic scaling. *Magnetic Resonance in Medicine*, *52*(5), 953-964.

Wu, F. (2016). Heat-based tumor ablation: role of the immune response. *Advances in Experimental Medicine and Biology, 880*, 131-153.

Xi, Z. J., Wang, X. Y., Ye, K. & Wang, X. (2023). Performance evaluation of focused microwave brain hyperthermia guided by microwave-induced thermoacoustic tomography. *IEEE Journal of Electromagnetics RF and Microwaves in Medicine and Biology, 7*(4), 383-391.

Xiao, F. C. & Yabe, H. (1998). Microwave imaging of perfectly conducting cylinders from real data by micro genetic algorithm coupled with deterministic method. *IEICE Transactions on Electronics*, *E81C*(12), 1784-1792.

Xu, L. F., Dong, K., Zhang, Y. M. & Li, H. X. (2020). Comparison and analysis of several medical waste treatment technologies. *IOP Conference Series: Earth and Environmental Science*, *615*(1), 012031.

Yamada, M., Iijima, Y., Seo, M., Hino, S., Sano, M., Sakagami, H., Horie, N. & Kaneko, T. (2023). Cancer chemotherapy-associated pigmentation of the oral mucosa. *In Vivo*, *37*(4), 1880-1885.

Yang, C. X., Xiao, X., Hu, M. & Wang, Z, X. (2022). Non-invasive microwave blood glucose monitoring based on the diffusion limited aggregation earlobe model. *International Journal of RF and Microwave Computer-Aided Engineering*, *32*(12), e23443.

Yilmaz, T., Foster, R. & Hao, Y. (2014). Broadband tissue mimicking phantoms and a patch resonator for evaluating noninvasive monitoring of blood glucose levels. *IEEE Transactions on Antennas and Propagation*, *62*(6), 3064-3075.

Yilmaz, T., Foster, R. & Hao, Y. (2019). Radio-frequency and microwave techniques for non-invasive measurement of blood glucose levels. *Diagnostics*, *9*(1), 6.

Zamani, A., Abbosh, A. M. & Mobashsher, A, T. (2016). Fast frequency-based multistatic microwave imaging algorithm with application to brain injury detection. *IEEE Transactions on Microwave Theory and Techniques*, *64*(2), 653-662.

Zhang, Q. L. & Chang, J. M. (2023). Interstitial granulomatous drug reaction due to chemotherapy. *Clinical Cosmetic and Investigational Dermatology*, *16*, 3625-3628.

Zhang, Q. Y., Tian, Y., Duan, J. X., Gao, Z. Z. Wang, W. L. & Yan, S. (2020). 915 MHz microwave-assisted laparoscopic partial splenectomy: a case series. *Journal of Minimal Access Surgery*, *16*(4), 441-444.

Chapter 3

Microwave Irradiation for Food Drying

Zheng Wang
Wenke Zhao
Yaning Zhang*
and Bingxi Li
School of Energy Science and Engineering, Harbin Institute of Technology, Harbin, China

Abstract

Food drying as an initial part of food safety processes can be further developed with the assistance of microwave irradiation. This chapter provides a comprehensive introduction of microwave irradiation drying technology, emphasizing its significant role in various industries, especially in food processing. It covers fundamental principles, applications, technological advancements, and the challenges and limitations of this technology. This chapter evaluates traditional drying techniques and introduces microwave drying as a more efficient fourth-generation solution. It delves into the interaction of materials with the microwave electric field, exploring practical applications in food preservation and addressing the challenges in traditional drying methods. The potential and future prospects of microwave drying technology are also discussed, highlighting its innovative applications in different food categories and its roles in enhancing food quality and preservation.

Keywords: biomass, properties, structural, physical, chemical, compositions

---

* Corresponding Author's Email: ynzhang@hit.edu.cn.

In: Characteristics and Applications of Microwave Irradiation
Editors: Yaning Zhang and Wenke Zhao
ISBN: 979-8-89530-170-8

1. Introduction

The application of microwave irradiation technology in drying spans a wide array of industries and materials, with its role in the food industry being particularly crucial. This technology is instrumental in the rapid drying of various foods such as vegetables, meats, and seafood, effectively preserving their nutritional values and flavors. This chapter aims to provide an in-depth analysis of microwave radiation drying technology, encompassing its fundamental principles, areas of application, the latest technological advancements, as well as the challenges and limitations it faces. In the context of food preservation, drying processes are commonly used but are considered energy-intensive, as highlighted in the references. (Ratti, 2001). Studies (Clary et al. 2005; Zhang et al. 2003; Min et al. 2005) indicate that traditional thermal drying methods like air, vacuum, and freeze drying are less efficient during phases of reduced drying rates and can lead to unnecessary accumulation of heat over prolonged periods (Mousa and Farid, 2002).

Given the constraints of traditional drying techniques, researchers such as Vega-Mercado consider microwave technology as a more advanced, fourth-generation drying solution. This chapter further explores critical aspects of microwave radiation drying, including an analysis of the interaction between materials and the microwave electric field, as evidenced by the measurement of dielectric constants and dielectric loss factors. Beginning with the basic principles of microwave drying, the section comprehensively presents its applications in the food industry, particularly focusing on the operational mechanisms of microwave drying (Vega-Mercado et al. 2001). It then examines the latest developments in microwave drying technology, discussing how innovative approaches are addressing the challenges faced by traditional drying methods.

Despite being an emerging technology with its own set of challenges and limitations, this chapter methodically assesses these challenges to fully grasp the practical applicability of microwave drying technology. After thorough research, the chapter seeks to uncover the future prospects of microwave drying technology and the innovative strategies to surmount existing obstacles. Additionally, this section will categorize and detail the microwave irradiation drying process for different types of foods, including meats, fruits, vegetables, and seafood. Examples such as cooking and pasteurization will be used to demonstrate the application of microwave irradiation in food drying scenarios.

2. Microwave Irradiation Drying

2.1. Mechanism of Microwave Irradiation Drying

Microwaves, as part of the electromagnetic spectrum, are harnessed for food heating through two primary mechanisms: dipole relaxation and ionic conduction (as shown in Figure 1). Water molecules in food, with their dipolar nature, attempt to align with the rapidly changing electric fields, generating heat. This dielectric heating is the main effect for microwave heating of food. Additionally, the electric field causes ionic migration in salty foods, which also generates heat, forming the second crucial mechanism for microwave and radio frequency energy heating. Integrating these mechanisms, microwave heating offers efficient and distinctive advantages in food processing, bringing innovative solutions to the industry.

With a foundational understanding of the principles of microwave radiation drying, this article delves into the specifics, applications, technological advances, and challenges of microwave drying. A comprehensive analysis reveals the promising applications of microwave drying technologies in food processing. Various methods include microwave-assisted freeze drying, microwave-assisted vacuum drying, microwave-enhanced spouted bed drying, and microwave-assisted vacuum frying dehydration, microwave-assisted air drying, each with unique characteristics, providing diverse and optimized options for food drying.

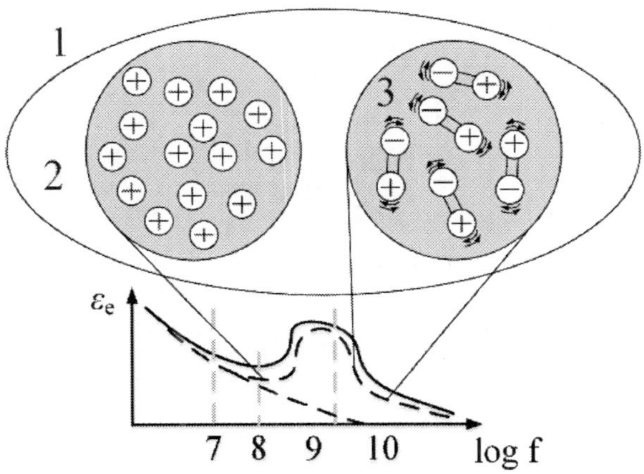

1 - Material boundary; 2 - Ionic conduction; 3 - Dipolar re-orientation.

Figure 1. Diagram of dipolar and ionic loss mechanisms (Metaxas, 1996).

2.2. Applications of Microwave Drying

2.2.1. Microwave-Assisted Freeze Drying

Microwave-assisted freeze drying (MFD) is a method that integrates microwave technology into the freeze-drying process. Similar in key aspects to traditional freeze drying, MFD uses microwaves to supply the necessary sublimation heat. This technique allows for uniform heating of the material under vacuum, significantly enhancing the sublimation rate. As shown in Figure 2, a classic microwave freeze dryer was proposed in 2010. During drying, a vacuum pump maintains a low-pressure environment of about 100 Pascal, while the cold trap temperature is set between -40 °C to -45 °C to facilitate vapor condensation. The microwave frequency is set at 2,450 MHz with an adjustable power range from 0 to 2,000 watts.

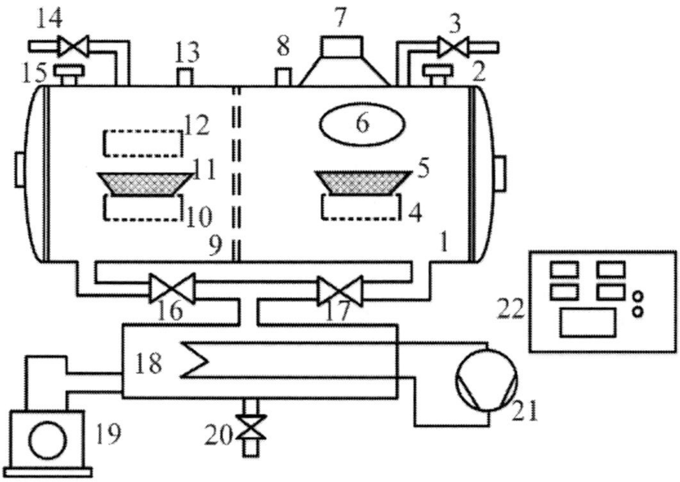

1-Microwave freeze drying chamber, 2-Optical fiber temperature sensor, 3-Vacuum breakage valve for MFD, 4- Sample supporting plate, 5-MFD Sample, 6,7-Microwave source, 8- Pressure sensor for MFD chamber, 9-Freeze drying chamber, 10,12-Heating plate, 11-FD sample, 13-Pressure sensor for FD chamber, 14- Vacuum breakage valve for FD, 15-Temperature sensor, 16-FD vacuum valve, 17- MFD vacuum valve, 18- Cold trap, 19-Vacuum pump, 20-Draining valve, 21-Refrigeration compressor, 22-Control system.

Figure 2. Diagram of a classic microwave freeze dryer (Jiang et al. 2010).

2.2.2. Microwave-Assisted Vacuum Drying

Microwave-assisted vacuum drying (MVD) technology harnesses the multifaceted advantages of microwave heating, such as rapid heating, high efficiency, excellent control, and optimal hygiene conditions (Min et al. 2000). This technique, by reducing the boiling point of water in a vacuum, enhances energy efficiency and minimizes scorch spots on the final product's surface. Another feature of MVD is its ability to increase product puffing, thereby improving the texture of the finished product (Zhang et al. 2007). As shown in Figure 3, within MVD experimental systems, microwave power can be finely adjusted and controlled via a programmable logic controller interface, ensuring a precise and efficient drying process. This system design not only improves product quality but also maintains consistency and stability in the processing phase (Cui et al. 2004).

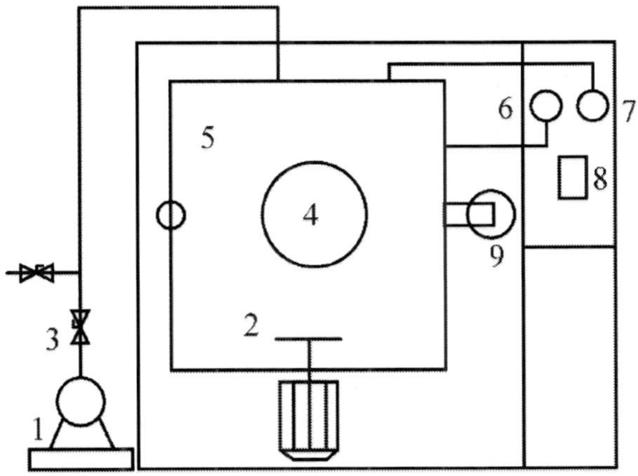

1-vacuum pump, 2-rotating tray, 3-valve, 4-observation window, 5-drying chamber, 6-infrared thermometer, 7-vacuum gauge, 8-magnetron power control unit, 9-magnetron.

Figure 3. Diagram of a classic microwave vacuum drying system.

2.2.3. Microwave-Enhanced Spouted Bed Drying

A key challenge in microwave drying is uneven heating due to irregular electromagnetic field distribution within the drying chamber. The combination of spouted bed drying technology is reckoned as an effective method to address this problem. A clasic microwave spouted bed dryer's schematic is depicted in Figure 4. This process involves injecting airflow into a bed of particles to maintain their suspension, thus achieving a more uniform and effective drying outcome (Lu et al. 2014).

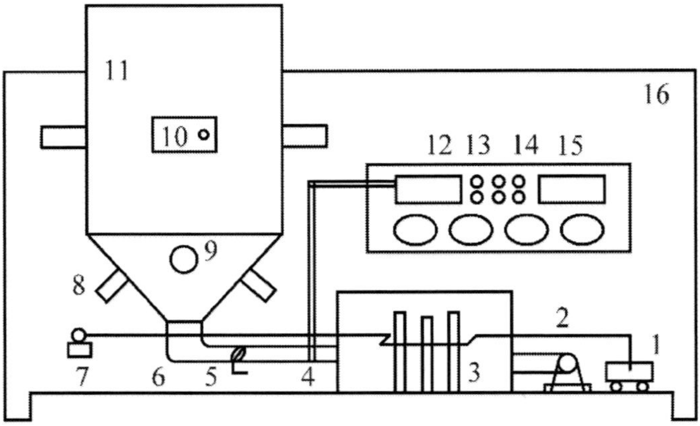

1-air compressor, 2-blower, 3-electric heating, 4-pressure detector, 5-valve, 6-nozzle, 7-peristaltic pump, 8-magnetron, 9-observation window, 10-operation door, 11-spouted bed, 12-air flow detector, 13-control box, 14-power regulator, 15-optical fiber measurement, 16-equipment support.

Figure 4. Diagram of a classic microwave spouted dryer.

2.2.4. Microwave-Assisted Vacuum Frying Dehydration

By employing Response Surface Methodology (RSM), optimization of vacuum-microwave drying time, frying temperature, and frying duration significantly affects key quality parameters of food, like moisture content, fat content, and breaking force. This optimization introduces an innovative technique in the food industry, combining vacuum frying with microwave-assisted drying. This method not only reduces fat content in fried products but also maintains quality, paving the way for healthier, better-tasting fried foods. However, further research is needed for industrial application, especially in model establishment, process optimization, hardware enhancement, and automated control systems.

Vacuum frying is a relatively new technique operating at pressures significantly lower than standard atmospheric pressure, ideally below 50 Torr (6.65 kPa). This low-pressure environment lowers the boiling points of oil and water, allowing for reduced frying temperatures, suitable for producing low-oil snacks from fruits, vegetables, and aquatic products, while enhancing various desired product qualities (Fan et al. 2005). A typical vacuum frying system, as shown in Figure 5, includes a 15-liter centrifuge capable of reaching up to 150 °C and 0.095 MPa vacuum, with a 5-liter oil capacity.

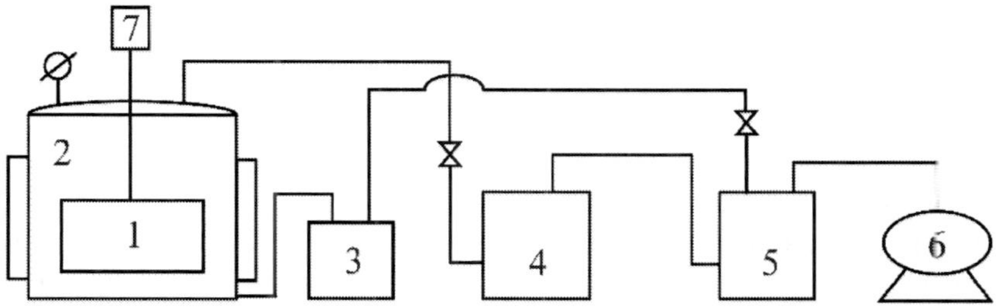

1-frying basket, 2-vacuum fryer, 3-store oilcan, 4-condenser, 5-segregator, 6-vacuum pump, 7-electric motor.

Figure 5. Diagram of a classic vacuum frying system.

2.2.5. Microwave-Assisted Air Drying

Microwave-assisted air drying (MAD) is a process that integrates hot air drying with microwave heating to enhance drying rates.

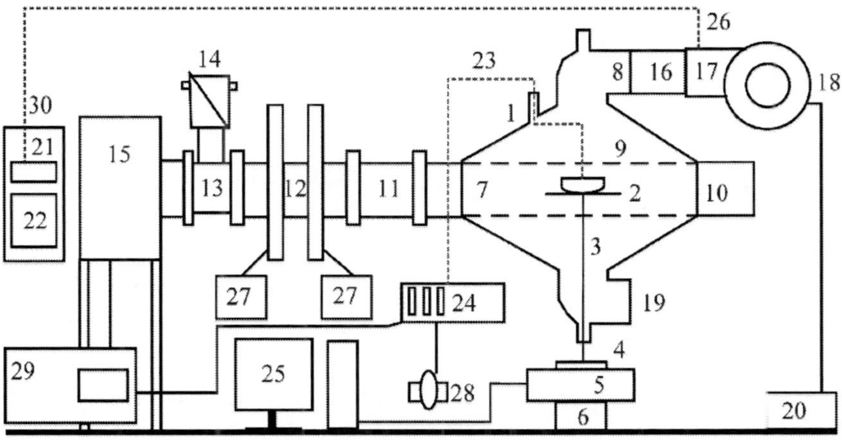

1-drying cavity, 2-polypropylene (pp) rotatable plate of ring shape covered with a Teflon mesh, 3-rotatable polypropylene (pp) stem, 4-roatable stainless steel plate, 5-rotatable electronic scale, 6-rotary system, 7-Microwave arrival in cavity (in Teflon), 8-Convective air arrival, 9-Protection metal grid against microwaves exit, 10-shunt carried out with aluminum foil, 11-waveguide with impedance adapter, 12-directing coupler, 13-waves circulator, 14-water trap, 15-magnetron, 16-air-flue pipe, 17-heating resistance, 18-ventilator, 19-opening for humid air exit, 20-air velocity variator, 21-PID regulator of temperature, 22-power electric meter, 23-fiber optical sensor, 24-data acquisition power station, 25-P.C., 26-PT100, 27-wattmeters, 28-electromagnetic ratemeter, 29-generator, 30-electronic box.

Figure 6. Diagram of a classic microwave–convective dryer (Koné et al. 2013).

As depicted in Figure 6, microwave heating can be utilized in three distinct stages of the drying process combined with air. Initially, microwave heating is applied at the start of the dehydration process for rapid internal heating. During the accelerated drying phase, a stable temperature profile is established to improve drying rates, forcing out steam due to enhanced drying efficiency, creating a porous structure known as "puffing" which further facilitates mass transfer of water vapor. In the final phase, when the drying rate slows down, moisture present at the core is expelled with the aid of microwave heating to remove bound water. This technique, combining the advantages of hot air drying and microwave heating, aims to expedite the drying process while maintaining product quality.

In food production, raw biomass with high moisture content often needs to be processed into low moisture content to meet the requirements for long-term storage and reduced volume and mass. If the biomass is put outside in a natural environment (in sunshine) or in a storage space, the (external) moisture contained in the biomass will be released quickly or gradually depending on the conditions, leading to unpredictable results. Therefore, using different types of microwave drying can meet the needs of various drying scenarios.

2.3. Advances in Microwave Drying Technologies

Microwave drying technology offers several advantages, including enhanced processing speed and the potential for uniform heating of materials. Typically, it achieves a more uniform heat distribution than conventional methods, reducing temperature gradients. This technology also boasts high energy efficiency, precise and rapid process control, minimal space usage, selective heating capability, and the potential for improved product quality, alongside beneficial chemical and physical effects (Mujumdar, 2006; Datta et al. 2014). However, one of its major challenges is the uneven temperature distribution caused by non-uniform microwave fields. As moisture content decreases, the efficiency of microwave energy absorption also drops, but product temperatures may continue to rise, occasionally causing overheating or charring (Bouraoui, 1994).

To overcome these limitations, researchers have explored other drying methods in combination with microwave drying. For example, studies show that integrating freeze drying with vacuum microwave drying can halve the drying time while maintaining good product quality (Duan et al. 2007). Similarly, the combination of hot air and microwave vacuum drying has been effective, indicating that integrating various drying technologies can effectively address some inherent challenges of microwave drying (Xu et al. 2004).

2.3.1. Microwave-assisted Freeze Drying

Microwave-assisted freeze drying (MFD) surpasses traditional freeze-drying methods in efficiency, featuring marked reductions in both drying time and energy consumption. This technique's primary benefits include speeding up the drying process, conserving energy, elevating the quality of the final product, and its versatile adaptability for a wide array of dried goods. As a highly effective dehydration method, MFD is particularly beneficial for various food types, such as seafood, solid soups, fruits, and vegetables, making it an invaluable asset in contemporary food processing (Ang et al. 1997).

2.3.2. Microwave-assisted Vacuum Drying

Beyond the conventional advantages of microwave radiation drying, Microwave-assisted vacuum drying (MVD) also enhances the sensory qualities of aquatic products, such as Bighead Carp fillets. By fine-tuning the microwave power and vacuum levels, it's possible to increase the puffiness and crispness of the product, thereby improving the sensory experience of the end product. The use of MVD is not confined to vegetables but can also be extended to other food items like banana and potato chips. Furthermore, the integration of MVD with other drying technologies is being extensively researched, broadening its application spectrum in the food processing industry.

2.3.3. Microwave-enhanced Spouted Bed Drying

Microwave-enhanced spouted bed drying (MSD) combines microwave technology with spouted bed drying, bringing numerous benefits to food processing. Particularly effective for producing healthier puffed snacks like potato chunks, MSD optimizes the puffing ratio, breaking strength, and rehydration rate during drying, offering a healthier alternative to traditional fried potato chips. MSD stands out for its faster drying rates and energy efficiency compared to other microwave drying techniques, especially microwave-assisted freeze drying (MFD). This feature not only boosts production efficiency but also positions MSD as an energy-saving drying option. Additionally, MSD enhances product puffiness, accelerates drying, and better preserves nutritional content, making it an advanced, practical technology in the food industry, ideal for creating healthy snacks and other puffed foods.

2.3.4. Microwave Vacuum Frying Dehydration Drying

Microwave Vacuum Frying Dehydration Drying is emerging as a potential food processing technique with multiple advantages. It significantly reduces fat content in food by incorporating microwave drying as a pre-treatment step before vacuum frying, paving the way for healthier fried products. For instance, in processing foods like potato chips, extending the microwave drying time results in reduced oil content during vacuum frying. Moreover, this method aids in maintaining the overall quality of food, enhancing color and texture while minimizing oil absorption. Experiments, such as with potato chips, show that products treated through vacuum microwave drying maintain a lower oil content and exhibit superior color and texture after vacuum frying. Furthermore, an increase in vacuum microwave drying time leads to a reduction in both moisture and fat content, thereby improving the overall quality of fried products. These benefits make the combination of vacuum microwave drying and vacuum frying a noteworthy innovation in the food processing industry.

2.3.5. Microwave-Assisted Air Drying

Microwave-assisted air drying technology showcases multiple advantages in the field of food drying. It significantly reduces the drying time of foods like fruits and vegetables compared to traditional hot air drying, thereby enhancing production efficiency. Moreover, it excels in preserving product quality, maintaining the color, texture, and other sensory qualities of food. This technology also improves the rehydration capacity of dried foods, making it easier to absorb moisture while retaining a pleasant texture. In terms of energy efficiency, microwave vacuum drying is more energy-efficient than microwave convective drying. Additionally, microwave-assisted air drying reduces energy consumption while improving product quality, demonstrating its environmentally friendly benefits. Practical applications across various food materials have proven its broad applicability, offering an efficient and energy-saving drying solution for the food industry.

2.4. Challenges and Limitations in Microwave Drying

Microwave drying technology in food processing has its limitations. A significant issue is the potential for overheating or burning, particularly at the corners or edges of food during the later drying stages. This challenge often arises from the difficulty in precisely controlling the final temperature in microwave drying, contrasting with hot air drying where product temperature doesn't exceed that of the hot air. At 2.45 GHz microwave frequency, limited penetration depth can pose challenges in large-scale operations, whereas Radio Frequency (RF) heating at 10–300 MHz offers deeper penetration, more suitable for large-scale applications.

Additionally, rapid mass transfer in microwave drying can lead to texture changes in food, known as "puffing." Whether this effect is desirable depends on the final product's requirements. Therefore, combining microwave with traditional drying methods can be an effective strategy to both enhance drying speed and maintain product quality. This combined approach enables better temperature control, preventing burning, and fully utilizing the rapid drying benefits of microwave technology.

Freeze drying, highly effective in dehydration, is perceived as costly and time-intensive, restricting production capacity and resulting in higher capital and energy expenditures. In microwave-assisted freeze drying, thermal runaway issues may arise due to inadequate interaction between microwave energy and ice, mainly stemming from partial melting in the frozen regions. Additionally, plasma discharge or arcing in industrial settings could degrade product quality or damage food items, with occurrences linked to the environmental pressure (Zhang et al. 2006). To mitigate arcing risks, it's advisable to use lower microwave powers in low-pressure environments. Regulating environmental pressure between low and medium levels can control plasma discharges effectively. Thus, environmental pressure is a crucial factor in managing and preventing plasma discharges, making microwave-assisted freeze drying a complex control challenge that necessitates meticulous operation and monitoring (Duan et al. 2010).

The process of removing moisture in the final stages of drying fruits and vegetables with high water content can be time-consuming. Microwave-assisted drying, as the final phase of hot air drying, effectively addresses this issue by demonstrating high thermal efficiency. Hot air drying in its later stages cannot speed up the diffusion process, hence not aiding in further moisture loss. While microwave-assisted hot air drying is efficient, it has its limitations. To achieve effective spouting of materials, sufficient hot air velocity is necessary. The use of high-speed hot air results in difficulty in efficiently recovering and utilizing thermal and residual heat. This loss of thermal energy is a notable downside of this technology, potentially increasing overall energy consumption and influencing the energy efficiency of the drying process.

3. Types of drying food with microwave radiation

3.1. Drying of Meat Products

In the production process of dried meat and its products, traditional hot air drying methods are widely used, but they come with significant limitations. Firstly, the energy consumption of hot air drying is relatively high, leading to increased production costs and environmental impact. Secondly, for heat-sensitive foods, high temperatures may adversely affect their quality, resulting in a loss of flavor, texture, and nutritional value. Consequently, finding more efficient and quality-assured drying methods has become an important research direction in the field of food processing (Aykın-Dinçer et al. 2020).

In recent years, microwave-assisted drying technology has garnered attention for its unique advantages. Research by Elif et al. (2021) has shown that introducing microwave heating can significantly enhance drying efficiency and reduce drying time. As shown in Figure 7, the microwave drying is a technique that uses microwave radiation to heat and dehydrate food. Its principle involves microwave penetration into the food, causing water molecules inside to vibrate and generate heat, leading to rapid moisture evaporation. In the production of beef jerky, microwave-assisted hot air drying has demonstrated its unique advantages. By increasing the air temperature and microwave power, the drying process can be completed in a shorter time, and the moisture content, water activity (aw value), pH value, and TBARS value (an index measuring the degree of oxidation) of beef jerky can be improved. This not only enhances production efficiency but also helps improve the product's quality and shelf life.

Furthermore, studies by Nayar et al. (2014) have compared the effects of hot air oven and microwave drying methods on dehydrating goat meat. The results showed that although there were no significant differences in yield, water-holding capacity, and dehydration and rehydration ratios between the two methods, there were notable differences in other aspects such as pH value, water activity, and fat content. Microwave drying excelled in texture parameters and microbial quality, possibly due to its more uniform heating of food, reducing the opportunity for microbial growth while maintaining the food's structure and texture. However, in terms of flavor and overall acceptability, microwave drying was inferior to hot air ovens. This could be due to microwave drying causing excessive dehydration, affecting the sensory experience, such as producing a burnt taste and lower sensory scores, which are particularly important in market acceptance since consumers often prioritize taste and texture when choosing food.

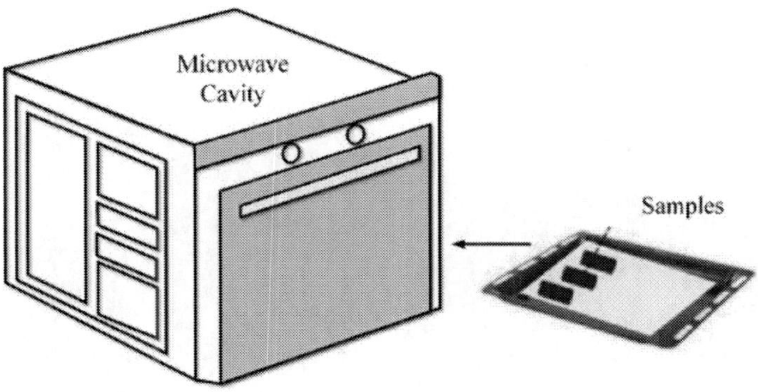

Figure 7. Microwave oven used for drying meat production.

Overall, these studies reveal the potential of microwave-assisted hot air drying technology in enhancing the quality of dried meat products, particularly beef jerky. They also highlight the challenges in applying microwave drying technology, especially in maintaining the flavor and sensory quality of food. For different types of meat, like beef and goat, specific adjustments to the drying parameters might be required due to their unique characteristics, to achieve optimal drying effects and product quality.

3.2. Drying of Cereals

Up to now, significant advancements and research efforts have been devoted to the development of microwave-assisted spouted bed drying techniques for food products. For example, studies have explored the effect of spouted bed and microwave-enhanced spouted bed drying on the efficiency of drying parboiled wheat. It was discovered that utilizing microwave-assisted spouted bed drying at microwave powers of 3.5 W/g and 7.5 W/g can decrease the drying duration by a minimum of 60% and 85%, respectively, in contrast to solely relying on spouted bed drying (Kahyaoglu et al. 2012; Sahin et al. 2013). Illustratively, Figure 8 demonstrates that peas processed through a spouted bed attain a more robust and plumper form when compared to those dried under the sun. Furthermore, investigative work on the microwave-assisted spouted bed drying of lettuce cubes has been carried out. This method outperformed traditional hot air flow drying, conventional spouted bed drying, and microwave vacuum drying in aspects such as rehydration rate, chlorophyll retention, color, and overall sensory quality. Notably, the duration required for microwave hot air spouted fluidized drying was substantially less than that for vacuum freeze drying, while still delivering comparable outcomes in terms of rehydration rates and chlorophyll preservation (Feng et al. 2012).

Figure 8. Comparison of pea between (a) sun drying and (b) spouted bed drying.

3.3. Drying of Fruit Products

3.3.1. Banana

Jiang et al. (2010) undertook a detailed investigation into the microwave freeze-drying process of banana slices with different maturity levels and microwave intensities, examining the physicochemical transformations at various drying phases. Their research demonstrated that the drying duration and significant changes in parameters such as starch and reducing sugar contents, as well as the structure and color of the banana slices, were considerably affected by the microwave power and the ripeness of the fruit. The initial phase of drying saw the most critical transformations, while the secondary drying phase witnessed the most notable shift in expansion ratio, leading to the conclusion that the initial stage is pivotal in determining the quality of the dried banana slices. When comparing with traditional hot air convective drying, the integration of microwave assistance with hot air drying markedly shortened the drying time by 64% during the falling rate period (Maskan, 2001). Drouzas and Schubert (Drouzas et al. 1996) delved into the empirical aspects of microwave vacuum drying for banana slices, conducting their experiments at 150 W of microwave power under a vacuum below 2500 Pa. Their findings showed that drying could be achieved within 30 minutes at temperatures not surpassing 70 °C, yielding a high-quality product on par with freeze-dried counterparts. The final dried products retained desirable textures and flavors without significant shrinkage or color changes, underscoring the efficacy of microwave vacuum drying in preserving quality. The experiments were rigorously controlled for temperature and pressure, with the structural integrity of the dried products before and after processing being a key determinant of quality. Sensory assessments, including evaluations of taste, aroma, and rehydration capabilities, confirmed the high quality of the dehydrated products produced by this method. Kiranoudis et al. (1997) have engaged in similar studies within this domain, further enriching the body of knowledge on microwave-assisted drying techniques.

3.3.2. Kiwifruit

In the drying process of kiwifruit, the use of microwave heating alone led to more pronounced shrinkage compared to conventional hot air drying. However, the shrinkage of kiwifruit was less significant when a combined hot air-microwave drying method was employed, in comparison to either method used independently. Moreover, kiwifruit is dried using this combined hot air-microwave technique exhibited superior rehydration capabilities than those dried solely with either microwave or hot air methods. This suggests that the synergistic effect of combining hot air and microwave drying not only mitigates shrinkage but also enhances the fruit's ability to reabsorb water, preserving more of its original qualities (Maskan, 2001).

3.3.3. Juice

Microwave vacuum drying shows great potential for producing high-quality dehydrated materials. In the study by Drouzas et al. (1996), significant drying efficiency was achieved in microwave vacuum drying of model fruit gels simulating orange juice. They managed to reduce the moisture content from 38.4% to less than 3% in less than 4 minutes, whereas traditional air drying methods require over 8 hours to achieve a moisture reduction to 10%. This substantial efficiency improvement suggests that microwave vacuum drying not only accelerates the drying process but also could match freeze-drying in preserving the sensory properties and rehydration capacity of food products, thereby producing high-quality dehydrated juice products.

3.3.4 Apple

Advancements in the field of Microwave Freeze Drying (MFD) of apple slices have been notable, as evidenced by the research of Funebo and Ohlsson (1996) and Huang et al. (2011). As shown in Figure 9, the pore structure appears more expanded in the sample treated with microwave irradiation. These studies contribute significantly to our understanding of modern drying techniques. Additionally, there has been intriguing research in the area of drying mixed-ingredient extruded reconstituted chips, particularly those combining potatoes and apples, which could emerge as a novel snack option. The drying speed in these studies was calculated by dividing the airflow rate by the spouted bed's larger end cross-sectional area. The experiments involved drying chopped apples with 24% moisture content at various microwave power densities (0-6.1W/g), successfully reducing the moisture content to around 5%. A remarkable improvement in temperature uniformity of the chopped apples was achieved using a combined drying method, compared to microwave drying on a static bed.

Figure 9. Micrographs of dried apple chips processed with (a) microwave freeze drying (b) freeze drying.

3.4. Drying Vegetable Products

In their investigation, Hu et al. (2006) assessed the effects of various drying techniques, such as vacuum microwave drying (MVD), freeze drying, hot air drying, and a combined approach of hot air and vacuum microwave drying, on the quality of granulated fruits and vegetables. This comparative analysis focused on several quality metrics, such as vitamin C and chlorophyll content, shrinkage, rehydration ability, color, texture, and microstructure. The results revealed that the quality of products subjected to vacuum microwave drying was marginally inferior to that of freeze-dried goods yet significantly superior to those dried using conventional hot air methods. This study is corroborated by additional MVD research, including works by Hu et al. (2006) and Zhang et al. (2003), further elucidating the advantages of microwave vacuum drying in maintaining the quality of dried food products.

3.4.1. Potato Chip

Exploring the potential of microwave drying technology in potato chip production offers a promising avenue for enhancing efficiency and quality in snack food processing. In the study of Zhang et al. (2011), the quality of modified vegetable chips was compared with air-dried (AD) and microwave-vacuum dried (MVD) samples. The findings indicated that MVD was the fastest drying method, taking less than an hour. Among the products, chips processed through MFD (presumably a typo for MVD) exhibited superior quality in terms of color, texture, and sensory evaluation, while AD products were the least ideal. Wang et al. (2010) explored the impact of vacuum and microwave freezing on potato chips' microstructure and quality, finding calcium ion treatment and bleaching effective in preventing shape changes during microwave freezing. Further, Wang et al. (2010; 2010) investigated the effects of adding substances through soaking to alter the dielectric properties of the material, thereby enhancing drying speed, and found significant results.

In the study by Song et al. (2007), an experimental analysis of vacuum-microwave drying on potato slices was conducted. The results revealed a decrease in moisture and fat content in the potato slices with extended MVD duration. Additionally, frying temperature and duration inversely affected the moisture content, while increasing the fat content of the potato slices. Response Surface Methodology (RSM) showed significant effects of MVD duration, frying temperature, and time on the moisture content, fat content, and breaking strength of the potato slices. In another experiment, the material was placed on a rotating tray to ensure uniform microwave exposure. Higher microwave power and lower vacuum levels were found to increase drying rate, though vacuum pressure had less impact on drying rate than microwave power.

Yan et al. (2010) developed an innovative Spouted Bed Microwave Drying method aimed at producing healthier puffed potato snacks, presenting a possible healthier alternative to traditional fried potato chips. Their research evaluated the impact of Spouted Bed Microwave Drying in comparison with conventional microwave and hot air drying techniques on the puffing attributes of potato segments. The integrated Microwave Enhanced Spouted Bed Drying (MWSB) method achieved favorable outcomes in terms of puffing ratios, structural integrity, and rehydration capabilities. The study highlighted that the puffing qualities of the potato segments were significantly affected by factors such as the pre-microwave treatment's humidity, the microwave's power level, and the initial size of the potato pieces. These studies collectively highlight the potential of microwave drying technologies to revolutionize the efficiency and quality of potato chip production, offering significant advancements in food processing.

3.4.2. Edamame

In the realm of drying technology research within the modern food industry, the drying process of edamame (young soybeans) presents a unique case study. Zhang et al. (2007) have empirically demonstrated that Microwave Vacuum Drying (MVD) is a faster drying process compared to Air Drying (AD) or Freeze Drying (FD). This is demonstrated by the curves that plot moisture content against time for edamame dried using FD, AD, AD-MVD, and MVD methods. At any given time point, MVD consistently shows a quicker drying process than either AD or FD. Notably, MVD reaches the final moisture content in just 40 minutes, whereas traditional air-dried samples require 9 hours, and FD takes a substantially longer time of 18 hours.

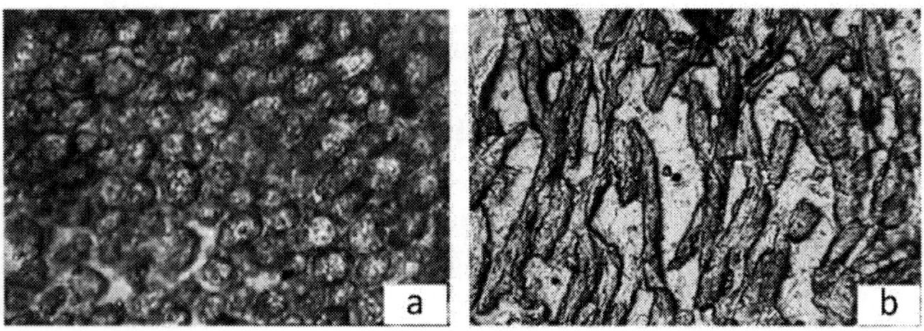

Figure 10. Micrographs of edamame processed with (a) air drying (b) vacuum microwave and convective air drying.

Hu et al. (2006; 2007) carried out a comprehensive examination of edamame drying using hot air and microwave vacuum methods. Their findings suggest that a hybrid approach, combining both hot air and microwave vacuum drying, yields superior drying results compared to using either microwave vacuum drying or hot air drying alone. Figure 10 illustrates that the edamame cells, having undergone microwave irradiation treatment, exhibit minimal shrinkage and remain intact without any signs of damage (Hu et al. 2006). As microwave power increases and the load mass decreases, the drying time significantly reduces. Although high vacuum facilitates the evaporation and volatilization of moisture, it may lead to electrical arcing, potentially causing product overheating. The optimal drying conditions for edamame are: (1) hot air drying at 70°C for 20 minutes, followed by (2) microwave vacuum drying with a power intensity of 9.33 W/g and a vacuum pressure of 95 kPa (gauge pressure), lasting for 15 minutes. In conclusion, these investigations not only shed light on how various drying techniques affect the drying efficacy of edamame but also offer important insights for refining drying procedures within the food industry.

3.4.3. Carrot

In theoretical aspect, Cui et al. (2004) formulated a model grounded in the principles of energy conservation. This model integrates the concepts of sensible heat, latent heat, and the heat produced by microwave energy, aiming to predict moisture content changes in carrots during microwave drying under varying power and vacuum pressure conditions. In experimental aspect, Lin et al. (1998) utilized vacuum microwave drying for carrots, finding that this method produced slices with favorable color, texture, flavor, and overall preference, both dry and rehydrated, compared to freeze-drying. Yan et al. (2010) examined various combined microwave drying techniques for carrots, including microwave-assisted vacuum drying (MVD), microwave-assisted freeze drying (MFD), and microwave-enhanced spouted bed drying (MSD), focusing on drying rate, uniformity, color, rehydration ratio, nutrient retention, and energy consumption. Mayer-Miebach et al. (2005) explored the thermal processing of carrots, particularly the stability and isomerization of lycopene during heat treatment, assessing its impact on antioxidant potential by heating carrot homogenates at different temperatures. Salehi et al. (2023) studied the effects of ultrasound and microwave pretreatments on carrot drying, aiming to enhance the process and the quality of the dried product, noting improvements in drying rate, time, color, texture, and overall sensory appeal. Figure 11 shows the process of drying carrot slices with microwave irradiation.

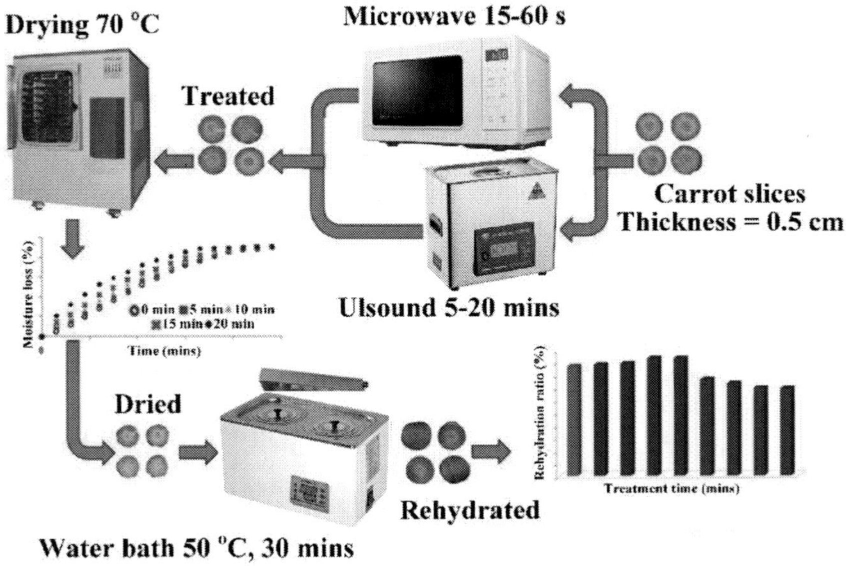

Figure 11. Process of drying carrot slices with microwave irradiation.

In summary, microwave drying pretreatments enhance the drying efficiency and quality of carrots, leading to better sensory attributes and a more appealing final product.

3.4.4. Cabbage

Xu et al. (2004) explored the dehydration of wild cabbage by integrating hot-air and microwave vacuum drying methods. Their findings highlighted a substantial reduction in drying duration, along with a notable preservation of nutrients and chlorophyll, leading to a product with an enhanced porous texture. Moreover, the microwave drying process demonstrated potent bactericidal effects, contributing to the high quality of the dried produce. The primary goal of this research was to optimize the drying process for wild cabbage, leveraging microwave-vacuum technology to achieve superior quality in the dried product.

3.4.5. Garlic Slices

Cui et al. (2003) initiated the drying process of garlic slices with microwave-vacuum drying until the moisture level was approximately 10% (on a wet basis). Subsequently, these slices underwent further drying through traditional hot-air methods at 45°C, aiming for a final moisture content below 5% (wet basis). Their research assesses and contrasts the spiciness, color, texture, and rehydration capacity of garlic slices prepared by this method against those processed by freeze drying and standard hot-air drying techniques. The findings suggest that garlic slices dried using a hybrid microwave-vacuum and air drying technique achieved a quality comparable to freeze-dried garlic slices and markedly superior to those dried by traditional hot-air methods.

3.5. Drying of Seafood

In the realm of seafood preservation, Duan et al. (2008; 2008)'s work has been pivotal. They investigated the impact of ultrasound-enhanced osmotic pre-treatment on sea cucumbers, utilizing microwave freeze-drying and nano-silver coating to extend shelf life and enhance quality. Duan et al. (2010) developed a method for processing sea cucumbers with MVD, while also exploring the integration of MVD with other drying methods (Duan et al. 2007; Xu, et al. 2004). Their evaluation showed that nano-silver, due to its extensive antimicrobial properties, significantly reduced microbial counts in microwave freeze-dried sea cucumbers without affecting the drying efficiency or sensory quality.

Zhang et al. (2007) concentrated on the parameters of microwave vacuum heating for the treatment of bighead carp slices. They found that the right moisture content and microwave power in pre-dehydrated fish slices could improve puffiness and crispness, enhancing the sensory quality of the final product. Higher vacuum levels were found to amplify this effect, improve crispness, increase energy efficiency in microwave heating, and reduce surface scorching.

In conclusion, these findings mark significant progress in seafood processing, leading to improvements in energy efficiency and product quality, thus advancing food preservation methods.

4. Applications of microwave irradiation in food drying

4.1. Microwave Cooking

Microwave cooking relies on the emission of microwave energy to water molecules in food, leading to accelerated molecular vibration and heat generation. This heating method is uniform as microwaves penetrate food, heating it evenly from inside out. It's particularly suited for quickly reheating pre-cooked foods, defrosting frozen items, and heating liquids like soups or beverages. Compared to traditional cooking methods, microwave heating is faster and better preserves the nutrients and flavors of food. The key advantage in cooking is its ability to heat food rapidly and uniformly, maintaining nutritional and flavor integrity, which is especially convenient in today's fast-paced lifestyle. However, its main drawback is the potential for uneven heating, especially when food thickness and water content vary, sometimes affecting the overall quality and texture of the food (Chandrasekaran et al. 2013).

Besides heating, microwave ovens are also used for defrosting, penetrating food to heat water molecules, allowing for more uniform and faster thawing than traditional methods. Additionally, microwaves can be used for baking small quantities of pastries or bread, although they may not produce the browning effect of traditional ovens (Sumnu, 2001). Furthermore, microwaves heat water to generate steam for cooking, suitable for vegetables and fish, and are used for poaching eggs and fish, illustrating their versatility in cooking methods (Ulusoy et al. 2019).

4.2. Microwave Pasteurization

Microwave pasteurization of tomato puree is an effective technique, maintaining stability, sensory acceptability, and safety within FDA microbial reduction standards during refrigeration (Pérez-Tejeda et al. 2016). Factors influencing microwave pasteurization include input power, flow rate, and feed temperature (Gentry and Roberts, 2005). The process typically involves four stages: preheating, microwave heating, holding, and cooling. Compared to traditional hot water pasteurization, microwaved carrots showed better color retention with no significant difference in texture and pectin methylesterase (PME) activity, indicating less thermal damage and improved quality uniformity (Peng et al. 2017).

Microwave technology showcases unique advantages in milk pasteurization, preserving nutritional integrity without the thermal damage associated with high-temperature, short-time (HTST) methods. It upholds the proteins, fats, and other vital nutrients within milk. The rapid processing cycle of microwave treatment not only conserves energy but also enhances efficiency by completing pasteurization swiftly. It also achieves microbial inactivation quickly, even with high microbial loads, offering an efficient and reliable option for the dairy industry (Dehghan et al. 2012; Malinowska-Pańczyk et al. 2019).

For ready-to-eat meals, while microwave sterilization provides significant advantages in food drying processes, it also presents challenges such as engineering complexity and high costs. Precise design and control are essential for uniform heating, involving complex machinery and detailed temperature monitoring to avoid overheating or underheating. The high initial investment can be a barrier for small-scale producers, necessitating a cost-benefit analysis to assess the feasibility of microwave technology. Additionally, as a relatively new technique, there may be a lack of experience in operating and maintaining these systems, underlining the need for substantial training and education in these technologies, including safety procedures and food processing standards (Tang et al. 2018).

Conclusion

In this chapter, the multifaceted advantages and innovative applications of microwave irradiation drying technology for food processing have been illuminated. The mechanisms, various applications, advancements, and challenges of microwave drying in food preservation have been explored. Importantly, efficient drying, superior quality preservation, and energy savings are offered by this technology compared to traditional methods. Its effectiveness has been observed across different food categories, from meats to vegetables, enhancing drying efficiency while maintaining nutritional and sensory qualities.

Despite its benefits, challenges such as uneven heating and potential overheating are faced by this technology, especially at later drying stages. The importance of integrating microwave drying with traditional methods to enhance control and uniformity is underscored by these issues. Additionally, the need for ongoing research and development is highlighted, particularly in large-scale operations and in overcoming technical limitations.

In conclusion, microwave irradiation drying is emerging as a promising, energy-efficient, and quality-preserving technique in the food industry. Its potential to revolutionize food processing and preservation is substantial, though continued innovation and optimization are required to realize its full potential.

Acknowledgments

The following financial supports are acknowledged: Financial support was provided by the National Natural Science Foundation of China (52076049, 52476005), Heilongjiang Province "Double First-class" Discipline Collaborative Innovation Achievement Project (LJGXCG2023-080), Heilongjiang Provincial Key R&D Program (2023ZX02C05), and Heilongjiang Provincial Key R&D Program "Unveiling the Leader" Project (2023ZXJ02C04).

References

Ang, T. K., Ford, J. D., & Pei, D. C. T. (1977). Microwave freeze-drying of food: A theoretical investigation. *International Journal of Heat and Mass Transfer*, 20(5), 517-526.

Aykın-Dinçer, E., Atlı, B., Çakmak, Ö., Canavar, S., & Çalışkan, A. (2021). Drying kinetics and quality characteristics of microwave-assisted hot air dried beef chips. *Journal of Microwave Power and Electromagnetic Energy*, 55(3), 219-235.

Aykın-Dinçer, E., Kılıç-Büyükkurt, Ö., & Erbaş, M. (2020). Influence of drying techniques and temperatures on drying kinetics and quality characteristics of beef slices. *Heat and Mass Transfer*, 56, 315-320.

Bouraoui, M., Richard, P., & Durance, T. (1994). Microwave and convective drying of potato slices. *Journal of Food Process Engineering*, 17(3), 353-363.

Chandrasekaran, S., Ramanathan, S., & Basak, T. (2013). Microwave food processing—A review. *Food Research International*, 52(1), 243-261.

Clary, C. D., Wang, S., & Petrucci, V. E. (2005). Fixed and incremental levels of microwave power application on drying grapes under vacuum. *Journal of Food Science*, 70(5), E344-E349.

Cui, Z. W., Xu, S. Y., & Sun, D. W. (2003). Dehydration of garlic slices by combined microwave-vacuum and air drying. *Drying Technology*, 21(7), 1173-1184.

Cui, Z. W., Xu, S. Y., & Sun, D. W. (2004). Microwave–vacuum drying kinetics of carrot slices. *Journal of Food Engineering*, 65(2), 157-164.

Datta, A. K., Sumnu, G., & Raghavan, G. S. V. (2014). Dielectric properties of foods. In Engineering Properties of Foods (pp. 523-588). *CRC Press.*

Dehghan, A., Jamalian, J., Farahnaky, A., Mesbahi, G., & Moosavi-Nasab, M. (2012). The effect of microwave pasteurization on some physical and chemical characteristics of milk. *International Journal of Food Engineering*, 8(1), 1-12.

Drouzas, A. E., & Schubert, H. (1996). Microwave application in vacuum drying of fruits. *Journal of Food Engineering*, 28(2), 203-209.

Drouzas, A. E., Tsami, E., & Saravacos, G. D. (1999). Microwave/vacuum drying of model fruit gels. *Journal of Food Engineering*, 39(2), 117-122.

Duan, X., Zhang, M., & Mujumdar, A. S. (2007). Study on a combination drying technique of sea cucumber. *Drying Technology*, 25(12), 2011-2019.

Duan, X., Zhang, M., Li, X., & Mujumdar, A. S. (2008). Microwave freeze drying of sea cucumber coated with nanoscale silver. *Drying Technology*, 26(4), 413-419.

Duan, X., Zhang, M., Li, X., & Mujumdar, A. S. (2008). Ultrasonically enhanced osmotic pretreatment of sea cucumber prior to microwave freeze drying. *Drying Technology*, 26(4), 420-426.

Duan, X., Zhang, M., Mujumdar, A. S., & Wang, R. (2010). Trends in microwave-assisted freeze drying of foods. *Drying Technology*, 28(4), 444-453.

Duan, X., Zhang, M., Mujumdar, A. S., Huang, L. L., & Wang, H. (2010). A novel dielectric drying method of sea cucumber. *International Journal of Food Science & Technology*, 45(12), 2538-2545.

Fan, L. P., Zhang, M., & Mujumdar, A. S. (2005). Vacuum frying of carrot chips. *Drying Technology*, 23(3), 645-656.

Feng, Y. F., Zhang, M., Jiang, H., & Sun, J. C. (2012). Microwave-assisted spouted bed drying of lettuce cubes. *Drying Technology*, 30(13), 1482-1490.

Funebo, T., & Ohlsson, T. (1998). Microwave-assisted air dehydration of apple and mushroom. *Journal of Food Engineering*, 38(3), 353-367.

Gentry, T. S., & Roberts, J. S. (2005). Design and evaluation of a continuous flow microwave pasteurization system for apple cider. *LWT - Food Science and Technology*, 38(3), 227-238.

Hu, Q. G., Zhang, M., Mujumdar, A. S., Du, W. H., & Sun, J. C. (2006). Effects of different drying methods on the quality changes of granular edamame. *Drying Technology*, 24(8), 1025-1032.

Hu, Q. G., Zhang, M., Mujumdar, A. S., Xiao, G. N., & Jin-cai, S. (2006). Drying of edamames by hot air and vacuum microwave combination. *Journal of Food Engineering*, 77(4), 977-982.

Hu, Q. G., Zhang, M., Mujumdar, A. S., Xiao, G. N., & Sun, J. C. (2007). Performance evaluation of vacuum microwave drying of edamame in deep-bed drying. *Drying Technology*, 25(4), 731-736.

Huang, L. L., Zhang, M., Mujumdar, A. S., & Lim, R. X. (2011). Comparison of four drying methods for re-structured mixed potato with apple chips. *Journal of Food Engineering*, 103(3), 279-284.

Jiang, H., Zhang, M., & Mujumdar, A. S. (2010). Microwave freeze-drying characteristics of banana crisps. *Drying Technology*, 28(12), 1377-1384.

Jiang, H., Zhang, M., & Mujumdar, A. S. (2010). Microwave freeze-drying characteristics of banana crisps. *Drying Technology*, 28(12), 1377-1384.

Kahyaoglu, L. N., Sahin, S., & Sumnu, G. (2012). Spouted bed and microwave-assisted spouted bed drying of parboiled wheat. *Food and Bioproducts Processing*, 90(2), 301-308.

Kiranoudis, C. T., Tsami, E., Maroulis, Z. B., & Marinos-Kouris, D. (1997). Drying kinetics of some fruits. *Drying Technology*, 15(5), 1399-1418.

Koné, K. Y., Druon, C., Gnimpieba, E. Z., Delmotte, M., Duquenoy, A., & Laguerre, J. C. (2013). Power density control in microwave assisted air drying to improve quality of food. *Journal of Food Engineering*, 119(4), 750-757.

Lin, T. M., Durance, T. D., & Scaman, C. H. (1998). Characterization of vacuum microwave, air, and freeze dried carrot slices. *Food Research International*, 31(2), 111-117.

Lu, Y., Zhang, M., Liu, H., Mujumdar, A. S., Sun, J., & Zheng, D. (2014). Optimization of potato cube drying in a microwave-assisted pulsed spouted bed. *Drying technology*, 32(8), 960-968.

Malinowska-Pańczyk, E., Królik, K., Skorupska, K., Puta, M., Martysiak-Żurowska, D., & Kiełbratowska, B. (2019). Microwave heat treatment application to pasteurization of human milk. *Innovative Food Science & Emerging Technologies*, 52, 42-48.

Maskan, M. (2001). Kinetics of colour change of kiwifruits during hot air and microwave drying. *Journal of Food Engineering*, 48(2), 169-175.

Mayer-Miebach, E., Behsnilian, D., Regier, M., & Schuchmann, H. P. (2005). Thermal processing of carrots: Lycopene stability and isomerisation with regard to antioxidant potential. *Food Research International*, 38(8-9), 1103-1108.

Metaxas, A. C. (1996). Foundations of electroheat: A unified approach. *Fuel and Energy Abstracts*, 37(3), 193.

Min, Z., Chunli, L., & Xiaolin, D. (2000). Application of various drying methods in selenium-enriched vegetable dehydration. *Drying Technology*, 18(4-5), 1139-1146.

Min, Z., Chunli, L., & Xiaolin, D. (2005). Effects of heating conditions on the thermal denaturation of white mushroom suitable for dehydration. *Drying Technology*, 23(5), 1119-1125.

Mousa, N., & Farid, M. (2002). Microwave vacuum drying of banana slices. *Drying Technology*, 20(10), 2055-2066.

Mujumdar, A. S. (2006). Handbook of Industrial Drying. *CRC Press.*

Nayar, R., Mendiratta, S. K., Prabhakaran, P., Chand, S., & Beura, C. K. (2014). Comparison of hot air oven dried and microwave dried extended and dehydrated goat meat cubes. *Research Opinion in Animal & Veterinary Sciences*, 4(6), 336-342.

Peng, J., Tang, J., Luan, D., Liu, F., Tang, Z., Li, F., & Zhang, W. (2017). Microwave pasteurization of pre-packaged carrots. *Journal of Food Engineering*, 202, 56-64.

Pérez-Tejeda, G., Vergara-Balderas, F. T., López-Malo, A., Rojas-Laguna, R., Abraham-Juárez, M. D. R., & Sosa-Morales, M. E. (2016). Pasteurization treatments for tomato puree using conventional or microwave processes. *Journal of Microwave Power and Electromagnetic Energy*, 50(1), 35-42.

Ratti, C. (2001). Hot air and freeze-drying of high-value foods: A review. *Journal of Food Engineering*, 49(4), 311-319.

Sahin, S., Sumnu, G., & Tunaboyu, F. (2013). Usage of solar-assisted spouted bed drier in drying of pea. *Food and Bioproducts Processing*, 91(3), 271-278.

Salehi, F., Goharpour, K., & Kamran, H. R. (2023). Effects of ultrasound and microwave pretreatments of carrot slices before drying on the color indexes and drying rate. *Ultrasonics Sonochemistry*, 101, 106671.

Song, X. J., Zhang, M., & Mujumdar, A. S. (2007). Optimization of vacuum microwave pre-drying and vacuum frying conditions to produce fried potato chips. *Drying Technology*, 25(2), 2027-2034.

Song, X. J., Zhang, M., Mujumdar, A. S., & Fan, L. (2009). Drying characteristics and kinetics of vacuum microwave–dried potato slices. *Drying Technology*, 27(9), 969-974.

Sumnu, G. (2001). A review on microwave baking of foods. *International Journal of Food Science & Technology*, 36(2), 117-127.

Tang, J., Hong, Y. K., Inanoglu, S., & Liu, F. (2018). Microwave pasteurization for ready-to-eat meals. *Current Opinion in Food Science*, 23, 133-141.

Ulusoy, Ş., Üçok Alakavuk, D., Mol, S., & Coşansu, S. (2019). Effect of microwave cooking on foodborne pathogens in fish. *Journal of Food Processing and Preservation*, 43(8), e14045.

Vega-Mercado, H., Góngora-Nieto, M. M., & Barbosa-Cánovas, G. V. (2001). Advances in dehydration of foods. *Journal of Food Engineering*, 49(4), 271-289.

Wang, R., Zhang, M., & Mujumdar, A. S. (2010). Effect of food ingredient on microwave freeze drying of instant vegetable soup. *LWT - Food Science and Technology*, 43(7), 1144-1150.

Wang, R., Zhang, M., & Mujumdar, A. S. (2010). Effect of osmotic dehydration on microwave freeze-drying characteristics and quality of potato chips. *Drying Technology*, 28(6), 798-806.

Wang, R., Zhang, M., & Mujumdar, A. S. (2010). Effects of vacuum and microwave freeze drying on microstructure and quality of potato slices. *Journal of Food Engineering*, 101(2), 131-139.

Yan, W. Q., Zhang, M., Huang, L. L., Tang, J., Mujumdar, A. S., & Sun, J. C. (2010). Study on optimization of puffing characteristics of potato cubes by spouted bed drying enhanced with microwave. *Journal of the Science of Food and Agriculture*, 90(8), 1300-1307.

Yan, W. Q., Zhang, M., Huang, L. L., Tang, J., Mujumdar, A. S., & Sun, J. C. (2010). Studies on different combined microwave drying of carrot pieces. *International Journal of Food Science & Technology*, 45(10), 2141-2148.

Yanyang, X., Min, Z., Mujumdar, A. S., Le-qun, Z., & Jin-cai, S. (2004). Studies on hot air and microwave vacuum drying of wild cabbage. *Drying Technology*, 22(9), 2201-2209.

Zhang, F., Zhang, M., & Mujumdar, A. S. (2011). Drying characteristics and quality of restructured wild cabbage chips processed using different drying methods. *Drying Technology*, 29(6), 682-688.

Zhang, J., Zhang, M., Shan, L., & Fang, Z. (2007). Microwave-vacuum heating parameters for processing savory crisp bighead carp (Hypophthalmichthys nobilis) slices. *Journal of Food Engineering*, 79(3), 885-891.

Zhang, M., Li, C., Ding, X., & Cao, C. W. (2003). Optimization for preservation of selenium in sweet pepper under low-vacuum dehydration. *Drying Technology*, 21(3), 569-579.

Zhang, M., Tang, J., Mujumdar, A. S., & Wang, S. (2006). Trends in microwave-related drying of fruits and vegetables. *Trends in Food Science & Technology*, 17(10), 524-534.

Chapter 4

Microwave Irradiation for Heating Organic Wastes

Longfei Cui
Wenke Zhao
Yaning Zhang*
and Bingxi Li
School of Energy Science and Engineering, Harbin Institute of Technology, Harbin, China

Abstract

Biomass and plastic are two important organic wastes, and they can be converted into energy products by pyrolysis technology. Using organic waste can protect the environment, reduce the use of fossil energy and reduce CO_2 carbon dioxide emissions. Microwave-assisted pyrolysis is a useful technology which can convert organic wastes to high valued products. Microwave irradiation heating process affects the degree of pyrolysis. This chapter comprehensively introduces the microwave irradiation heating technology for organic waste (biomass and plastic), microwave absorbent, organic waste mixed with microwave absorbent. The effects of microwave power, feedstock load and microwave absorbent/organic waste mass ratio on microweave irradiation heating processes are discussed. These can help regulate the microwave irradiation heating process.

* Corresponding Author's Email: ynzhang@hit.edu.cn.

In: Characteristics and Applications of Microwave Irradiation
Editors: Yaning Zhang and Wenke Zhao
ISBN: 979-8-89530-170-8

Keywords: microwave irradiation, heating, organic wastes, biomass, plastics, microwave absorbent

1. Microwave Irradiation for Heating Biomass

With economic development and technological advances (Kamari et al. 2023), demand for energy is gradually increasing (Depren et al. 2022). The use of fossil fuels will pollute the environment (Ji et al. 2022). Renewable energy has been investigated by many scholars (Amjith et al. 2022), such as solar energy (Caporale et al. 2023), wind energy, geothermal power, and biomass (Rajpoot et al. 2022; Su et al. 2022), and renenwable energy application has developed rapidly (Amalina et al. 2023).

Among the renewable resources, biomass has many advantages, such as abundantly available, zero SO_x emissions (Yana et al. 2022), lower NO_x emissions, environmentally friendly (Gadkari et al. 2017), sustainable (Sinha et al. 2023), strong renewable and plentiful, and it is considered to be part of alternative to fossil fuels (Gao et al. 2022), and it has been recognized as a renewable carbon source.

Biomass can be converted to energy products through a pyrolysis process (Sun et al. 2022), and the products include biogas, bio-oil, and biochar (Iturbides et al. 2022). Wang et al. (2022a) used corn straw as feedstock to prepare biochar which was used as constructed wetland substrate and treated city tail water, and this was also an effective method to achieve high-value utilization of corn straw. Shi et al. (2023b) used three commercial-ready catalysts for catalytic fast pyrolysis of sawdust for bio-oil production.

Microwave irradiation technology causes polar molecules to rotate at high speeds in space, resulting in rapid and uniform heating (Pritam et al. 2023). Due to its advantages of environmental protection (Guo et al. 2015), cleanliness and greatly reduced heating time (Binner et al. 2014), microwave heating plays an increasingly important role in the fields of drying (Pickles et al. 2014), carbonization (Wang et al. 2020), liquefaction, pyrolysis (Abdelsayed et al. 2019) and gasification (Uhm et al. 2014).

Traditional heating technology and microwave irradiation heating process were also used as heating methods for pyrolysis process (Li et al. 2023). Recently, microwave irradiation has been used as a heating technology for pyrolysis of biomass and plastics (Zhang et al. 2022). Microwave irradiation heating technology has many advatanges (Zhao et al. 2011), such as, fast, selective, and internal heating (Zou et al. 2022). Zou et al. (2022) used microwave-assisted pyrolysis method to investigate the co-pyrolysis performance of plastics and biomass with biochar as catalyst, and the results showed that the quality of syngas and bio-oil was improved by the use of biochar. Zhang et al. (2022) enhanced the microwave-assisted pyrolysis process of biomass using needle metal (iron wire), and the results showed that the interaction between the metal and microwave effectively improved the degree of biomass pyrolysis.

During the microwave-assisted pyrolysis process, the heating process of feedstock affected the pyrolysis degree (Li et al. 2023). Heating performance under microwave irradiation is different for different raw materials, which is related to the dielectric parameters of the raw materials themselves (Vos et al. 2003). The tangent value of the dielectric loss angle tan δ indicates the ability of the feedstock to absorb microwave (Mushtaq et al. 2014). With the increase of tanδ, the microwave absorption ability increases (Larhed et al. 2002). Table 1 shows the tanδ values of different biomass at room temperature and 2.45 GHz, and the tanδ value affect the microwave heating process of feedstocks (Namazi et al. 2015).

Microwave irradiation heating technology for biomass has been investigated, and microwave power affected the heating degree (Cui et al. 2023b). Figure 1 shows the microwave heating process of corn straw with particle size of 0.075 mm, biomass mass of 22 g, and reactor volume of 150 mL at microwave irradiation power from 400 W to 600 W.

Table 1. tanδ of different biomass at room temperature and 2.45 GHz

Feedstock	tanδ	Source
Pine Bark	0.18	Mushtaq et al. 2014
Oil palm fiber	0.08	Salemaa et al. 2013
Water	0.12	Larhed et al. 2002
Aspen Bark	0.22	Mushtaq et al. 2014
Pine wood	0.19	Vos et al. 2003
Hemicelluloses	0.062	Namazi et al. 2015
Oil palm shell	0.12	Salemaa et al. 2013
Fir plywood	0.01-0.05	Mushtaq et al. 2014
Particle board	0.1-1.0	Mushtaq et al. 2014
Acid washed sludge	0.019	Namazi et al. 2015
Pulp mill sludge	0.08	Namazi et al. 2015
Cellulose	0.035	Namazi et al. 2015
Lignin	0.052	Namazi et al. 2015
Sludge	0.035	Namazi et al. 2015

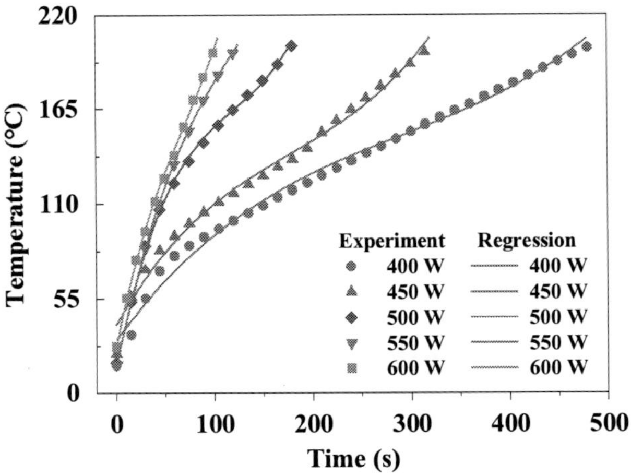

Figure 1. Heating performances at different microwave powers (Cui et al. 2023b).

From Figure 1, the temperature of heating process was strongly correlated with microwave irradiation time, and some equations can be used to describe the effects of time, power, and temperature.

$$T_{400\,\mathrm{W}} = 30.62 + 0.78t - 1.91\times10^{-3}t^2 + 2.18\times10^{-6}t^3 \quad (0 \le t \le 480) \tag{1}$$

$$T_{450\,\mathrm{W}} = 39.48 + 1.01t - 3.81\times10^{-3}t^2 + 7.11\times10^{-6}t^3 \quad (0 \le t \le 320) \tag{2}$$

$$T_{500\,\mathrm{W}} = 15.42 + 2.73t - 1.85\times10^{-2}t^2 + 5.11\times10^{-5}t^3 \quad (0 \le t \le 180) \tag{3}$$

$$T_{550\,\mathrm{W}} = 15.53 + 2.67t - 1.39\times10^{-2}t^2 + 3.60\times10^{-5}t^3 \quad (0 \le t \le 125) \tag{4}$$

$$T_{600\,\mathrm{W}} = 26.74 + 2.83t - 2.19\times10^{-2}t^2 + 1.07\times10^{-4}t^3 \quad (0 \le t \le 105) \tag{5}$$

where:

T is corn straw temperature (°C),
t is the heating time of microwave irridation heating process (s).

R^2 values of equations (1) - (5) were 0.990, 0.991, 0.997, 0.999, and 0.998, respectively, and the equations can be well used to describe the relationship of heating time and temperature.

Table 2. Microwave heating performances of corn straw at different particle sizes (Cui et al. 2023b)

Biomass mass (g)	Biomass size (mm)	Microwave power (W)	Average heating rate (°C/min)
22	0.075	600	101.46
22	0.125	600	61.07
22	0.200	600	52.70
22	0.355	600	37.34

Table 3. Microwave heating performances at different corn straw masses (Cui et al. 2023b)

Corn straw mass (g)	Corn straw size (mm)	Microwave power (W)	Average heating rate (°C/min)
10	0.075	600	36.31
16	0.075	600	46.24
22	0.075	600	101.46
28	0.075	600	92.95
34	0.075	600	46.01

During microwave irradiation heaing process, heating rate is also affected by biomass particle size and biomass mass. Table 2 shows the microwave heating process of corn straw affected by the biomass particle size, and the heating rate decreases gradually as the biomass zise increases (Cui et al. 2023b).

Table 3 shows the microwave heating process of corn straw affected by the biomass mass, and the heating rate increases and then decreases as the biomass mass increases (Cui et al. 2023b).

Huang et al. (2016) investigated the microwave-assisted heating process of different biomass feedstocks, and the heating rates are shown in Table 4.

When the particle size, reactor volume, and biomass mass were 0.058 mm, 150 mL, and 30 g, respectively, the microwave irradiation heating processes of wheat straw affect by microwave power are shown in Figure 2 (Cui et al. 2023a).

The microwave heating time required to increase the temperature of wheat straw pellets from room temperature to 252.2 °C was 9.5 min with the microwave power of 250 W, and the average heating rate was 23.41 °C/min. When the power was 300 and 350 W, the average heating rate were 30.55 and 56.18 °C/min, respectively.

Table 4. Microwave heating performances of different biomass feedstocks (Huang et al. 2016)

Biomass feedstock	Microwave power (W)	Average heating rate (°C/min)
Bamboo leaves	300	51 ± 8
	400	76 ± 15
	500	98 ± 10
Rice straw	300	52 ± 8
	400	89 ± 7
	500	115 ± 9
Rice husk	300	48 ± 10
	400	94 ± 8
	500	140 ± 14
Sugarcane peel	300	64 ± 18
	400	84 ± 5
	500	120 ± 19
Corn stover	300	58 ± 6
	400	98 ± 15
	500	132 ± 15
Sugarcane bagasse	300	69 ± 13
	400	103 ± 4
	500	133 ± 13
Coffee grounds	300	53 ± 13
	400	86 ± 14
	500	124 ± 12

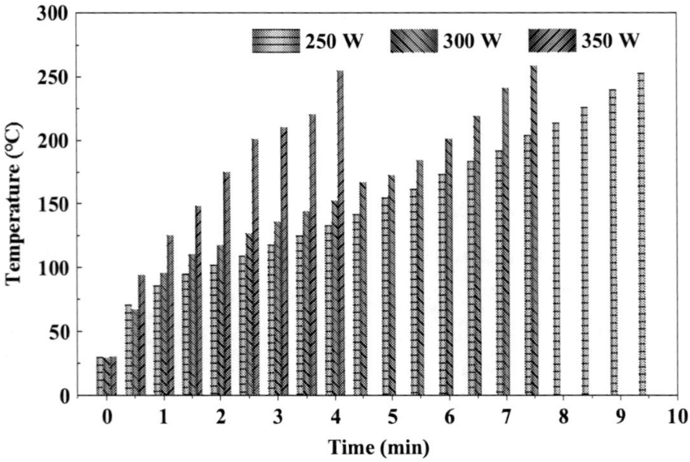

Figure 2. Wheat straw temperature with various microwave powers (Cui et al. 2023a).

Table 5. Performances of wheat straw during microwave heating process with various biomass masses and particle sizes (Cui et al. 2023a)

Biomass mass (g)	Biomass size (mm)	Microwave power (W)	Average heating rate (oC/min)
10	0.058	350	5.96
30	0.058	350	56.18
50	0.058	350	14.17
30	0.106	350	25.87
30	0.270	350	20.59

Table 5 shows the microwave heating process of wheat straw affected by biomass mass and particle size (Cui et al. 2023a). From Table 5, the average heating rate of wheat straw increased with the biomass mass increased from 10 g to 30 g, and when the biomass mass increased from 30 g to 50 g, the average heating rate decreased. The average heating rate decreased with increasing particle size.

2. Microwave Irradiation for Heating of Plastics

Plastics are increasingly appearing and play an important role in our daily life (Maqsood et al. 2021), they are widely used in construction, packaging, automobile and the other fields. The global production of plastics has gradually increased in recent years (Lopez et al. 2017). The waste generated by the use of plastics accounts for a large proportion of the inorganic component of municipal solid waste (Alabi et al. 2019; Williams et al. 2007), and the overuse and improper disposal of plastic waste has a negative impact on the ecosystem (Schirmer et al. 2001). Disposal of plastic waste is a serious problem (Ragaert et al. 2017). Plastics are usually composed of polymers that are difficult to biodegrade (Jambeck et al. 2015), so how to dispose of plastic waste efficiently and without pollution is an issue that deserves in-depth study (Alabi et al. 2019).

In recent years, many scholars have researched and carried out studies on plastics pyrolysis technology (Fan et al. 2023), through which waste plastics can be converted into high-quality pyrolysis oil and pyrolysis gas (Cui et al. 2023d). Utilizing microwave irradiation heating technology as the heat source of pyrolysis process is a relatively new type of pyrolysis (Wang et al. 2022b). Fan et al. (2023) used microwave-assisted pyrolysis method to prepare aviation oil from polystyrene plastics with iron-based microwabe absorbent, and the results showed that the oil yield varied in the range of 78.00 ~ 97.67 wt.%, and microwave absorbent type showed no significant difference for the oil yield. Cui et al. (2023d) used microwave-assisted pyrolysis method to prepare high-quality oil from polypropylene plastic with SiC, and the experiment results indicated that the optimal oil yield was 79.3 wt.% at pyrolysis temperature, mass ratio of SiC and polypropylene, mass of polypropylene particle, and microwave power of 450 °C, 2.5, 30 g, and 500 W, respectively.

The heating performances of different plastics under microwave irradiation are different (Imenokhoyev et al. 2013; Somani et al. 2017), and they are related to the dielectric parameter of the feedstock itself. The tangent value of the dielectric loss angle tanδ indicates the ability of the feedstock to absorb microwave (Sarker et al. 2011). With the increase of tanδ, the microwave absorption ability increases.

Table 6. tanδ of different plastics at room temperature and 2.45 GHz

Feedstock	tanδ	Source
Polypropylene	0.0003-0.0004	Zhang et al. 2017
Polystyrene	0.0002-0.0003	Zhang et al. 2017
Polyethylene	0.0001-0.0002	Zhang et al. 2017

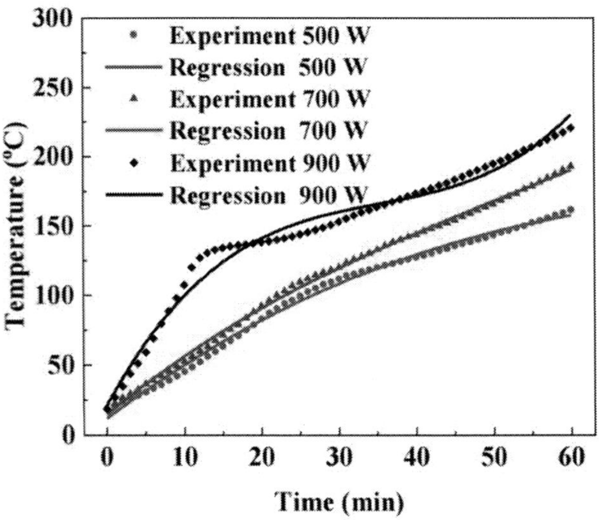

Figure 3. Heating performances of HDPE at different microwave powers (Fu et al. 2021).

Table 6 shows the tanδ values of different plastics at room temperature and 2.45 GHz, and the tanδ value affects the microwave irradiation heating performances of different plastic (Larhed et al. 2002).

Figure 3 illustrates the microwave irradiation heating process of high-density polyethylene plastic particles (Fu et al. 2021). From Figure 3, when the heating time, mass of high-density polyethylene (HDPE) plastic, and reactor chamber volume were fixed at 60 min, 100 mL, and 30 g, respectively, the temperatures of plastic at different microwave irradiation powers increased from room temperature to 161.5, 192.8, and 220.2 °C, and the average heating rates were 2.38, 2.89, and 3.36 °C/min, respectively.

From Figure 3, the temperature of HDPE of heating process was strongly correlated with microwave irradiation time, and equations (6) - (8) can be used to describe the relevance of time, power, and temperature.

$$T = 11.93 + 4.14t - 3.354 \times 10^{-2} t^2 + 8.30 \times 10^{-5} t^3 \quad \text{(500 W)} \tag{6}$$

$$T = 14.77 + 4.561t - 4.396 \times 10^{-2} t^2 + 2.77 \times 10^{-4} t^3 \quad \text{(700 W)} \tag{7}$$

$$T = 21.18 + 10.18t - 0.2587t^2 + 2.449 \times 10^{-3} t^3 \quad \text{(900 W)} \tag{8}$$

where:

T is HDPE temperature (°C),
t is the heating time of microwave heating process (min).

During the microwave irradiation heating process, the heating rate is also affected by plastic mass and reactor chamber volume. Table 7 shows the microwave irradiation performances of HDPE affected by the plastic mass, and reactor chamber volume (Fu et al. 2021).

Table 7. Average heating rates of HDPE at various masses and reactor volumes

Plastic mass (g)	Volume (mL)	Microwave power (W)	Average heating rate (°C/min)
10	100	900	2.22
20	100	900	2.29
30	100	900	3.36
30	150	900	4.49
30	200	900	5.50

From Table 7, the average heating rate increased with increasing plastic mass and decreased with increasing reaction chamber volume.

Figure 4 shows the microwave irradiation heating performances of low-density polyethylene (HDPE) plastic particles (Hong et al. 2021). From Figure 4, when the heating time, plastic mass, and reactor chamber volume were fixed at 60 min, 100 mL, and 30 g, respectively, the temperatures of plastic at different microwave irradiation powers (500, 700, and 900 W) were increased from room temperature to 159.8, 259.1, and 314.4 °C, and the average heating rates were 2.34, 3.96 and 4.89 °C/min, respectively.

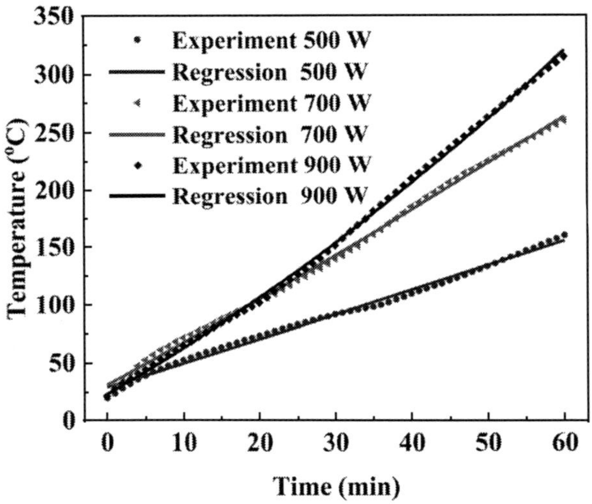

Figure 4. Heating performances of low density polyethylene at different microwave powers (Hong et al. 2021).

From Figure 4, the temperature of LDPE of heating process was strongly correlated with microwave irradiation time, and equations (9) - (11) can be used to describe the relevance of time, power, and temperature.

$$T = 28.92 + 2.0332t + 1.17 \times 10^{-2} t^2 \qquad (500\ \mathrm{W}) \tag{9}$$

$$T = 31.07 + 3.5753t + 4.82 \times 10^{-3} t^2 \qquad (700\ \mathrm{W}) \tag{10}$$

$$T = 22.32 + 3.7292t - 2.031 \times 10^{-2} t^2 \qquad (900\ \mathrm{W}) \tag{11}$$

where:

T	is the temperature of LDPE (°C),
t	is the heating time of microwave heating process (min).

Table 8 shows the microwave irradiation performances of LDPE affected by the plastic mass, and reactor chamber volume (Hong et al. 2021). From Table 8, the average heating rate increased with increasing plastic mass, and the average heating rate decreased with increasing of reactor chamber volume.

Table 8. Microwave heating performances of HDPE at different masses and reactor volumes (Hong et al. 2021)

Plastic mass (g)	Volume (mL)	Microwave power (W)	Average heating rate (°C/min)
10	100	900	2.03
20	100	900	4.28
30	100	900	4.89
30	150	900	3.68
30	200	900	2.76

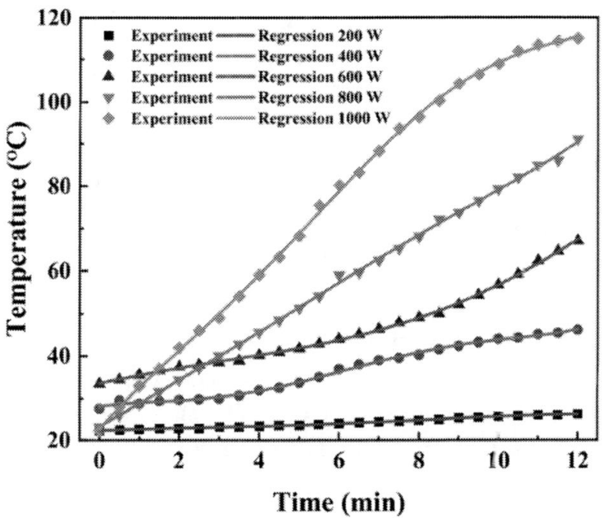

Figure 5. Heating performances of PP plastic at different microwave powers (Cui et al. 2023e).

The performance of microwave irradiation heating of polypropylene (PP) plastic particles in a fluidized bed reactor is shown in Figure 5 (Cui et al. 2023e). From Figure 5, when the heating time, plastic size, and fluidization velocity were fixed at 15 min, 96-109 μm, and 2.36 × 10^{-3} m/s, the average heating rates of PP plastic particles with various microwave irradiation powers (200 - 1000 W) were 0.3, 1.5, 2.8, 7.4 and 8.1 °C/min, respectively.

Table 9 shows the performances of microwave irradiation heating process of PP particles affected by particle size and fluidization velocity in a fluilized bed reactor (Cui et al. 2023e). From Table 9, the average heating rate decreased as the fluidization velocity increased and increased as the particle size increased.

Table 9. Performances of microwave heating polypropylene particles with various sizes and fluidization velocities (Cui et al. 2023e)

PP size (μm)	Fluidization velocity (×10^{-3} m/s)	Microwave power (W)	Average heating rate (°C/min)
96-109	1.18	600	6.70
96-109	2.36	600	2.80
96-109	3.54	600	1.00
96-109	4.72	600	0.86
96-109	5.90	600	0.64
53-63	2.36	600	0.30
63-74	2.36	600	0.33
74-86	2.36	600	1.18
86-96	2.36	600	1.91

3. Microwave Irradiation for Heating of Microwave Absorbent

Biomass and plastics have a weak microwave absorption capacity, and they cannot reach the set pyrolysis temperature during microwave-assisted pyrolysis (Qiu et al. 2023). Therefore, in the microwave-assisted pyrolysis process of biomass and plastics (Jafari et al. 2017), it is necessary to add some microwave absorbents with strong microwave absorption ability to strengthen the microwave heating process (Lin et al. 2017).

The main microwave absorbents commonly used are SiC, biochar, and activated carbon, etc (Qiu et al. 2023), and Table 10 shows the tanδ values of different microwave absorbents at room temperature and 2.45 GHz. During the microwave irradiation heating process, when the biomass and plastics are heated by microwave irradiation with microwave absorbents, the average heating rate is usually high (Mushtaq et al. 2014).

Table 10. tanδ of different microwave absorbents at room temperature and 2.45 GHz

Microwave absorbent	tanδ	Source
SiC	0.02-1.05	Zhang et al. 2017
Kernel shell activated carbon	0.40	Omar et al. 2011
Biochar	0.20	Peng et al. 2011
Activated carbon	0.62	Ramasamy and Moghtaderi, 2010
Wet fruit bunch char	0.30	Omar rt al., 2011
Dried fruit bunch char	0.13	Omar rt al., 2011
Graphite	0.10-0.17	Sanchez-Coronado et al. 2003
Carbon	0.28	Salema and Ani, 2012
Palm shell char	0.08	Salemaa et al. 2013

Figure 6 illustrates the performances of microwave irradiation heating process of SiC particles in a fluidized bed reactor at different microwave irradiation powers with SiC size of 96-109 μm, heating time of 15 min and fluidization velocity of 2.36×10^{-3} m/s (Cui et al. 2023e). When the microwave power and heating time were 200 W and 15 min, respectively, the temperature of SiC increased from room temperature to 330.5 °C, and the average heating rate was 20.0 °C/min, and when the microwave power was 1000 W, the average heating rate was 63.2 °C/min.

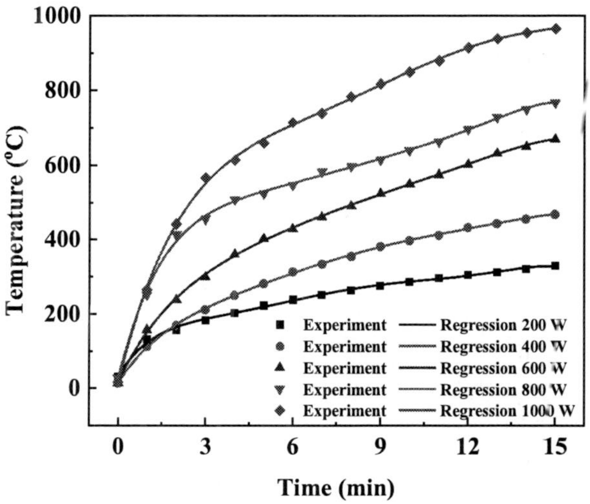

Figure 6. Performances of microwave heating process of SiC in fluidized bed reactor at different microwave powers (Cui et al. 2023e).

Table 11. Microwave heating performances of SiC particles at different particle sizes and fluidization velocities (Cui et al. 2023e)

SiC size (μm)	Fluidization velocity ($\times 10^{-3}$ m/s)	Microwave power (W)	Average heating rate (°C/min)
53-63	2.36	600	22.7
96-109	2.36	600	43.4
96-109	1.18	600	55.4
96-109	5.90	600	15.9

Table 11 illustrates the average heating rate of SiC particles at different particle sizes and fluidization velocities during microwave heating process, and with the increased of fluidization, the averaging heating rate decreased.

During the microwave-assisted pyrolysis process, some scholars used activated carbon as microwave absorber, and Shi et al. (2023a) investigated the microwave irridiation heating process of activated carbon as affected by microwave irradiation power, mass, and reactor chamber volume. Figure 7 shows the temperatures of activated carbon as affected by microwave irradiation powers (400-600 W). As the microwave irradiation power decreased from 600 W to 400 W, the average heating rates were 5.9, 3.0, 2.8, 2.2, and 2.0 °C/s, respectively (Shi et al. 2023a).

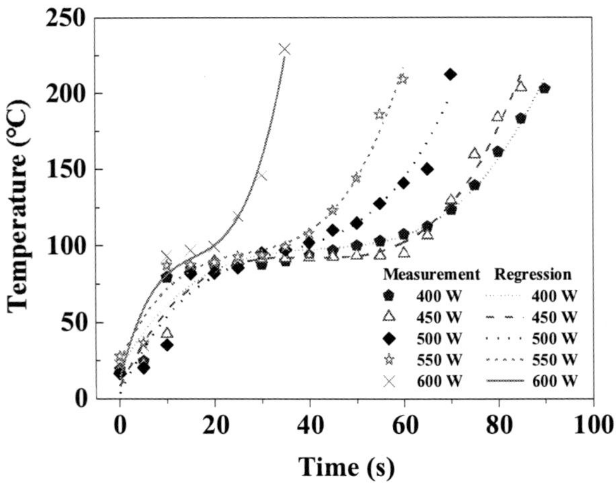

Figure 7. Microwave irradiation heating performances of activated carbon at different powers (Shi et al. 2023a).

From Figure 7, the temperature of activated carbon of heating process was strongly correlated with microwave irradiation time, and some equations (12) - (16) can be used to describe the relevance of heating time, power, and temperature.

$$T = 20.159 + 5.163t - 0.118t^2 + 9.369\times10^{-4}t^3 \quad (400\ W) \tag{12}$$

$$T = 8.097 + 6.391t - 0.161t^2 + 1.340\times10^{-3}t^3 \quad (450\ W) \tag{13}$$

$$T = 3.333 + 6.542t - 0.168t^2 + 1.640\times10^{-3}t^3 \quad (500\ W) \tag{14}$$

$$T = 24.385 + 6.724t - 0.228t^2 + 2.820\times10^{-3}t^3 \quad (550\ W) \tag{15}$$

$$T = 12.955 + 12.073t - 0.670t^2 + 1.421\times10^{-2}t^3 \quad (600\ W) \tag{16}$$

where:

T is the activated carbon temperature (°C),
t is the microwave irradiation heating time (s).

The R^2 values of the fitting equations were 0.973, 0.972, 0.963, 0.973, and 0.978, respectively. Table 12 shows the microwave irradiation heating performances of activated carbon affected by reactor volume and mass in microwave irradiation heating device (Shi et al. 2023a).

Table 12. Microwave heating performances of activated carbon at different masses (Shi et al. 2023a)

Activated carbon mass (g)	Reactor volume (mL)	Microwave power (W)	Average heating rate (°C/s)
10	150	500	2.8
15	150	500	3.1
20	150	500	3.5
25	150	500	3.55
10	50	500	7.6
10	100	500	3.3
10	200	500	2.6
10	250	500	1.2

Figure 8 shows the microwave irradiation heating process affected by microwave power (300, 350, 400, 450, 500 W) with SiC mass of 30 g, reactor volume of 50 mL (Ke et al. 2022). From Figure 8, the average heating rate increased with the microwave irradiation power increasing, and when the powers were 300, 350, 400, 450, 500 W, the average heating rates were 0.89, 3.04, 3.95, 5.67, and 7.08 °C/s, respectively.

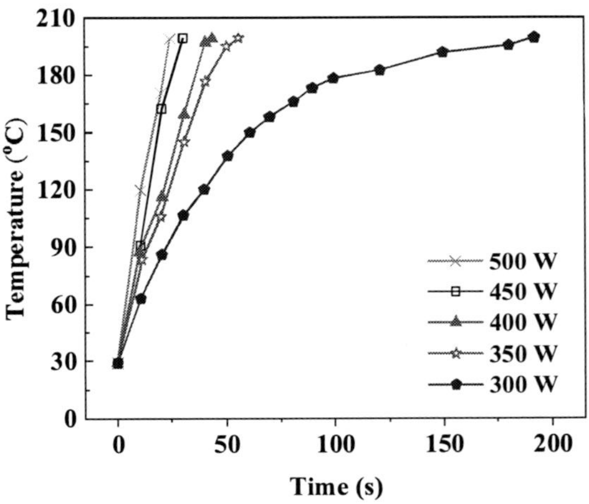

Figure 8. Microwave irradiation heating process of SiC at different powers (Ke et al. 2022).

Table 13 shows the microwave irradiation heating process of SiC affected by SiC mass in a microwave heating device, and from the results, the average heating rate increased and then decreased, as the mass of SiC increased (Ke et al. 2022).

Table 13. Average heating rates of SiC at different masses in fixed bed (Ke et al. 2022)

SiC mass (g)	Reactor volume (mL)	Microwave power (W)	Average heating rate (°C/s)
20	50	300	0.36
25	50	300	0.86
30	50	300	0.63
35	50	300	0.52
40	50	300	0.40

4. Microwave Irradiation for Heating Organic Waste with Microwave Absorbent

Plastics and biomass have a weak microwave absorption capacity, and the microwave irradiation heating process of biomass and plastics without microwave absorbers does not allow for deep pyrolysis of biomass. Therefore, microwave absorbent is used in the microwave irradiation heating and pyrolysis process of biomass and plastics for the production of energy products (Folorunso et al. 2012; Jeong et al. 2019).

4.1. Microwave Irradiation for Heating Biomass with Microwave Absorbent

Some scholars studied the heating process of microwave heating of biomass with the addition of microwave absorbent. SiC is mainly chosen as microwave absorbent for microwave heating process of biomass (Jafari et al. 2017; Lin et al. 2017).

Figure 9 shows the comparison of microwave irradiation heating processes of corn straw with and without SiC at different powers (400, 450, 500, 550, and 600 W) (Cui et al. 2023c), and the average heating rates are shown in Table 14. From Table 14 and Figure 9, the average heating rate of the mixture of SiC and biomass is higher than that of biomass, and with the increase of SiC, the microwave irradiation heating degree can be deepened.

Table 14. Microwave heating performances of corn straw at different SiC masses (Cui et al. 2023c)

Microwave power (W)	Biomass mass (g)	SiC mass (g)	Average heating rate (°C/min)
400	22	0	23.06
400	22	22	83.68
450	22	0	33.19
450	22	22	98.45
500	22	0	61.35
500	22	22	123.18
550	22	0	88.70
550	22	22	151.71
600	22	0	104.46
600	22	22	236.88

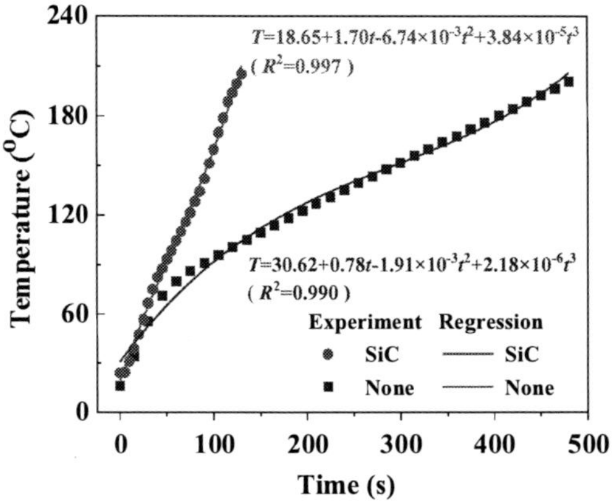

(a) 400 W

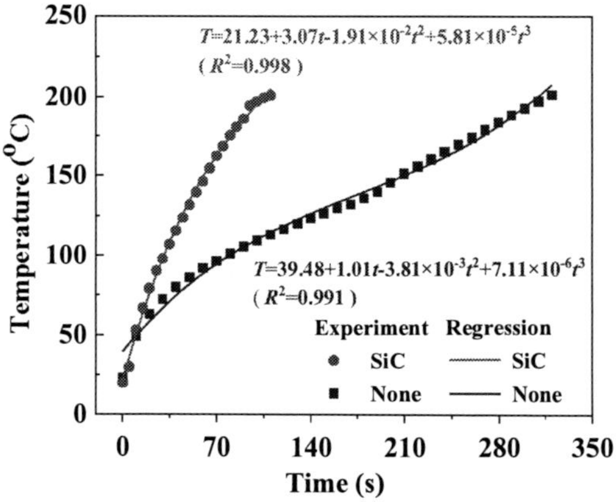

(b) 450 W

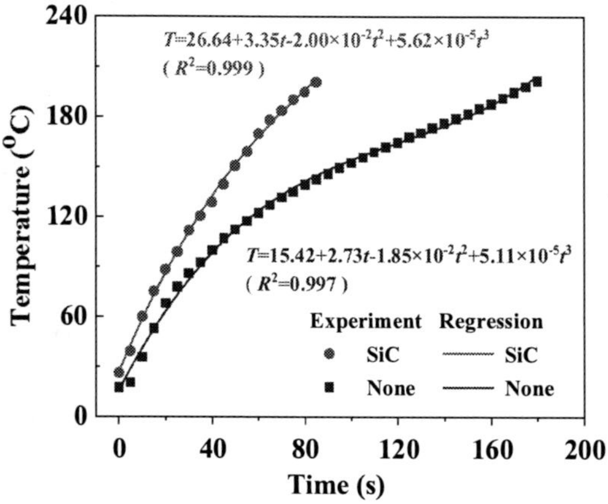

(c) 500 W

Figure 9. (Continued).

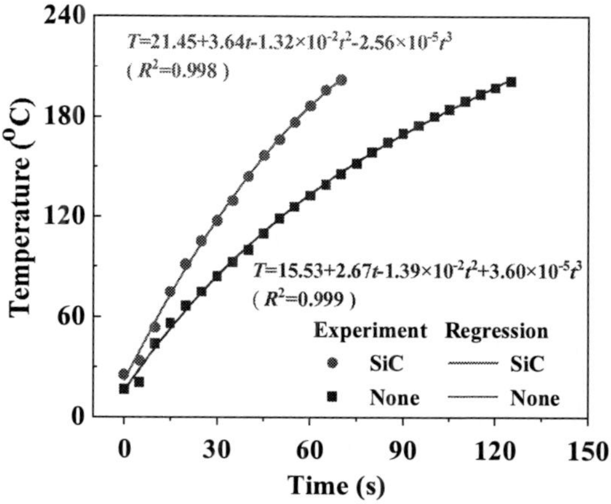

(d) 550 W

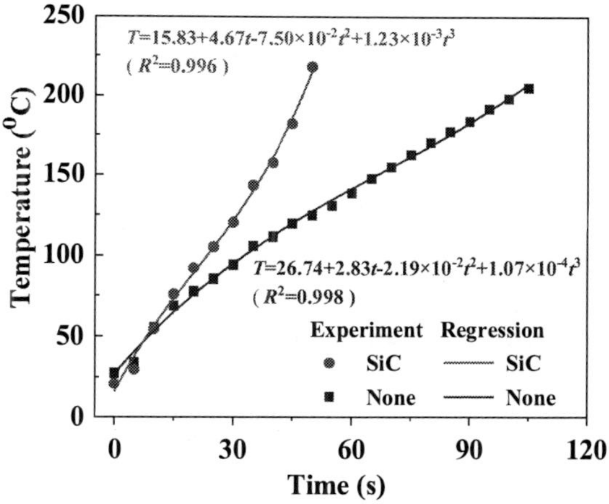

(e) 600 W

Figure 9. Comparison of microwave heating process of corn straw particles with and without SiC with various microwave powers (Cui et al. 2023c).

From Figure 9, the temperature of heating process of corn straw and SiC was strongly correlated with microwave power, and microwave irradiation time. When the mass of corn straw, mass of SiC, and microwave power were 22 g, 22 g and 400 W, respectively, formulae (17) to (18) can be used to describe the relationship of heating time and temperature.

$$T = 2.18\times10^{-6}t^3 - 1.91\times10^{-3}t^2 + 0.78t + 30.62 \text{ (400 W, without SiC)} \quad (17)$$

$$T = 3.84\times10^{-5}t^3 - 6.74\times10^{-3}t^2 + 1.70t + 18.65 \text{ (400 W, with SiC)} \quad (18)$$

where:

T = Temperature, °C,
t = Heating time of microwave irradiation proces, s.

The values of R^2 of formulae (17) and (18) were 0.990 and 0.997, respectively, and the formulae can be well used to describe the relationship of heating time and temperature with the microwave power of 400 W. When the microwave irradiation powers were 450, 500, 550, and 600 W, the formulae are shown in Figure 9, and the formulae can be used to describe the relationship of heating time, microwave irradiation power, and temperature, respectively.

The microwave irradiation heating process was affected by the mass ratio of biomass and SiC, and Table 15 shows the heating performances of corn straw at different SiC masses (Cui et al. 2023c).

Table 15. Average heating rates of corn straw at different SiC masses (Cui et al. 2023c)

Biomass mass (g)	Microwave power (W)	SiC mass (g)	Average heating rate (°C/min)
22	600	0	101.49
22	600	5.5	130.35
22	600	11	154.20
22	600	16.5	204.00
22	600	22	236.88

4.2. Microwave Irradiation for Heating Plastic with Microwave Absorbent

Some scholars studied the microwave-assisted heating process of microwave absorbent and plastics for energy recovery and energy production (Kulas et al. 2021).

Figure 10 shows the microwave irradiation heating performances of mixture particles of SiC and PP plastic in a fluidized bed reactor, and the particle sizes of PP and SiC were 96-109 μm and 74-86 μm, respectively (Cui et al. 2023e). In Figure 10, the heating time, mass ratio of SiC and PP, and fluidization velocity were fixed at 30 min, 8, and 3.54×10^{-3} m/s, respectively. When the microwave irradiation powers were 400, 600, 800, and 1000 W, the temperature of the mixture of SiC and PP increased to 600 °C rapidly.

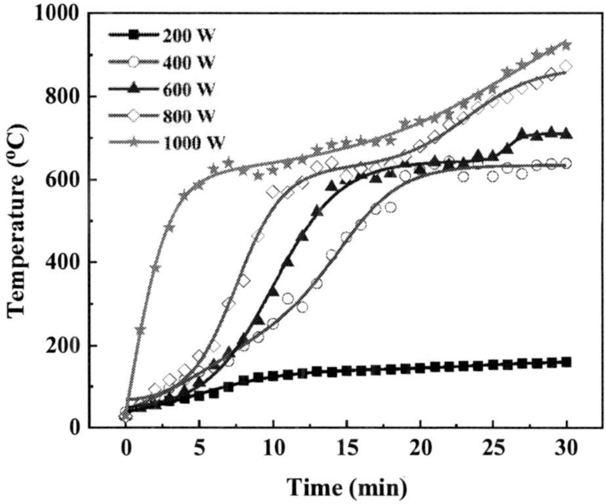

Figure 10. Heating process of PP with SiC addition in a fluidized bed reactor with various microwave powers (Cui et al. 2023e).

Reddy et al. (2022) studied microwave heating process of SiC and different plastic wastes at different mass ratios. Figure 11 shows the microwave irradiation heating performances of the mixture of different plastics and SiC at the mass ratio of 1, microwave power of 100 W, plastic size of 0.7 mm, and SiC size of 0.7 mm.

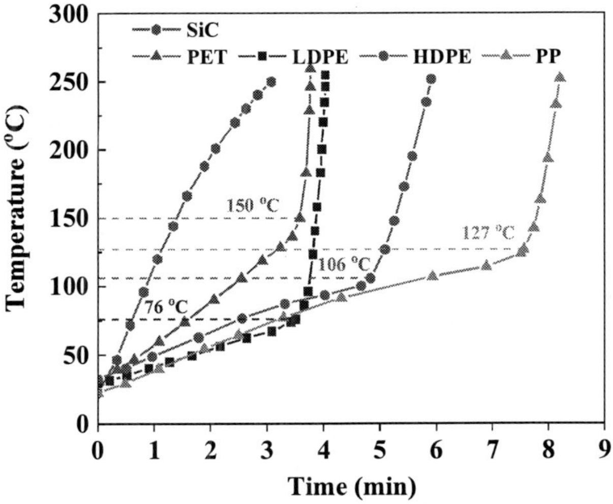

(a) Temperature of the mixture of SiC and different plastic

Figure 11. (Continued).

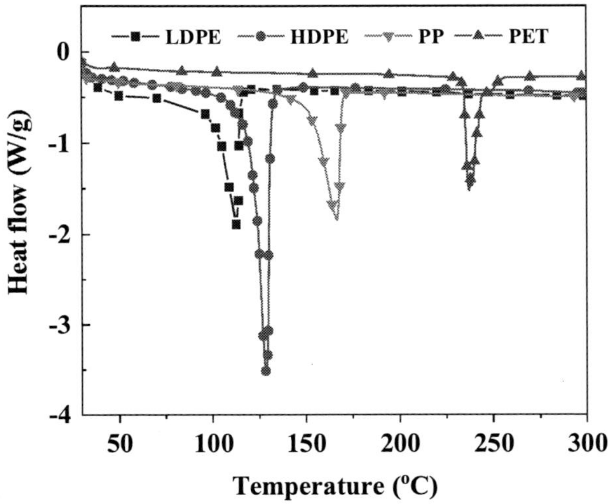

(b) DSC thermograms for different mixtures

Figure 11. Microwave heating process of SiC and different plastics by physical mixing (Reddy et al. 2022).

Figure 12 (a) shows the microwave irradiation heating performances of PP films embedded with SiC at different mass ratios, and Figure 12 (b) compares the heating process of the mixture of PP and SiC between physical mixing and hot-pressed film method.

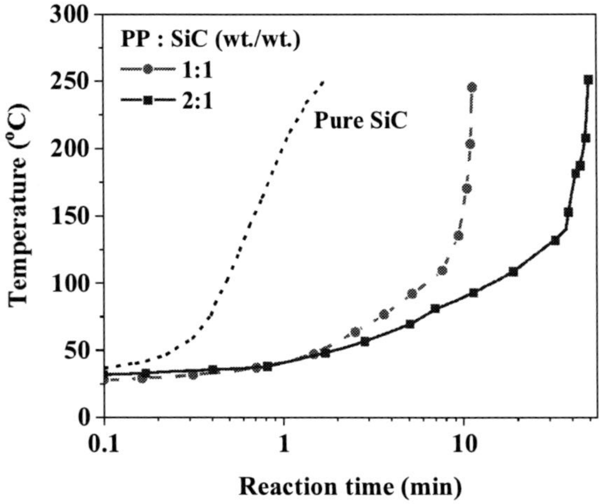

(a) Heating process of mixtures of SiC and PP with different mass ratios

Figure 12. (Continued).

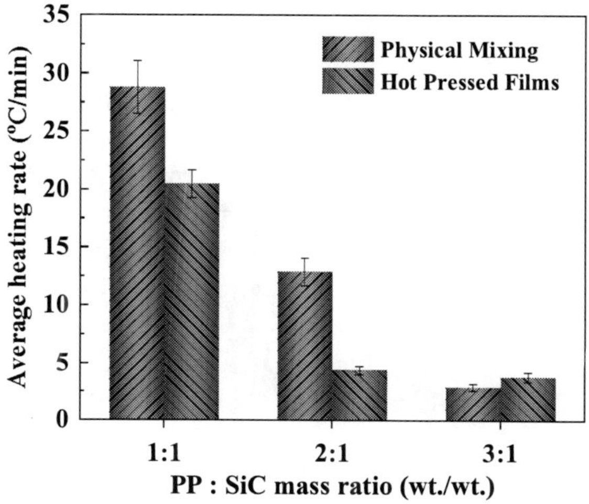

(b) Comparison of heating rates between physical mixing and hot-pressed film

Figure 12. Microwave heating performances of PP films embedded with SiC (Reddy et al. 2022).

Conclusion

Microwave irradiation for heating organic wastes is a useful technology, and this chapter gives a comprehensive introduction to microwave irradiation heating technology for organic wastes (biomass and plastics), microwave absorbent, and the mixture of organic wastes and microwave absorbent. Microwave-assisted pyrolysis technology can convert organic wastes to high valued products, which uses microwave heating as a heating method, and the microwave irradiation heating process affects the pyrolysis degree. In this chapter, the factors affecting the heating performance of microwave irradiation process, such as microwave power, organic waste quality, microwave adsorbent/organic waste mass ratio, are discussed. The microwave absorption of biomass and plastics is generally weak, and for the microwave-assisted pyrolysis process, the microwave absorbent is necessary. Mixtures of microwave absorbers with organic wastes have higher average heat-up rates than biomass and plastics. Some equations were introduced for regulating the microwave irradiation heating process.

Acknowledgments

The following financial supports are acknowledged: Financial support was provided by the National Natural Science Foundation of China (52076049, 52476005), Heilongjiang Province "Double First-class" Discipline Collaborative Innovation Achievement Project (LJGXCG2023-080), Heilongjiang Provincial Key R&D Program (2023ZX02C05), and Heilongjiang Provincial Key R&D Program "Unveiling the Leader" Project (2023ZXJ02C04).

References

Abdelsayed, V., Shekhawat, D., Smith, M. W., Link, D. & Stiegman, A. E. (2018). Microwave assisted pyrolysis of Mississippi coal: a comparative study with conventional pyrolysis. *Fuel*, *217*, 656-667.

Alabi, O. A., Ologbonjaye, K. I., Awosolu, O. & Alalade, O. E. (2019). Public and environmental health effects of plastic wastes disposal: a review. *Journal of Toxicology Risk Assessment*, *5*, 534-547.

Amalina, F., Krishnan, S., Zularisam, A. W. & Nasrullah, M. (2023). Effect of process parameters on bio-oil yield from lignocellulosic biomass through microwave-assisted pyrolysis technology for sustainable energy resources: Current status. *Journal of Analytical and Applied Pyrolysis*, *171*, 105958.

Amjith, L. R. & Bavanish, B. (2022). A review on biomass and wind as renewable energy for sustainable environment. *Chemosphere*, *293*, 133579.

Binner, E., Mediero-Munoyerro, M., Huddle, T., Kingman, S., Dodds, C., Dimitrakis, G., Robinson, J. & Lester, E. (2014). Factors affecting the microwave cooking of coals and the implications on microwave cavity design. F*uel Processing Technology*, *125*, 8-17.

Caporale, G. M., Spagnolo, N. & Almajali Awon. (2023). Connectedness between fossil and renewable energy stock indices: The impact of the COP policies. *Economic Modelling*, *123*, 106273.

Cui, L. F., Liu, C. Y., Liu, H., Zhao, W. K. & Zhang, Y. N. (2023a). Exergy transfer analysis of biomass and microwave based on experimental heating process. *Sustainability*, *15*, 388.

Cui, L. F., Qin, N., Li, H., Mostafa, E., Liu, H., Zhao, W. K. & Zhang, Y. N. (2023b). Heating performances of corn straw particles in a microwave chamber. *Energy Sources, Part A: Recovery, Utilization, and Environmental Effects*, *45*(3), 7186-7197.

Cui, L. F., Zhao, W. K., Mostafa, E. & Zhang, Y. N. (2023c). Heating performances of corn straw particle with/without SiC particle in a microwave chamber. *Environmental Science and Pollution Research*, https://doi.org/10.1007/s11356-023-30375-1.

Cui, Y. L., Zhang, Y. N., Cui, L. F., Liu, Y. F., Li, B. X. & Liu, W. (2023d). Microwave-assisted pyrolysis of polypropylene plastic for liquid oil production. *Journal of Cleaner Production*, *411*, 137303.

Cui, Y. L., Zhang, Y. N., Cui, L. F., Zhao, W. K., & Ahmad, F. (2023e). Microwave heating of silicon carbide and polypropylene particles in a fluidized bed reactor. *Applied Thermal Engineering*, *232*, 121009.

Depren, S. K., Kartal, M. T., Çoban, N. & Depren, Ö. (2022). Energy consumption and environmental degradation nexus: A systematic review and meta-analysis of fossil fuel and renewable energy consumption. *Ecological Informatics*, *70*, 101747.

Fan, S. H., Zhang, Y. N., Cui, L. F., Xiong, Q. G. & Maqsood, T. (2023). Conversion of polystyrene plastic into aviation fuel through microwave-assisted pyrolysis as affected by iron-based microwave absorbents. *Acs Sustainable Chemistry and Engineering*, *11*, 1054-1066.

Folorunso, O., Dodds, C., Dimitrakis, G. & Kingman, S. (2012). Continuous energy efficient exfoliation of vermiculite through microwave heating. *International Journal of Mineral Processing*, *114*, 69-79.

Fu, W. M., Dai, J. N., Zhang, Y. N., Guang, M. M., Liu, Y. F., & Li, B. X. (2021). Heating performances of high density polyethylene (HDPE) plastic particles in a microwave chamber. *Sustainable Energy Technologies and Assessments*, *48*, 101581.

Gadkari, S., Fidalgo, B. & Gu, S. (2017). Numerical investigation of microwave-assisted pyrolysis of lignin. *Fuel Processing Technology*, *156*, 473-484.

Gao Y. Z., Chen, B., Wu, D. X., Dai, Z. F., Wang, C. L. & Zhang, X. S. (2022). Comparative study of various solar power generation systems integrated with nanofluid-flat heat pipe. *Applied Energy*, *327*, 120052.

Guo, X., Ren, J., Xie, C. J., Lin, J. Y. & Li, Z. (2015). A comparison study on the deoxygenation of coal mine methane over coal gangue and coke under microwave heating conditions. *Energy Conversion Management*, *100*, 45-55.

Hong, K., Fu, W. M., Guang, M. M., Zhang, Y. N. & Li, B. X. (2021). Microwave heating performances of low density polyethylene (LDPE) plastic particles. *Journal of Analytical and Applied Pyrolysis*, *160*, 105356.

Huang, Y. F., Chiueh, P. T., Kuan, W. H. & Lo, S. L. (2016). Microwave pyrolysis of lignocellulosic biomass: Heating performance and reaction kinetics. *Energy*, *100*, 137-144.

Imenokhoyev, I., Windsheimer, H., Waitz, R., Kintsel, N. & Linn, H. (2013). Microwave heating technology: potentials and limits. *CFI-Ceramic Forum International*, *90*, 41-49.

Iturbides, R. D. L. C., Jauregui, H. U. & Polaert, I. (2022). Recent technological innovations on continuous microwave assisted biomass pyrolysis and perspectives for industrial scale applications. *Bioresource Technology Reports*, *19*, 101202.

Jafari, H., Kalantari, D. & Azadbakht, M. (2017). Semi-industrial continuous band microwave dryer for energy and exergy analyses, mathematical modeling of paddy drying and its qualitative study. *Energy*, *138*, 1016-1029.

Jambeck, J. R., Geyer, R., Wilcox, C., Siegler, T. R., Perryman, M., Andrady, A., Narayan, R. & Law, K. L. (2015). Plastic waste inputs from land into the ocean. Science, 347, 768-771.

Jeong, C. H., Ahn, S. H. & Lee, W. S. (2019). Four-kilowatt homogeneous microwave heating system using a power-controlled phase-shifting mode for improved heating uniformity. *Electronics Letters*, *55*, 465-467.

Ji, Z., Niu, D., Li, W., Wu, G., Yang, X. & Sun, L. (2022). Improving the energy efficiency of China: An analysis considering clean energy and fossil energy resources. *Energy*, *259*, 124950.

Kamari, M. L., Maleki, A., Daneshpour, R., Rosen, M. A., Pourfayaz, F. & Nazari M. A. (2023). Exergy, energy and environmental evaluation of a biomass-assisted integrated plant for multigeneration fed by various biomass sources. *Energy*, *263*, 125649.

Ke, C. F., Liu, T., Zhang, Y. N. & Xiong, Q. G. (2022). Energy absorption performances of silicon carbide particles during microwave heating process. *Chemical Engineering and Processing: Process Intensification*, *172*, 108796.

Kulas, D. G., Zolghadr, A. & Shonnard, D. (2021). Micropyrolysis of polyethylene and polypropylene prior to bioconversion: The effect of reactor temperature and vapor residence time on product distribution. ACS Sustainable Chemistry and Engineering, 9, 14443-14450.

Larhed, M., Moberg, C. & Hallberg, A. (2002). Microwave-accelerated homogeneous catalysis in organic chemistry. *Accounts of Chemical Research*, *35*, 717-727.

Li, L. Z., Cai, D. Q., Zhang, L. J., Zhang, Y., Zhao, Z. Y., Zhang, Z. L., Sun, J. F., Tan, Y. D. & Zou, G. F. (2023). Synergistic effects during pyrolysis of binary mixtures of biomass components using microwave-assisted heating coupled with iron base tip-metal. *Renewable Energy*, *203*, 312-322.

Lin, B. Q., Li, H., Chen, Z. W., Zheng, C. S. Hong, Y. D., Wang, Z. (2017). Sensitivity analysis on the microwave heating of coal: a coupled electromagnetic and heat transfer model. *Applied Thermal Engineering*, *126*, 949-962.

Lopez, G., Artetxe, M., Amutio, M., Bilbao, J. & Olazar, M., (2017). Thermochemical routes for the valorization of waste polyolefinic plastics to produce fuels and chemicals. Renewable and Sustainable Energy Reviews, 73, 346-368.

Maqsood, T., Dai, J. Z., Zhang, Y. N., Guang, M. M. & Li, B. X. (2021). Pyrolysis of plastic species: a review of resources and products. *Journal of Analytical and Applied Pyrolysis*, *159*, 105295.

Mushtaq, F., Mat, R. & Ani, F. N. (2014). A review on microwave assisted pyrolysis of coal and biomass for fuel production. *Renewable and Sustainable Energy Review*, *39*, 555-574.

Namazi, A. B., Allen, D. G., Jia, C. Q. (2015). Probing microwave heating of lignocellulosic biomasses. *Journal of Analytical and Applied Pyrolysis*, *112*, 121-128.

Omar, R., Idris, A., Yunus, R., Khalid, K. & Isma, M. I. A. (2011) Characterization of empty fruit bunch for microwave-assisted pyrolysis. *Fuel*, *90*, 1536-1544.

Peng, Z., Hwang, J., Bell, W., Andriese, M. & Xie, S. (2011). Microwave dielectric properties of pyrolyzed carbon. 2nd International Symposium on High-Temperature Metallurgical Processing. February 27-March 3 2011, California, USA, pp. 77-83.

Pickles, C. A., Gao, F. & Kelebek, S. (2014). Microwave drying of a low-rank sub-bituminous coal. *Minerals Enginnering*, *62*, 31-42.

Pritam, K., Puppala, H., Palla, S., Suriapparao, D. V. & Basak, T. (2023). A two-step hybrid multi-criteria approach to analyze the significance of parameters affecting microwave-assisted pyrolysis. *Process Safety and Environmenral Protection*, *171*, 975-985.

Qiu, H. P., Wu,Y. C., Chen, H. X., Wang, R. Y., Yu, J. C. & Lin,Y. R., (2023). Influence of SiC on the thermal energy transfer and storage characteristics of microwave-absorbing concrete containing magnetite and/or carbonyl iron powder. *Construction amd Building Materials*, *366*, 130191.

Ragaert, K., Delva, L. & Geem, K. V. (2017). Mechanical and chemical recycling of solid plastic waste. *Waste Management*, *69*, 24-58.

Rajpoot, L., Tagade, A., Deshpande, G., Verma, K., Geed, S. R., Patle, D. S. & Sawarkar, A. N. (2022). An overview of pyrolysis of de-oiled cakes for the production of biochar, bio-oil, and pyro-gas: Current status, challenges, and future perspective. *Bioresource Technology Reports*, *19*, 101205.

Ramasamy, S. & Moghtaderi, B. (2010). Dielectric properties of typical Australian wood-based biomass materials at microwave frequency. *Energy and Fuels*, *24*, 4534-4548.

Reddy, B. R., Malhotr, A., Najmi, S., Baker-Fales, M., Coasey, K., Mackay, M. & Vlachos, D. G. (2022). Microwave assisted heating of plastic waste: Effect of plastic/susceptor (SiC) contacting patterns. *Chemical Engineering & Processing: Process Intensification*, *182*, 109202.

Salema, A. A. & Ani, F. N. (2012). Microwave-assisted pyrolysis of oil palm shell biomass using an overhead stirrer. *Journal of Analytical and Applied Pyrolysis*, *96*, 162-172.

Salemaa, A. A., Yeowb, Y. K., Ishaquec, K., Anid, F. N., Afzala, M. T. & Hassane, A. (2013). Dielectric properties and microwave heating of oil palm biomass and biochar. *Industrial Crops and Products*, *50*, 366-374.

Sanchez-Coronado, J. & Chung, D. D. L. (2003). Thermomechanical behavior of a graphite foam. Carbon, 41, 1175-1180.

Sarker, M. (2011). Converting plastic waste into hydrocarbon fuel materials. Energy Engineering, 108, 35-43.

Schirmer, J., Kim, J. S. & Klemm, E. (2001). Catalytic degradation of polyethylene using thermal gravimetric analysis and a cycled-spheres-reactor. *Journal of Analytical and Applied Pyrolysis*, *60*, 205-217.

Shi, C., Shi, H. Q., Li, H., Liu, H., Mostafa, E., Zhao, W. K. & Zhang, Y. N. (2023a). Efficient heating of activated carbon in microwave field. *C-Journal of Carbon Research*, *9*, 48.

Shi, Z. Y., Jin, Y. H., Svanberg, R., Han, T., Minidis, A. B. E., Ann-Sofi, K. D., Kjeldsen, C., Jönsson, P. G. & Yang, W. H. (2023b). Continuous catalytic pyrolysis of biomass using a fluidized bed with commercial-ready catalysts for scale-up. *Energy*, *273*, 127288.

Sinha, A. A., Saini, G., Sanjay, Shukla, A. K., Ansari, M. Z., Dwivedi, G. & Choudhary, T. (2023). A novel comparison of energy-exergy, and sustainability analysis for biomass-fueled solid oxide fuel cell integrated gas turbine hybrid configuration. *Energy Conversion and Management*, *283*, 116923.

Somani, A., Nandi, T. K., Pal, S. K. & Majumder, A. K. (2017). Pre-treatment of rocks prior to comminution - a critical review of present practices. *International Journal of Mining Science and Technology*, *27*, 339-348.

Su, W., Lu, Z. F., She, X. H., Zhou, J. M., Wang, F., Sun, B. & Zhang, X. S. (2022). Liquid desiccant regeneration for advanced air conditioning: A comprehensive review on desiccant materials, regenerators, systems and improvement technologies. *Applied Energy*, *308*, 118394.

Sun, J. M., Luo, J., Lin, J. H., Ma, R., Sun, S. C., Fang, L. & Li, H. W. (2022). Study of co-pyrolysis endpoint and product conversion of plastic and biomass using microwave thermogravimetric technology. *Energy*, *247*, 123547.

Uhm, H. S., Na, Y. H., Hong, Y. C., Shin, D. H., Cho, C. H. & Park, Y. K. (2014). High-efficiency gasification of low-grade coal by microwave steam plasma, *Energy Fuels*, *28*, 4402-4408.

Vos, B., Mosman, J., Zhang, Y., Poels, E. & Bliek, A. (2003). Impregnated carbon as a susceptor material for low loss oxides in dielectric heating. *Journal of Materials Science*, *38*, 173-182.

Wang, H. X., Teng, H. W., Wang, X. Y., Xu, J. L. & Sheng L. X. (2022a). Physicochemical modification of corn straw biochar to improve performance and its application of constructed wetland substrate to treat city tail water. *Journal of Environmental Management*, *310*, 114758.

Wang, W. Z., Wang, C., Huang, Y., Lu, H. Y., Chen, J., Shi, J. W. & Jin, H. (2022b). Heat, electricity, and fuel gas ploy-generation system on an island based on plastic waste gasification in supercritical water. ACS Sustainable Chemistry and Engineering, 10, 13786-13791.

Wang, X. W., Wang, X. J., Muhoza, B., Feng, T. T., Xia, S. Q. & Zhang, X. M. (2020). Microwave combined with conduction heating effects on the tenderness, water distribution, and microstructure of pork belly. *Innovtive Food Science and Emerging Technologies*, *62*, 102344.

Williams, P. T. & Slaney, E., (2007). Analysis of products from the pyrolysis and liquefaction of single plastics and waste plastic mixtures. *Resources Conservation and Recycling*, *51*, 754-769.

Yana, S., Nizar, M., Irhamni, & Mulyati, D. (2022). Biomass waste as a renewable energy in developing bio-based economies in Indonesia: A review. *Renewable and Sustainable Energy Reviews*, *160*, 112268.

Zhang, L. J., Tan, Y. D., Cai, D. Q., Sun, J. F., Zhang, Y., Li, L. Z., Zhang, Q., Zou, G. F., Song, Z. L. & Bai, Y. H. (2022). Enhanced pyrolysis of woody biomass under interaction of microwave and needle-shape metal and its production properties. *Energy*, *249*, 123667.

Zhang, Y. N., Chen, P., Liu, S. Y., Fan, L. L., Zhou, N., Min, M., Cheng Y. L., Peng, P., Anderson, E., Wang, Y. P., Wan, Y. Q., Liu, Y. H., Li, B. X. & Ruan, R. (2017). Chapter 6 - Microwave-assisted pyrolysis of biomass for bio-oil production. Intech Open, London, UK, 133-148.

Zhao, X. Q., Zhang, J., Song, Z. L., Liu, H. Z., Li, L. Z. & Ma, C. Y. (2011). Microwave pyrolysis of straw bale and energy balance analysis. *Journal of Analytical and Applied Pyrolysis*, *92*, 43-49.

Zou, R. G., Wang, C. X., Qian, M., Huo, E., Kong, X., Wang, Y. P., Dai, L. L., Wang, L., Zhang, X. S., Mateo, Wendy. C., Ruan, R. & Lei, H. W. (2022). Catalytic co-pyrolysis of solid wastes (low-density polyethylene and lignocellulosic biomass) over microwave assisted biochar for bio-oil upgrading and hydrogen production. *Journal of Cleaner Production*, *374*, 133971.

Chapter 5

Microwave Irradiation for Pyrolysis of Organic Wastes for Oil Production

Sichen Fan
Wenke Zhao
Yaning Zhang*
and Bingxi Li
School of Energy Science and Engineering, Harbin Institute of Technology, Harbin, China

Abstract

As microwave irradiation technology advances rapidly, the application for oil production via pyrolysis has garnered significant attention. This chapter introduces the use of microwave radiation for the pyrolysis of biomass, and assesses the impacts of biomass type, microwave power, and pyrolysis temperature on oil yields and components. The effects of plastic types, reaction conditions, microwave absorbent, and catalyst on oil yields and components during microwave radiation for the pyrolysis of plastics were evaluated in detail. To elucidate the function of microwave absorbents, the impacts of different types of microwave absorbents and the ratio of plastics to catalysts on the yields and components of microwave radiation for pyrolysis of plastics were reviewed. Additionally, to comprehend the mechanism of catalyst action, the effects of the types of catalysts, the ratios of plastics to catalysts, and the catalytic mode were also studied from the perspectives of two target factors: oil yields and components. This chapter provides data to support the screening of organic wastes, microwave absorbents, catalysts, and

* Corresponding Author's Email: ynzhang@hit.edu.cn.

In: Characteristics and Applications of Microwave Irradiation
Editors: Yaning Zhang and Wenke Zhao
ISBN: 979-8-89530-170-8

suitable reaction parameters for oil production through microwave irradiation for pyrolysis.

Keywords: microwave irradiation, oil production, biomass, plastics, microwave absorbent, catalyst

1. Introduction

Oil is a crucial chemical with extensive applications in both industrial production and daily life. It is a type of liquid fuel obtained by high-temperature decomposition of organic solid wastes such as biomass, coal, plastic, etc. It serves as a fundamental raw material for the manufacture of synthetic rubber, synthetic resin, synthetic coatings, and petrochemical products (Wei et al. 2014). Through different processing methods, oil can be made into various chemical products, such as styrene and polymers, which find widespread use in the fields of synthetic fibers, rubber, and various other applications (Wang et al. 2014). Moreover, oil plays a vital role in the energy industry, where it is utilized as a fuel to power industrial boilers, power plants, and other equipment, providing heat and electricity (Schüppel & Gräbner, 2024). Oil can also be used as a kerosene substitute for refining lubricants and coal chemical products, significantly aiding the replacement of traditional petroleum energy. In addition, oil has many uses in daily life and consumer products. It can be used as a raw material to make cosmetic products such as perfumes and spices and to produce daily necessities such as detergents, pigments, and dyes (Mostafazadeh et al. 2018). Oil can also be used in paving asphalt roads and producing waterproof materials to ensure the smoothness of roads and the waterproof performance of buildings. As science and technology continue to advance and environmental awareness improves, the demand for oil continues to increase. Through technological development and innovation, oil will increasingly provide convenience and support for human production and life (Lu et al. 2009).

The composition of oil is primarily determined by the raw materials used, the operating methods, and the reaction conditions. It encompasses several chemical categories, such as ethers, esters, aldehydes, ketones, phenols, organic acids, and alcohols (Zhang et al. 2007). For instance, oil derived from woody biomass primarily includes six categories: simple oxygenated compounds (non-aromatic), complex oxygenated compounds, sugars and sugar derivatives, furans, aromatic compounds, and high molecular substances

(Mohan et al. 2006). Simple oxygenated compounds include acids, esters, alcohols, ketones, and aldehydes. Complex oxygenated compounds include 2-hydroxyacetaldehyde, 1-hydroxy-2-propanone, etc. Common sugars and sugar derivatives include levoglucosan, xylose, arabinose, glucose, fructose, and sorbitol (Sakulkit et al. 2020). Furan compounds primarily include furan, furfural, 2-furanone, butyrolactone, and methyl-2-furanone (Fukuda, 2015). Aromatic substances, which are compounds with benzene rings, are divided into oxygen-free aromatic compounds, phenolic aromatic compounds, catechol aromatic compounds, guaiacol aromatic compounds, and dimethoxyphenol aromatic compounds. The main components include toluene, benzene, xylene, phenol, methylphenol, dimethylphenol, ethylphenol, and guaiacol. Polymer substances are mostly dimers, trimers, and oligomers. The oil derived from plastics is a complex mixture consisting of numerous hydrocarbons and hundreds of chemical compounds. Oil obtained from the pyrolysis of polypropylene (PP) exhibits a diverse range of chemical compounds and a wide distribution of products. The compositions typically range from C5 to C38, encompassing monocyclic aromatic hydrocarbons (MAHs), polycyclic aromatic hydrocarbons (PAHs), alkanes, alkenes, alcohols, and esters (Chang, 2023). Like the oil compositions from PP, the oil compositions from high-density polyethylene (HDPE) and low-density polyethylene (LDPE) also mainly contain MAHs, PAHs, alkanes, and alkenes. Although the random chain scissions both occur during the pyrolysis of HDPE and PP, the alkane and olefin contents obtained from HDPE are generally higher than those from PP, due to the linear and unbranched carbon chain structures in HDPE (Miandad et al. 2019). In contrast, the polymer chain of polystyrene (PS) contains thermally unstable vinyl functional groups that distinguish it from HDPE and PP, facilitating chain end breakage and depolymerization to generate styrene monomer, resulting in high MAHs content (Sharuddin et al. 2016). The pyrolysis oil also contains PAHs, such as indene, naphthalene, and methylnaphthalene, but the concentrations of other compounds (i.e., alkanes and olefins) are relatively low or even absent. Compared with HDPE and PP, the pyrolysis oil produced from PS has the simplest chemical composition and the narrowest product distribution, consisting mainly of C6-C14 aromatic hydrocarbons. The abundant styrene monomer in oil extracted from PS is a promising source of chemical precursors for the plastics industry.

The oil color is usually yellow to black, depending on the chemical composition and the presence of carbon particles. The density is approximately 1.2 g/cm^3, which exceeds that of crude oil. However, its energy density is less than half that of crude oil, and its energy content amounts to roughly 30-41% of crude oil (Arjharn et al. 2022). The higher heating value (HHV) of the oil is influenced by the type of raw material and the oxygen content. Generally, the HHV of oil derived from polyolefin plastic can reach up to 40 MJ/kg (Zevenhoven et al. 1997), while the HHV of oil derived from biomass is around 20 MJ/kg (Günther et al. 2012). The maximum moisture content of the oil can reach 30% to 45%. The moisture in the oil mainly comes from the surface water carried by the material and the dehydration reaction during the thermal cracking process. The moisture in the oil helps to reduce the viscosity and improve its stability, but the existence of moisture reduces the HHV of the oil. The viscosity of the oil ranges from 25 cP to 1 000 cP, depending on the water content (Ahmad et al. 2014). The low pH value of the oil is primarily due to the presence of organic acids. Therefore, it is advisable to use acid corrosion-resistant materials for oil collection and storage, such as stainless steel or polyolefin compounds (Wang et al. 2022). Due to the presence of various oxygenated compounds, bio-oil exhibits undesirable properties, including strong corrosiveness, high viscosity, poor stability, and low HHV (Chen et al. 2024). The strong acidity of the oil indicates significant corrosiveness, necessitating strict processing, transportation, and storage conditions (Yang et al. 2015). Oil can also easily undergo gelatin during storage, resulting in suboptimal atomization and combustion characteristics. The moisture in the oil poses challenges in its removal and contributes to issues such as corrosiveness, cold filter plugging, and catalyst deactivation (Lujaji et al. 2016). The high oxygen content of the oil imparts high polarity, making it immiscible with conventional petroleum. Moreover, the presence of acids, olefins, and phenols can induce the formation of macromolecular products through polymerization, resulting in increased viscosity and reduced fluidity. Therefore, to facilitate further utilization, it is essential to modify bio-oil (Huber et al. 2006).

The preparation and upgrading of oil have consistently been research focal points for scholars worldwide. Currently, numerous technologies including hydrothermal liquefaction, pyrolysis, Fischer-Tropsch synthesis, and transesterification are frequently employed (Moya et al. 2017; Okolie et al. 2021). Pyrolysis technology involves the conversion of raw materials into three-phase products consisting of low-molecular-weight substances under oxygen-free or low-oxygen conditions. This technology enables the conversion of low-energy-density biomass, which is challenging to process using conventional methods, into high-energy-density gas, liquid, and solid products through a cost-effective and continuous production process. It reduces the volume of biomass, enhances the convenience of storage and transportation, and enables the extraction of high-value-added liquid chemicals (Zhang et al. 2023). The continuous growth of global energy demand and the improvement of environmental protection awareness have led to a gradual increase in the emphasis on oil policy (Guo et al. 2015). Over the past decade, a series of measures have been implemented to promote the development of the oil industry. These measures include improving oil quality standards, increasing the development of alternative oil energy sources, encouraging market competition, strengthening quality supervision, and issuing oil price control policies. These efforts have yielded remarkable results, actively contributing to the reduction of oil consumption, environmental pollution, and the promotion of sustainable energy development.

This chapter presents an exploration into the microwave irradiation process for the pyrolysis of organic wastes (i.e., biomass and plastic), focusing on various influencing factors. The effect of biomass or plastic types, microwave power, pyrolysis temperature, microwave absorbent types, plastic to microwave absorbent ratio, catalyst type, plastic to catalyst ratio, and catalytic mode are presented, and the yield and component of oil are analyzed.

2. Microwave Irradiation for Pyrolysis of Biomass

2.1. Effect of Biomass Type

The process of utilizing microwave technology for biomass oil production represents a novel approach to biomass conversion. The high heating rate and conversion efficiency of microwave irradiation for biomass pyrolysis have garnered significant attention in research. Given the distinct chemical composition and content of various types of biomass, the pyrolysis reaction characteristics, product yield, and components also exhibit variations. Currently, the raw materials used for microwave-assisted pyrolysis (MAP) of biomass to produce oil can be mainly divided into three categories: lignocellulosic biomass, algal biomass, and sludge biomass. Lignocellulosic biomass primarily comprises cellulose, hemicellulose, and lignin, with the three components accounting for approximately 40-60%, 20-41%, and 10-25% of the biomass, respectively (Huang et al. 2012). Different types of raw materials will result in varying proportions of the three components, leading to differences in oil yield. Table 1 lists oil yields from microwave irradiation for pyrolysis of different biomass types. Overall, the oil yields from MAP of biomass are in the range of 9-49.79 wt.%. Within lignocellulosic biomass, the oil yields from woody biomass (32.92 wt.% and 32.63 wt.%, Halim & Swithenbank, 2016) exceed those from straw biomass (26.1 wt.%, Suriapparao et al. 2020). What's more, sludge biomass (49.79 wt.%, Tian et al. 2011) and algal biomass (35.02 wt.%, Ferrera-Lorenzo et al. 2014) can also obtain pyrolysis oil with relatively high yields through precise control of conditions.

Table 1. Oil yield from microwave irradiation for pyrolysis of different biomass types

Biomass type	Microwave absorbent	Oil yield (wt.%)	Reference
Chlorella vulgaris (CV)	/	17.76	Chen et al. 2023
Tea oil seed residue (TR)	/	21.03	Chen et al. 2023
CV+TR	/	18.17-21.13	Chen et al. 2023
Macroalgae	Char	35.02	Ferrera-Lorenzo et al. 2014
Microalgae	Char	15.60	Beneroso et al. 2013
Mississippi coal	/	57.00	Abdelsayed et al. 2018
Bituminous coal		14.00	Gasner et al. 2007
Kraft lignin	Char	38.00	Farag et al. 2016
Rubber wood	/	32.92	Halim & Swithenbank, 2016
Malaysian wood pellets	/	36.23	Halim & Swithenbank, 2016
Sewage sludge	Graphite	49.79	Tian et al. 2011
Dyeing sludge	/	8	Liu et al. 2018
Empty oil palm fruit bunch	Activated carbon	21.9 ± 1.4	Francis Prashanth et al. 2020
Kitchen waste	Graphite	34.15	Suriapparao & Vinu, 2015a
Garden waste	Graphite	31.83	Suriapparao & Vinu, 2015a
KW + GW + paraffin oil (PO) + cellulose (Cl)	Silica beads	41.09	Suriapparao & Vinu, 2015a
Prosopis jouliflora	Fly ash	39.87	Suriapparao et al. 2015
Rice straw	biochar	26.1	Suriapparao et al. 2020
Sugarcane bagasse	biochar	29.3	Suriapparao et al. 2020

Due to the presence of nitrogen and oxygen elements in biomass, the components of bio-oil obtained through microwave irradiation for pyrolysis exhibit notable diversity. Apart from various hydrocarbon compounds, bio-oil obtained through microwave irradiation for pyrolysis also contains aromatic oxygenates and aliphatic oxygenates. Furthermore, it can be also subdivided into acids, aldehydes, ketones, esters, and other compounds. The classification of oil components depends on the research objective and oil application. Table

2 and Table 3 list the compositions of oil obtained from MAP of rice straw and sugarcane bagasse by Suriapparao et al. (2020). The content of oxygen compounds in oil is significantly higher than that of hydrocarbon compounds. The highest contents were observed in aromatic and aliphatic oxygenates, which were 58.76 area% and 18.81 area% in pyrolysis oil from rice straw, and 76.82 area% and 13.11 area% in pyrolysis oil from sugarcane bagasse, respectively. The main oxygen compounds from rice straw are phenol, 4-pethyl- (5.16 area%), 2-furanmethanol (5.11 area%), phenol (4.91 area%), 2-methoxy-4-vinylphenol (4.75 area%), phenol, 2-methoxy-(4.35 area%). Among the hydrocarbon compounds, the content of aromatic hydrocarbons exceeds that of aliphatic hydrocarbons, and this phenomenon is most obvious in the oil from MAP of sugarcane bagasse (6.99 area% vs 0.50 area%). The primary aromatic hydrocarbon compositions in the oil obtained through MAP of rice straw were benzene, ethyl- (2.05 area%) and naphthalene (1.46 area%). Additionally, other components mainly consisit of nitrogen-containing compounds, including 1-pyrazole, 3, 5-dimethyl- (2.72 area%), 1, 4-dimethylpyrazole (2.29 area%).

Table 2. Oil components from microwave irradiation for pyrolysis of different biomass types. (Data from (Suriapparao et al. 2020))

Components (area%)	Rice straw	Sugarcane bagasse
Monoaromatic Hydrocarbons	3.70	4.59
Polyaromatic Hydrocarbons	2.94	2.40
Aromatic Oxygenates	58.76	76.82
Aliphatic Oxygenates	18.81	13.11
Aliphatic Hydrocarbons	6.33	0.50
Other	9.46	2.58

Table 3. Oil compositions from microwave irradiation for pyrolysis of rice straw. (Data from (Suriapparao et al. 2020))

Compound Name	Selectivity (%)
Monoaromatic Hydrocarbons	3.70
Benzene, Ethyl-	2.05
Benzene, 1,4-Dimethyl-	0.71
Benzene, 1,2-Dimethyl-	0.54
D-Limonene	0.41
Polyaromatic Hydrocarbons	2.94
Naphthalene	1.46
Naphthalene, 1-Methyl-	0.74

Table 3. (Continued)

Compound Name	Selectivity (%)
Phenanthrene	0.41
1,4-Dimethylnaphthalene	0.33
Aromatic Oxygenates	58.76
Phenol, 4-Methyl-	5.16
2-Furanmethanol	5.11
Phenol	4.91
2-Methoxy-4-Vinylphenol	4.75
Phenol, 2-Methoxy-	4.35
Phenol, 3-Ethyl-	3.58
2,3-Dihydro-Benzofuran	2.86
Phenol, 2-Methyl-	2.74
Guaiacol, 4-Ethyl-	2.44
Phenol, 3,4-Dimethyl-	1.92
1,2-Benzenediol	1.73
Phenol, 2,6-Dimethyl-	1.73
3-Oxaspiro[5.5]Undecane-1,5-Dione, 4-Methyl-3-Phenyl-	1.72
2-(Acetyloxy)Ethyl Acetate	1.34
2-Furanmethanol, Tetrahydro-	1.25
Creosol	1.23
Phenol, 2,6-Dimethoxy-	1.14
Phenol, 2,4-Dimethyl-	1.10
2-Ethylphenol	0.81
1,2-Benzenediol, 3-Methyl-	0.75
Phenol, 2-Ethyl-5-Methyl-	0.71
2h-1-Benzopyran-2-One, 3,4-Dihydro-6-Hydroxy-	0.60
Phenol, 2,6-Dimethoxy-4-(2-Propenyl)-	0.59
Oxalic Acid, Isohexyl 2-Isopropylphenyl Ester	0.57
Phenol, 3-Ethyl-5-Methyl-	0.54
9,10-Epoxy-4,8-Ethanocyclohepta[C]Furan-1,3-Dione, Hexahydro-	0.51
4-Ethylcatechol	0.50
1-Naphthalenol, 3-Methyl-	0.49
Phenol, 2,3,5-Trimethyl-	0.46
1,2-Benzenediol, 4-Methyl-	0.44
Benzofuran, 4,7-Dimethyl-	0.44
1,5-Naphthalenediol, 5,6,7,8-Tetrahydro-	0.39
Trans-Isoeugenol	0.37
2-(2-Formylbenzyl)Phenol	0.32
2-Propanone, (4-Hydroxy-3-Methoxyphenyl)-	0.32
1-Inden-1-One, 2,3-Dihydro-	0.31
5-Benzoylpentanoic Acid	0.31
3,5-Dimethoxy-4-Hydroxytoluene	0.30
Aliphatic Oxygenates	18.81
3-Methyl-2-Cyclopenten-1-One	3.36
2-Cyclopenten-1-One, 2-Hydroxy-3-Methyl-	2.10
Ethanone, 1-(2-Furanyl)-	2.09

Compound Name	Selectivity (%)
2-Cyclopenten-1-One, 2-Methyl-	1.62
2-Cyclopenten-1-One, 2,3-Dimethyl-	1.57
2-Cyclopenten-1-One, 3-Ethyl-2-Hydroxy-	0.99
1-Octanol, 2-Butyl-	0.85
Acetic Acid, 3,7,11,15-Tetramethyl-Hexadecyl Ester	0.81
2-Cyclopenten-1-One, 3-Ethyl-2-Hydroxy-	0.78
7-OxabicycloHept-3-En-2-One, 5-Ethenyl-1,3,4,5,6-Pentamethyl-	0.60
1-Dodecanol	0.52
Cyclohexanone, 4-Methylene-	0.48
9-Octadecenoic Acid -, Methyl Ester	0.44
Cyclohexanone	0.42
Hexanoic Acid	0.41
Cyclohexanone, 3-Ethenyl-	0.40
3,4-Dimethyl-2-Cyclopenten-1-One	0.37
Hexadecanoic Acid, Methyl Ester	0.34
2-Cyclopenten-1-One, 2,3,4-Trimethyl-	0.33
2-Myristynoic Acid	0.33
Aliphatic Hydrocarbons	6.33
BicycloOctane, 2-Methyl-	0.85
Tetradecane	0.77
1,3-Cyclohexadiene, 1-Methyl-4-(1-Methylethyl)-	0.71
Pentadecane	0.70
Tridecane	0.63
Decane, 2-Methyl-	0.51
Nonadecane	0.48
Dodecane, 2-Methyl-6-Propyl-	0.36
1-Methyl-4-(1,5-Dimethyl-4-Hexenylidene)-1-Cyclohexene	0.34
Eicosane	0.34
2,4-Hexadiene, 2,3-Dimethyl-	0.32
6,6-Diethylhoctadecane	0.31
Others	9.46
1-Pyrazole, 3,5-Dimethyl-	2.72
1,4-Dimethylpyrazole	2.29
Acetamide, N-Tert-Butyl-2-(1h-Indol-3-Yl)-2-Oxo-	0.62
1-Indole-3-Carboxaldehyde	0.49
1-Indole, 3-Methyl-	0.46
1-Ethyl-3,5-Dimethyl-1h-Pyrazole-4-Carbaldehyde -3-HydroxyHydrazone	0.41
1-Indole	0.41
6-(Ethylamino)-1,3,5-Triazine-2,4-Diol	0.40
1-Pyrazole-4-Carboxylic Acid, 3-Methyl-	0.35
5-Tetrazol-5-One, 1,4-Dihydro-1,4-Dimethyl-	0.35
1-Pyrrole, 1-Pentyl-	0.34
(2-3-Anilino-2-Propenal	0.31
[Diphenylcarbamyl]TerephthalicDihydroxamic Acid	0.31

2.2. Effect of Microwave Power

The reaction conditions during the process of microwave irradiation for biomass pyrolysis significantly impact the quality and quantity of the products. The oil yield and component changes vary under different reaction conditions influenced by factors such as microwave power, pyrolysis temperature, and residence time. Microwave power is a distinctive reaction condition in the MAP process, directly impacting the heating rate of biomass. Higher microwave power results in increased microwave absorption by both the biomass and microwave absorbent, converting them into heat energy. The process reduces the temperature gradient between the internal and external surfaces of the raw material, ultimately resulting in a higher final heating rate. Figure 1 shows the effect of microwave power on the oil yield from microwave irradiation for pyrolysis of biomass. In general, insufficient heat supply from low microwave power leads to inadequate cracking and consequently, low oil yield. Conversely, excessive microwave power inhibits the formation of oil. It is obtained from Figure 1 that the oil yields produced from MAP of sewage sludge (Tian et al. 2011; Dai et al. 2014), prosopis jouliflora (Suriapparao et al. 2015), oil plam shell (Mushtaq et al. 2015), food waste (Liu et al. 2014), algae (Li et al. 2013), and microalgae (Du et al. 2011) initially exhibited an increase followed by a decrease as the microwave power increased. Typically, the appropriate microwave power range falls between 400-750 W. However, it is important to note that this trend may not apply universally to all studies. In the case of corn stover studied by Yu et al. (2019) during MAP, the oil yield showed minor fluctuations within a small range as the microwave power increased, eventually showing a gradual increase. Therefore, finding a critical value to maximize the oil yield is crucial when determining the optimal microwave power.

Table 4. Oil components from microwave irradiation for pyrolysis of prosopis jouliflora at different microwave powers (Data from (Suriapparao et al. 2015)

Components (%)	280 W	420 W	560 W	700 W
Total phenolics	~32	~42	~44	~45
Acids/alcohols/ketones	~37	~33	~23	~21
Furans	~17	~13.5	~15.5	~12.5
Cyclopentaneones	~12.5	~10	~8	~8.5
Aromatics	~1	~1.5	~2.5	~9

Microwave power significantly influences the composition of oil generated from microwave irradiation for biomass pyrolysis. Table 4 lists the oil components from microwave irradiation for pyrolysis of prosopis jouliflora at different microwave powers (Suriapparao et al. 2015). The presence of oxygen in prosopis jouliflora results in the formation of oil components such as phenolics, acid, alcohols, ketones, furans, cyclopentaneones, and aromatics. Phenolics and aromatics showed the tendency of increasing gradually with the improvement of microwave power, whereas the contents of acid, alcohols, ketones, and cyclopentaneones displayed contrasting trends. The main components consist of phenolics and acid/alcohols/ketones, accounting for 32-45% and 21-37%, respectively. Oil compositions from microwave irradiation for pyrolysis of sewage sludge at different microwave powers by Tian et al. (2011) are listed in Table 5. The main oil components from sewage sludge contained alkanes (12.8-21.7 area%), monoaromatic compounds (14,1-29.5 area%), and polycyclic aromatic hydrocarbons (5.3-7.2 area%) at the powers of 200-1200 W.

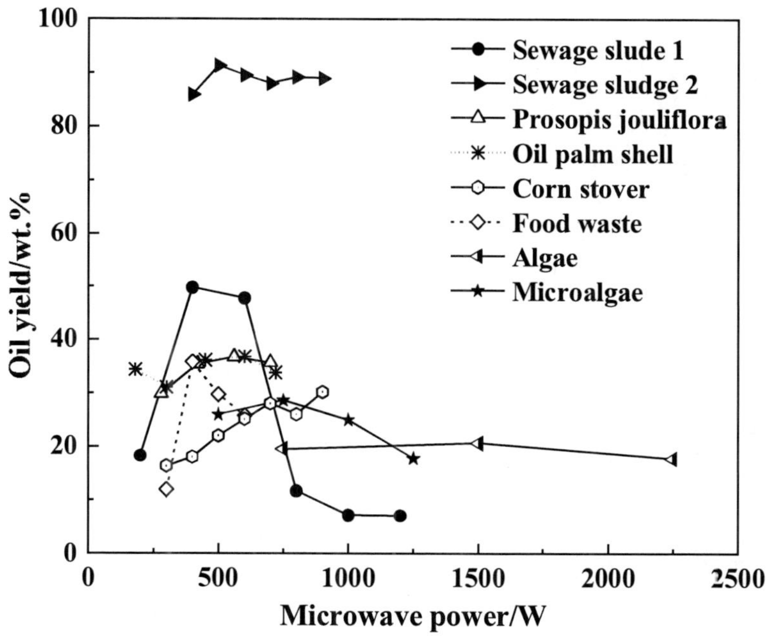

Figure 1. Oil yield from microwave irradiation for pyrolysis of biomass at different microwave powers: sewage sludge (Tian et al. 2011) (Dai et al. 2014), prosopis jouliflora (Suriapparao et al. 2015), oil palm shell (Mushtaq et al. 2015), corn stover (Yu et al. 2009), food waste (Liu et al. 2014), algae (Li et al. 2013), and microalgae (Du et al. 2011).

Table 5. Oil compositions from microwave irradiation for pyrolysis of sewage sludge at different microwave powers. (Data from (Tian et al. 2011))

Compositions (area%)	200 W	400 W	600 W	800 W	1000 W	1200 W
Alkanes	21.7	17.8	19.7	14.6	13.6	12.8
Heptane	/	/	/	/	0.1	0.2
Octane	/	/	/	/	/	0.5
Nonane	/	/	/	/	/	1.1
Decane	/	/	/	2.3	0.2	1.2
Undecane	/	0.1	0.4	2.4	2.2	1.1
Dodecane	/	/	0.3	2.5	1.4	0.9
Tridecane	/	0.8	2.0	2.2	1.8	1.2
Tetradecane	0.5	0.7	0.5	2.0	1.3	0.9
Pentadecane	1.2	1.1	1.8	1.2	2.4	2.3
Hexadecane	2.2	2.3	2.6	1.5	2.9	1.6
Heptadecane	3.9	1.3	1.5	0.3	1.3	0.9
Octacosane	2.3	2.2	1.4	0.1	/	0.4
Nonacosane	2.6	2.5	1.9	0.1	/	0.4
Eicosane	2.0	2.3	2.2	/	/	0.1
Heneicosane	1.9	0.8	2.2	/	/	/
Tricosane	1.7	2.4	1.9	/	/	/
Tetracosane	1.4	0.6	0.9	/	/	/
Pentacosane	1.1	0.3	0.1	/	/	/
Octacosane	0.6	0.2	/	/	/	/
Nonacosane	0.3	0.1	/	/	/	/
Triacontane	/	0.1	/	/	/	/
Monoaromatic compounds	26.5	29.5	27.7	16.5	14.1	14.3
Toluene	12.7	13.5	18.3	7.7	6.5	3.3
Xylene	1.0	2.6	3.2	/	/	/
Styrene	1.5	1.2	1.4	/	/	/
Benzene, propenyl-	2.9	0.9	0.3	1.2	0.8	1.6
Benzene, butenyl-	1.0	2.3	1.6	0.7	0.7	2.3
Ethylbenzene	1.7	1.2	/	1.5	1.2	3.7
Benzene, propyl-	1.8	0.3	0.2	1.8	1.4	/
Benzene, butyl-	0.6	1.4	1.2	/	0.9	0.8
Benzene, pentyl-	0.6	1.2	0.9	1.5	0.9	0.8
Benzene, hexyl-	1.1	1.6	/	0.3	0.2	0.2
Benzene, heptyl-	0.3	1.7	0.4	1.2	1.1	1.2
Benzene, nonyl-	1.3	1.6	0.2	0.6	0.4	0.4
Polycyclic aromatic hydrocarbons	7.2	6.0	5.5	5.3	6	5.5
Acridine, 9-methyl-	/	/	/	/	/	/
Indene	3.4	1.7	2.3	2.5	2.3	2.3
Naphthalene	1.2	0.5	0.9	0.4	0.7	0.6
Biphenyl	1.5	0.6	0.5	0.5	0.5	0.5
Fluorene	0.3	0.6	0.6	0.7	0.6	0.5
Phenanthrene	0.3	1.1	0.9	0.5	0.9	1.2
Acenaphthene	0.5	0.7	0.3	0.4	0.8	0.3
Benzofuran	/	0.8	/	0.3	0.2	0.1

It could be seen that the contents of alkanes, monoaromatic hydrocarbons, and polycyclic aromatic hydrocarbons gradually declined as the microwave power improved. Toluene (3.3-12.7 area%) reached the highest concentration among all the compositions. Additionally, certain polycyclic aromatic hydrocarbons and alkanes showed elevated levels, such as indene (1.7-3.4 area%), heptadecane (0.3-3.9 area%), and nonacosane (0.1-2.6 area%).

2.3. Effect of Pyrolysis Temperature

The pyrolysis temperature plays a crucial role in microwave-assisted pyrolysis reactions, which determines the degree of decomposition of raw materials. Inadequate pyrolysis temperatures result in incomplete biomass decomposition, whereas excessively high temperatures can accelerate the secondary decomposition of pyrolysis products, ultimately affecting the product yield. The changing trends of oil yields obtained from MAP of various biomass with the increase of pyrolysis temperature are summarized in Figure 2. Figure 2 illustrates that biomass decomposition was limited at lower pyrolysis temperatures, leading to decreased oil yields. As the pyrolysis temperature increased gradually, the highest oil yields were obtained at 500-700°C. However, further temperature elevation led to secondary cracking, resulting in a decline in oil yields. The highest oil yields reported were 49.79 wt.% from sewage sludge (Tian et al. 2011), 8 wt.% from dye sludge (Liu et al. 2018), 40.27 wt.% from camellia oleifera shell (Dai et al. 2019), 56.89 wt.% from stillingia oil (Dai et al. 2019), 41.72 wt.% from bamboo (Du et al. 2020), 17 wt.% from oil palm shell (Salema et al. 2012), and 22.97 wt.% from brown coal (Zhang et al. 2018). In the study of Liu et al. (2021) and Dominguez et al. (2003), the oil yields from MAP of sewage sludge showed the opposite trends as the pyrolysis temperature improved.

Pyrolysis temperature will also affect the oil components obtained from microwave irradiation for biomass pyrolysis. Dai et al. (2019) divided the oil components from MAP of camellia oleifera shell and stillingia oil into several categories, including acid, hydrocarbons, phenols, furans, aldehydes, esters, ketones, and alcohols at the pyrolysis temperatures of 400-600 °C. The acid (4.33-27.2%) and phenols (32-48%) accounted for the large proportions of oil components, and stillingia oil was prefferred to produce hydrocarbons (65-82%). The contents of acid (4.33-27.2%) and hydrocarbons increased gradually with the increase of pyrolysis temperatures, while the contents of phenols decreased.

Table 6. Oil components from microwave irradiation for pyrolysis of camellia oleifera shell/stillingia oil at different temperatures (Data from (Dai et al. 2019))

Components (%)	400 °C	500 °C	600 °C
Acid	4.33/~0.5	~10/0	27.2/0
Hydrocarbons	~8/~65	~6/~75	0/~82
Phenols	~48/~3	~37/~3.5	~32/~0.5
Furans	~4.5/0	~8/0	~12/0
Aldehydes	0/0	~2/0	~2/0
Esters	0/~4	0/~1.5	0/~1.2
Ketones	~4/0	~13/~1.2	~10/~1.6
Alcohols	~3/~1	~4.2/~1.3	~6/~1.2

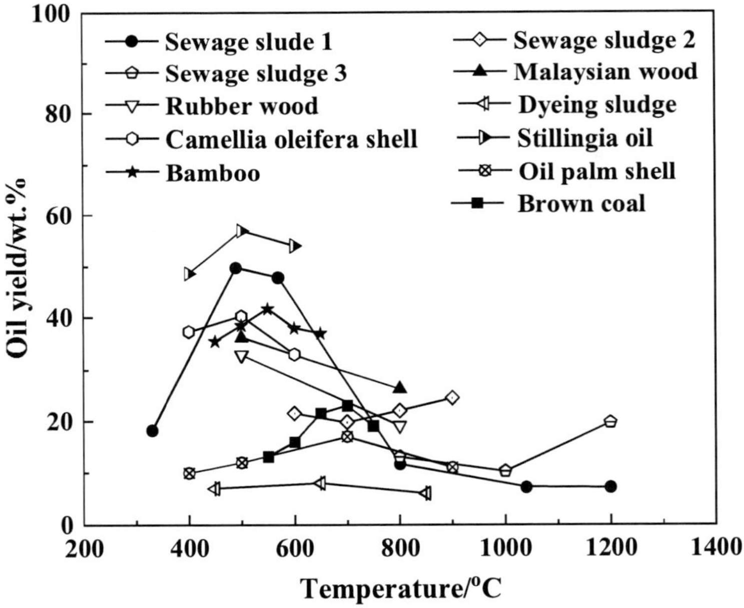

Figure 2. Oil yield from microwave irradiation for pyrolysis of biomass at different pyrolysis temperatures: sewage sludge (Tian et al. 2011) (Liu et al. 2021) (Dominguez et al. 2003), Malaysian wood (Halim & Swithenbank, 2016), rubber wood (Halim & Swithenbank, 2016), dyeing sludge (Liu et al. 2018), camellia oleifera shell (Dai et al. 2019), stillingia oil (Dai et al. 2019), bamboo (Du et al. 2020), oil palm shell (Salema & Ani, 2012), and brown coal (Zhang et al. 2018).

Table 7 lists oil compositions derived from microwave irradiation for pyrolysis of dyeing sludge at different temperatures by Liu et al. (2018). The primary constituents of the oil components were aniline (6.39-14.90), pyridine (3.20-15.82 area%), phenol (1.95-15.85 area%), pyridine, 3-methyl- (2.13-10.16 area%), pyridine, 2-methyl- (0-5.84 area%). High pyrolysis temperatures promoted the formation of aniline and phenol, whereas low pyrolysis temperatures increased the production of pyridine, pyridine, 3-methyl-, and pyridine, 2-methyl-.

Table 7. Oil compositions from microwave irradiation for pyrolysis of dyeing sludge at different temperatures (Data from (Liu et al. 2018))

Compositions (area%)	450 °C	650 °C	850 °C
Pyrrole	/	6.38	/
Pyridine	15.82	12.70	3.20
Pyrazine, methyl-	2.14	2.41	/
Dichloroacetic acid, allyl ester	/	/	/
Phenol	4.79	1.95	15.85
Aniline	6.39	23.48	14.90
Pyridine, 2-methyl-	4.48	5.84	/
Pyridine, 3-methyl-	7.31	10.26	2.13
2-Cyclopenten-1-one, 2-methyl-	4.72	0.74	/
2(1 H)-Pyrimidinethione, tetrahydro-1,3-dimethyl-	13.17	2.54	/
Phenol, 3-methyl-	4.45	0.66	5.88
Pyridine, 3,4-dimethyl-	1.83	4.24	/
Benzyl nitrile	/	1.05	3.33
1,8-Nonadien-3-ol	1.22	0.59	/
3-Ethyl-5-methylindolizidine	/	/	6.37
Cyclohexasiloxane, dodecamethyl-	/	/	/
1-Hydroxycyclododecanecarbonitrile	/	/	/
Cycloheptasiloxane, tetradecamethyl-	/	/	/
Octasiloxane, 1,1,3,3,5,5,7,7,9,9,11,11,13,13,15,15-hexadecamethyl-	/	/	/
cis-10-Heptadecenoic acid	/	/	/
Octadecanal, 2-bromo-	/	/	/
13-Docosenamide, (Z)-	4.27	/	/
1-Monolinoleoylglycerol trimethylsilyl ether	/	/	/

3. Microwave Irradiation for Pyrolysis of Plastics

3.1. Effect of Plastic Type

Given the surge in plastic production, the recycling of plastics has become increasingly essential (Geyer et al. 2017). Converting waste plastics into high-value-added products through microwave radiation for pyrolysis technology is a promising recycling approach that addresses the environmental pollution caused by waste plastics and contributes to environmental sustainability. This technology offers significant economic benefits and obvious environmental advantages, setting the stage for future developments in the plastic recycling industry. The high-value hydrocarbons produced by pyrolysis and recovery technology of waste plastic can serve as cleaner energy carriers or petrochemical raw materials, offering an effective solution for waste recycling. However, in daily life, various types of waste plastics, in the form of mixed plastics, are commonly encountered. So, the types of waste plastics have a huge impact on their recycling. Table 8 lists oil yields from microwave irradiation for pyrolysis of different plastic types. In general, polyolefin plastics could get a higher yield of oil through MAP, such as PS (76-98.78 wt.%), PP (44.8-79.3 wt.%), HDPE (40.2-83.9 wt.%), LDPE (46.5-68.44 wt.%), LLDPE (linear low-density polyethylene, 39.12-84.10 wt.%). Polyimide (PIP) and polycarbonate (PC) contain N and O elements, resulting in low oil yields through MAP compared to polyolefin plastics, which achieved yields of 45.94 wt.% (Suriapparao & Vinu, 2015b) and 39.78 wt.% (Fan et al. 2023a), respectively. The highest oil yield was achieved through MAP of PS (98.78 wt.%, Fan et al. 2023a), representing the global highest value for pyrolysis oil yield.

The elements in different types of plastics also directly determine the oil compositions obtained by MAP. The oil compositions resulting from microwave irradiation for pyrolysis of different plastic types are listed in Table 9 and Table 10 (Fan et al. 2023a). The oils from MAP of PS primarily consisted of aromatic hydrocarbons (73.39 area%), while aliphatic hydrocarbons were rich in the oils derived from PP (99.33 area%), HDPE (98.45 area%), and LDPE (100 area%). HDPE (48.44 area% and 28.11 area%) and LDPE (43.12 area% and 37.42 area%) mainly contained olefins and paraffins, while PP (58.55 area%) primarily consisted of cycloalkanes. The pyrolysis oils derived from MAP of HDPE and LDPE predominantly contained straight-chain olefins and paraffins with carbon numbers ranging from 8 to 18. The main olefins and paraffins from HDPE and LDPE were 1-decene (4.54 area% and 2.25 area%), 1-undecene (3.68 area% and 2.30 area%), 1-tetradecene (7.35 area% and 3.01 area%), decane (2.86 area% and 1.46 area% area%), undecane (2.77 area% and 1.72 area%), and tetradecane (1.57 area% and 2.54 area%). Due to the large amount of oxygen contained in PC, it mainly generated phenolic compounds (87.73 area%) thourhg MAP, including phenol (36.93 area%), phenol, m-tert-butyl- (18.17 area%), p-isopropenylphenol (10.94 area%), phenol, 4-(1-methylethyl)- (7.53 area%), and phenol, 4, 4'-(1-methylethylidene)bis- (4.86 area%).

Table 8. Oil yield from microwave irradiation for pyrolysis of different plastic types

Plastic type	Microwave absorbent	Oil yield (wt.%)	Reference
PS	Actived carbon	93.04	Prathiba et al. 2018
PS	SiC	98.78	Fan et al. 2023a
PS	Coconut sheath carbon	86.1	Rex et al. 2020
PS	Iron	80	Hussain et al. 2010
PS	Makarwal coal	76	Hussain et al. 2011
PP	Graphite	55.3	Suriapparao et al. 2018
PP	SiC	78.22	Fan et al. 2023a
PP	SiC	44.8	Zhou et al. 2021
PP	SiC	79.3	Cui et al. 2023
PP	Coconut sheath carbon	58.3	Rex et al. 2020
PP	Carbon	74.7	Undri et al. 2013
PP	Carbon	56.5	Undri et al. 2014b
HDPE	SiC	70.78a	Fan et al. 2023a
HDPE	SiC	40.2	Zhou et al. 2021
HDPE	Carbon	83.9	Undri et al. 2013
HDPE	Carbon	80.2	Undri et al. 2014b
LDPE	Graphite	46.50	Suriapparao & Vinu, 2015b
LDPE	SiC	68.44a	Fan et al. 2023a
LLDPE	SiC	84.1	Fan et al. 2021
LLDPE	Graphite	39.12	Suriapparao & Vinu, 2015b
PIP	Graphite	45.94	Suriapparao & Vinu, 2015b
PC	SiC	39.78	Fan et al. 2023a

Table 9. Oil components from microwave irradiation for pyrolysis of different plastic types (Data from (Fan et al. 2023a))

Component (area%)	PP	HDPE	LDPE	PS	PC
MAHs	/	/	/	56.75	1.85
PAHs	/	/	/	16.64	1.53
Paraffins	9.10	28.11	37.42	0.02	/
Cycloalkanes	58.55	13.47	13.18	/	/
Olefins	30.34	48.44	43.12	/	0.12
Cycloalkenes	1.33	8.43	6.28	0.11	/
Phenols	/	/	/	/	87.73
Aromatic hydrocarbons	/	/	/	73.39	3.38
Aliphatic hydrocarbons	99.33	98.45	100	0.13	0.12

Table 10. Oil compositions from microwave irradiation for pyrolysis of different plastic types (Data from (Fan et al. 2023a))

Compositions (area%)	PP	PS	HDPE	LDPE	PC
MAH	/	/	/	/	/
Methylbenzene	/	2.79	/	/	0.32
Styrene	/	44.34	/	/	0.58
Ethylbenzene	/	1.51	/	/	0.10
P-xylene	/	/	/	/	0.21
Alpha-methylstyrene	/	3.43	/	/	/
Benzene, (1, 3-dimethyl-3-butenyl)-	/	1.64	/	/	/
PAH	/	/	/	/	/
Bibenzyl	/	0.81	/	/	/
Benzene, 1, 1'-(1, 3-propanediyl) bis-	/	2.83	/	/	/
1,1'-Biphenyl, 3,3',4,4'-tetramethyl-	/	/	/	/	1.16
Cycloalkane	/	/	/	/	/
Cyclohexane, ethylidene-	/	/	1.18	/	/
Cyclohexane, ethyl-	/	/	0.84	0.63	/
Cycloheptane, methyl-	/	/	0.41	0.44	/
Cyclopentane, 1-ethyl-2-methyl-, cis-	/	/	0.50	0.27	/
Cyclopentane, butyl-	/	/	0.22	0.24	/
Cyclohexane, 1,2,4-trimethyl-	11.59	/	/	/	/
Tetramethylcyclopentane	3.13	/	/	/	/
Cyclohexane, 1, 2, 3-trimethyl-	2.28	/	/	/	/
Cyclohexane, 1-ethyl-2-propyl-	7.50	/	/	/	/
Cyclohexane, 1, 2-diethyl-3-methyl-	5.44	/	/	/	/
Cyclohexane, 2, 4-diethyl-1-methyl-	5.38	/	/	/	/
Cyclooctane, 1-methyl-3-propyl-	2.19	/	/	/	/
Cyclohexane, 1, 2, 4, 5-tetraethyl-	2.58	/	/	/	/
Cycloalkene	/	/	/	/	/
Cyclohexene, 1-methyl-	/	/	1.23	0.45	/
Paraffin	/	/	/	/	/
Nonane	/	/	2.85	1.35	/
Heptane, 2, 4-dimethyl-	/	/	2.77	/	/
Decane	/	/	2.86	1.46	/
Undecane	/	/	2.77	1.72	/
Dodecane	/	/	2.33	1.96	/
Tridecane	/	/	1.84	2.38	/
Tetradecane	/	/	1.57	2.54	/
Pentadecane	/	/	/	2.53	/
Hexadecane	/	/	1.18	2.44	/
Heptadecane	/	/	2.42	2.24	/
Octadecane	/	/	/	2.70	/
Olefin	/	/	/	/	/
1-Octene	/	/	2.79	/	/
1, 4-Hexadiene, 3, 3, 5-trimethyl-	2.35	/	/	/	/
2, 4-Dimethyl-1-heptene	7.54	/	/	/	0.12
1-Nonene	/	/	3.75	1.13	/

Compositions (area%)	PP	PS	HDPE	LDPE	PC
1-Decene	/	/	4.54	2.25	/
1-Undecene	/	/	3.68	2.3	/
1-Dodecene	/	/	/	2.48	/
3-Decene, 2, 2-dimethyl-, (E)-	3.10	/	/	/	/
2-Undecene, 4-methyl-	3.08	/	/	/	/
1-Tridecene	/	/	/	2.78	/
1-Tetradecene	/	/	7.35	3.01	/
1-Pentadecene	/	/	1.75	2.86	/
1-Hexadecene	/	/	/	2.65	/
Phenols	/	/	/	/	/
Phenol	/	/	/	/	36.93
Phenol, 4-methyl-	/	/	/	/	1.09
Phenol, 4-ethyl-	/	/	/	/	1.45
p-Isopropenylphenol	/	/	/	/	10.94
Phenol, 4-(1-methylethyl)-	/	/	/	/	7.53
Phenol, m-tert-butyl-	/	/	/	/	18.17
9, 10-Anthracenedione, 1, 8-dihydroxy-3-methyl-	/	/	/	/	1.15
Phenol, 4-(1-methyl-1-phenylethyl)-	/	/	/	/	1.54
Phenol, 4, 4'-(1-methylethylidene)bis-	/	/	/	/	4.86

3.2. Effect of Microwave Power

Microwave power in microwave irradiation for plastic pyrolysis can directly affect the heating rate of the pyrolysis reaction, thus impacting the quantity and quality of the pyrolysis products. Figure 3 illustrates the variation in oil yields from microwave irradiation for pyrolysis of plastics at different microwave powers. In the study of Suriapparao & Vinu (2015b), Cui et al. (2023), and Fan et al. (2022), the oil yields from MAP of PP and PS exhibited an initial increase followed by a subsequent decrease as the microwave power was raised. The optimal microwave power for achieving the maximum oil yield from PP and PS was found to be within the range of 450-650 W. The maximum oil yields were 48.16 wt.%, 79.3 wt.%, and 98.78 wt.%, respectively. As the microwave power increased, the oil yield from MAP of PS exhibited an initial fluctuating upward trend, followed by a gradual decline upon reaching the critical power value (2100 W) for the maximum oil yield. This critical value represents the optimum microwave power (Arshad et al. 2020). Conversely, a similar study conducted by Arshad et al. (2020) on MAP of PP and LDPE demonstrated a fluctuating upward trend in the oil yield.

The oil components obtained from microwave irradiation for pyrolysis of PP at different microwave powers are listed in Table 11 (Suriapparao & Vinu, 2015b). The main components contained alkenes (6.21-29.26%), cycloalkanes

(2.22-7.37%), alkynes (0-3%), alkanes (0.64-2.27%), and benzene derivates (0.2-2.86%) at different microwave powers ranging from 180 to 800 W. Additionally, there were small amounts of cycloalkenes (0-1%) and condensed ring compounds (0-1.2%). The contents of alkanes, cycloalkanes, alkenes, cycloalkenes, and benzene derivates all obtained the highest values at the microwave power of 450 W. Combined with the highest oil yield obtained at this microwave power in Figure 3, it can be easily inferred that 450W is the optimal microwave power for oil production by MAP of PP in this study. The variation trend of components with the change of microwave power was investigated, and the specific compositions are detailed in Table 12 (Fan et al. 2022). The monocyclic and polycyclic aromatic hydrocarbons were the main components, and they accounted for 56.751-82.464 area% and 3.218-16.638 area%, respectively. The contents of PAHs initially rose and then declined with the increase in microwave power, while the contents of MAHs showed the opposite changing tendency. It indicated that the gradual increase in microwave power resulted in secondary cracking, leading to more extensive fragmentation of the PS chains. The most abundant compositions were styrene (44.342-66.195 area%).

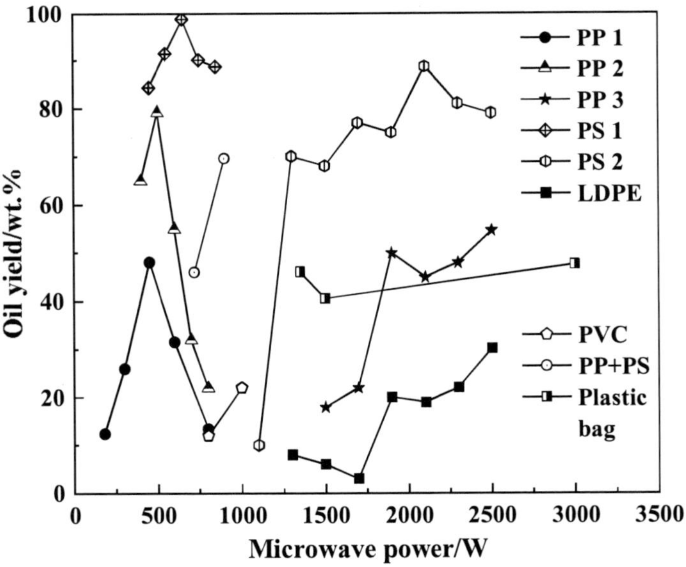

Figure 3. Oil yield from microwave irradiation for pyrolysis of plastics at different microwave powers: PP (Suriapparao & Vinu, 2015b) (Cui et al. 2023) (Arshad et al. 2020), PS (Fan et al. 2022) (Arshad et al. 2020), LDPE (Arshad et al. 2020), PVC (Dai et al. 2018), PP+PS (Rex et al. 2020), and plastic bag (Undri et al. 2014a).

Table 11. Oil components from microwave irradiation for pyrolysis of PP at different microwave powers (Data from (Suriapparao & Vinu, 2015b))

Components (%)	180 W	300 W	450 W	600 W	800 W
Alkanes	1.93	1.70	2.27	1.45	0.64
Cycloalkanes	2.22	3.95	7.37	7.05	4.83
Alkenes	6.50	15.46	29.26	19.04	6.21
Cycloalkenes	~0.30	~0.40	~1	~0.70	/
Alkynes	/	~3	/	~0.70	/
Benzene derivates	~0.30	~1.5	~2.86	~2.13	~0.2
Condensed ring compounds	~/	/	~0.40	/	~1.2
Others	~1	~0.50	~5	~1.50	~0.4

Table 12. Oil compositions from microwave irradiation for pyrolysis of PS at different microwave powers (Data from (Fan et al. 2022))

Compositions (area%)	450 W	650 W	850 W
Monocyclic aromatic hydrocarbons	82.464	56.751	73.551
Methylbenzene	6.146	2.791	4.412
Ethylbenzene	1.416	1.512	1.216
Styrene	66.195	44.342	60.115
Benzene, 2-propenyl-	0.175	0.098	0.216
Alpha-methylstyrene	7.271	3.435	5.447
Benzene, (1,3-dimethyl-3-butenyl)-	0.456	1.636	0.755
Polycyclic aromatic hydrocarbons	3.218	16.638	9.007
Bibenzyl	0.176	0.807	1.291
Benzene, 1,1'-(1-methyl-1,2-ethanediyl) bis-	0.061	0.326	0.492
Benzene, 1,1'-(1,3-propanediyl) bis-	1.022	2.834	1.301
Benzene, 1,1'-(1-butene-1,4-diyl) bis-, (Z)-	0.132	0.459	0.327
2,5-Diphenyl-1,5-hexadiene	0.041	0.702	0.528
Benzene, 1,1'-(1,4-pentadiene-1,5-diyl) bis-, (E, E)-	0.156	0.712	0.396
Tri-m-xylylene	0.146	0.635	0.171
1,1':3',1''-Terphenyl, 5'-phenyl-	0.077	0.779	0.303
Cycloalkenes	0.045	0.112	0
Triquinacene	0	0.043	0
Cyclohexene, 3-methylene-4-(1,2-propadienyl)-	0.045	0	0
3-[(1Z)-1,3-Butadienyl]-4-vinylcyclopentene	0	0.031	0
Olefins	0	0.016	0
1-Pentene, 2-methyl-	0	0.005	0
2,4-Dimethyl-1-heptene	0	0.011	0
Others	14.270	26.483	17.270
Benzene, (3-nitropropyl)-	8.790	13.358	9.536
1,2-Propanediol, 3-benzyloxy-1,2-diacetyl-	4.640	9.515	5.819

3.3. Effect of Pyrolysis Temperature

The pyrolysis temperature stands as one of the most critical parameters in the pyrolysis reaction. The distribution of pyrolysis temperatures determines the yields and properties of the resulting pyrolysis products. Varied final pyrolysis temperatures lead to different temperature rises and pyrolysis processes of the raw material, thereby influencing the yield distribution of different liquids, gases, and solids. Figure 4 shows the oil yield from microwave irradiation for pyrolysis of plastics at different pyrolysis temperatures. With the increase in pyrolysis temperatures, the oil yields exhibit an initial increase followed by a subsequent decrease.

Table 13. Oil compositions from microwave irradiation for pyrolysis of PS at different temperatures (Data from (Fan et al. 2022))

Main component (area%)	340 °C	460 °C	580 °C
Monocyclic aromatic hydrocarbons	80.566	56.751	76.001
Methylbenzene	3.327	2.791	10.192
Ethylbenzene	0.479	1.512	5.020
Styrene	73.588	44.342	44.121
Benzene, 2-propenyl-	0.164	0.098	0.421
Alpha-methylstyrene	2.054	3.435	12.032
Benzene, (1,3-dimethyl-3-butenyl)-	0.406	1.636	0.428
Polycyclic aromatic hydrocarbons	4.040	16.638	9.825
Bibenzyl	0.551	0.807	1.350
Benzene, 1,1'-(1-methyl-1,2-ethanediyl) bis-	0.254	0.326	0.401
Benzene, 1,1'-(1,3-propanediyl) bis-	0.375	2.834	0.691
Benzene, 1,1'-(1-butene-1,4-diyl) bis-, (Z)-	0.219	0.459	0.253
2,5-Diphenyl-1,5-hexadiene	0.438	0.702	0.683
Benzene, 1,1'-(1,4-pentadiene-1,5-diyl) bis-, (E, E)-	0.236	0.712	0.297
Tri-m-xylylene	0.142	0.635	0.117
1,1':3',1''-Terphenyl, 5'-phenyl-	0.080	0.779	0.289
Cycloalkenes	0.090	0.112	0.201
Triquinacene	0	0.043	0
1,2,3,3a-Tetrahydroazulene	0	0	0.201
3-[(1Z)-1,3-Butadienyl]-4-vinylcyclopentene	0	0.031	0
1,3,5,7-Cyclooctatetraene, 1-butyl-	0.090	0	0
Olefins	0	0.016	0
1-Pentene, 2-methyl-	0	0.005	0
2,4-Dimethyl-1-heptene	0	0.011	0
Others	15.304	26.483	13.973
Benzene, (3-nitropropyl)-	6.978	13.358	7.157
1,2-Propanediol, 3-benzyloxy-1,2-diacetyl-	6.532	9.515	5.449

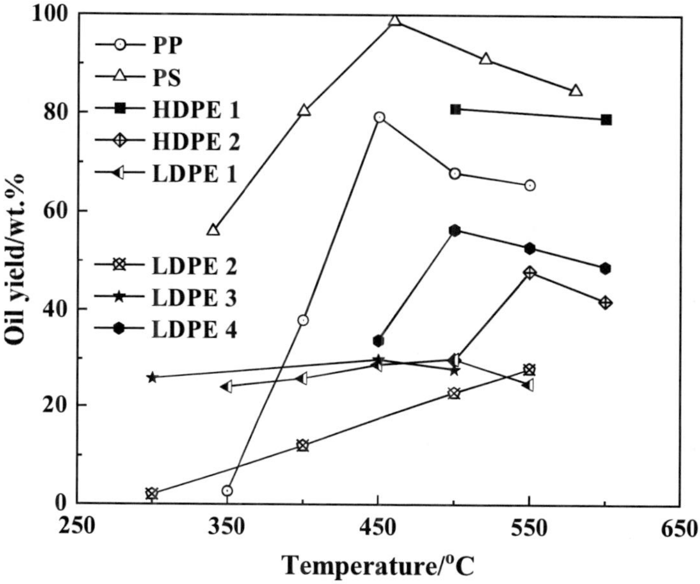

Figure 4. Oil yield from microwave irradiation for pyrolysis of plastics at different pyrolysis temperatures: PP (Cui et al. 2023), PS (Fan et al. 2022), HDPE (Ludlow-Palafox & A. Chase, 2001) (Zhou et al. 2021), and LDPE (Fan et al. 2017) (Juliastuti et al. 2017) (Zhang et al. 2015) (Ding et al. 2019).

This phenomenon can be attributed to the increased pyrolysis temperature, which facilitates the complete depolymerization and chain fragmentation of plastics, forming short chains of condensable liquid products. Furthermore, a further increase in pyrolysis temperature promotes the secondary pyrolysis of pyrolysis vapors, leading to a decrease in liquid products and an increase in small-molecule gas products. The optimum temperature for MAP of plastics is between 450 °C to 550 °C, and the highest oil yields were 79.3 wt.% from PP (Cui et al. 2023), 98.78 wt.% from PS (Fan et al. 2022), 48 wt.% from HDPE (Zhou et al. 2021), 56.53 wt.% from LDPE (Ding et al. 2019), respectively.

The oil compositions from microwave irradiation for pyrolysis of PS at different pyrolysis temperatures are summarized in Table 13 (Fan et al. 2022). The oil compositions produced from MAP of PS exhibited higher proportions of aromatic hydrocarbons and lower proportions of aliphatic hydrocarbons. The MAHs were in the range of 56.751-80.566 area% at different pyrolysis temperatures of 340-580 °C, exhibiting an initial decrease followed by an increase as the pyrolysis temperatures increased. In contrast to the trend observed for MAHs, the contents of PAHs showed the opposite changing trend. The contents of PAHs increased from 4.040 area% to 16.638 area% as the pyrolysis temperatures improved from 340 °C to 460 °C, and subsequently decreased to 9.825 area% as the pyrolysis temperatures further improved to 460°C. What's more, a minor quantity of unsaturated aliphatic hydrocarbon was generated, comprising 0.090-0.201 area% cycloalkenes and 0-0.016 olefins. These unsaturated aliphatic hydrocarbons were 1,2,3,3a-tetrahydroazulene (0.201 area%), 1,3,5,7-cyclooctatetraene, 1-butyl- (0.09 area%), triquinacene (0.043 area%), 3-[(1Z)-1,3-Butadienyl]-4-vinylcyclopentene (0.031 area%), 2,4-dimethyl-1-heptene (0.011 area%), and 1-Pentene, 2-methyl- (0.005 area%).

4. Microwave Irradiation for Pyrolysis of Plastics with Microwave Absorbent

4.1. Effect of Microwave Absorbent Type

Microwave heating is a selective process, and not all materials can absorb microwaves to reach the temperature required for pyrolysis. Therefore, a specific quantity of microwave absorbent needs to be added to make the material heat up to the temperature required for pyrolysis. In general, the absorbent can accelerate the pyrolysis process of the material without altering the composition of the resulting product. Moreover, microwave absorbents can lower the activation energy required for pyrolysis and enhance the overall efficiency of the pyrolysis process for plastics. The addition of different types of microwave absorbents has varying temperature effects on plastics. When the microwave absorbent is used, the microwave absorber particles remain heated through the input of microwave irradiation. Once the reactor temperature is stable, the microwave absorbent transfers heat to the plastic, initiating the pyrolysis process through indirect heating, specifically the heat conduction of the microwave absorbent. Table 14 lists oil yield from microwave irradiation for pyrolysis of plastic with different microwave absorbents. The solid product (carbon) via pyrolysis of biomass serves as a microwave absorbent with a high capacity for microwave absorption. The use of carbon as a microwave absorbent enables the rapid elevation of pyrolysis temperature, facilitate efficient pyrolysis, and stimulate the production of liquid and gas products. Rex et al. (2020) adopted coconut sheath carbon as the microwave absorbent to achieve the oil yield of 86.1 wt.% from MAP of PS. Activated carbon functions as both a microwave absorbent and a catalyst.

Table 14. Oil yield from microwave irradiation for pyrolysis of plastic with different microwave absorbents

Microwave absorbent	Biomass type	Oil yield (wt.%)	Reference
Actived carbon	PS	93.04	Prathiba et al. 2018
Activated carbon	PS+PP	59.05	Rex et al. 2020
Activated carbon	PP	29.8	Suriapparao & Vinu, 2015b
Rice husk activated carbon	PS	83.3	Rex et al. 2020
Rice husk activated carbon	PP	25.4	Rex et al. 2020
Corn husk activated carbon	PS	88.4	Rex et al. 2020
Corn husk activated carbon	PP	51.9	Rex et al. 2020
Coconut sheath activated carbon	PS	77.8	Rex et al. 2020
Coconut sheath activated carbon	PP	47.7	Rex et al. 2020
Carbon	PP	56.5	Undri et al. 2014b
Carbon	PP	74.7	Undri et al. 2013
Carbon	HDPE	80.2	Undri et al. 2014b
Carbon	HDPE	83.9	Undri et al. 2013
Rice husk carbon	PS	66.7	Rex et al. 2020
Rice husk carbon	PP	19	Rex et al. 2020
Rice husk carbon	PS+PP	69.55	Rex et al. 2020)
Corn husk carbon	PS	83.3	Rex et al. 2020
Corn husk carbon	PP	41.23	Rex et al. 2020
Corn husk carbon	PS+PP	77.40	Rex et al. 2020
Coconut sheath carbon	PS	86.1	Rex et al. 2020
Coconut sheath carbon	PP	58.3	Rex et al. 2020
Coconut sheath carbon	PS+PP	84.30	Rex et al. 2020
Lignin	PP	41.86	Suriapparao & Vinu. 2015b
Tire	PP	58.9	Undri et al. 2013
Tire	HDPE	49.7	Undri et al. 2013
Tire	HDPE	43.9	Undri et al. 2014b
Fly ash	PP	21.07	Suriapparao & Vinu, 2015b
Graphite	PP	48.16	Suriapparao & Vinu, 2015b
Graphite	PP	55.3	Suriapparao et al. 2018
Makarwal coal	PS	76	Hussain et al. 2011
SiC	PP	79.3	Cui et al. 2023
SiC	PP	31.89	Suriapparao & Vinu, 2015b
SiC	PP	44.8	Zhou et al. 2021
SiC	PP	78.22	Fan et al. 2023a
SiC	PS	98.78	Fan et al. 2023b
SiC	HDPE	40.2	Zhou et al. 2021
SiC	PC	39.78	Fan et al. 2023a
Fe	PS	97.67	Fan et al. 2023b
Fe_3O_4	PS	94.67	Fan et al. 2023b
FeS_2	PS	93.00	Fan et al. 2023b
Iron	PS	80	Hussain et al. 2010
Aluminum	PP	45.4	Suriapparao & Vinu, 2015b

The addition of activated carbon is beneficial to the generation of gaseous products. Rex et al. (2020) also found that compared with rice husk carbon and corn husk carbon, rice husk activated carbon and corn husk activated carbon can promote the oil yield from MAP of plastics. They respectively increased the oil yields from MAP of PS from 66.7 wt.% and 83.3 wt.% to 83.3 wt.% and 88.4 wt.%. Similarly, for MAP of PP, the oil yields improved from 19 wt.% and 41.23 wt.% to 25.4 wt.% and 51.9 wt.%. However, the effect was the opposite for coconut sheath carbon and coconut sheath-activated carbon. Silicon carbide, a non-reactive phase microwave absorbent, exhibits excellent oxidation and high-temperature resistance. Moreover, it effectively absorbs microwave energy to improve the efficiency of catalytic conversion. By employing SiC as a microwave absorbent, Fan et al. (2023b) achieved the highest pyrolysis oil yield of 98.78 wt.% from MAP of PS, while Cui et al. (2023) obtained a yield of 79.3 wt.% from PP. In addition, metallic microwave absorbents, such as Fe, Al, Fe_3O_4, etc., can be used to improve the pyrolysis conversion rate of plastics and promote the pyrolysis reaction process.

Table 15 lists oil components from microwave irradiation for pyrolysis of PS with different microwave absorbents (Fe, Al, Fe_3O_4, and FeS_2; Fan et al. 2023b). The oil components from MAP of PS with iron-based microwave absorbent mainly contained MAHs, PAHs, paraffins, cycloalkanes, olefins, and cycloalkenes. The total contents of aromatic hydrocarbons were in the range of 80.78-96.92 area%, while the total contents of aliphatic hydrocarbons ranged from 0area% to 2.56 area%. The utilization of FeS_2 as a microwave absorbent, as opposed to SiC, substantially increased the diversity of aliphatic hydrocarbons. The aliphatic hydrocarbon contents increased by 10 times, rising from 0.13 area% to 2. 6 area%.

Table 15. Oil components from microwave irradiation for pyrolysis of PS with different microwave absorbents (Data from (Fan et al. 2023b))

Component (area%)	Fe	Fe_3O_4	FeS_2	SiC
MAHs	93.93	74.16	48.36	56.75
PAHs	3.00	14.08	32.43	16.64
Paraffins	0	0	0.15	0
Cycloalkanes	0	0	0.05	0.11
Olefins	0	0	1.68	0
Cycloalkenes	0.07	0	0.69	0.02
Aromatic hydrocarbons	96.92	88.24	80.78	73.39
Aliphatic hydrocarbons	0.07	0	2.56	0.13

4.2. Effect of Plastic to Microwave Absorbent Ratio

The capacity of a material to absorb microwave is related to its dielectric constant. A high dielectric constant indicates a strong microwave absorption capacity, and the material is easy to be heated by microwave. Otherwise, materials with low microwave absorption capacity are difficult to heat efficiently. Plastic exhibits weak microwave absorption capacity, indicated by a very small medium loss angle tangent value, making it challenging to enhance the heating rate solely with plastic. Incorporating microwave absorbents with strong microwave absorption capacity in specific quantities is essential to enhance the microwave absorption performance of plastic (Bhattacharya & Basak, 2016). Figure 5 shows the oil yield from microwave irradiation for pyrolysis of plastics at different plastic to microwave absorbent ratios. When employing activated carbon, carbon, and SiC as microwave absorbents, the oil yields from MAP of plastic improved first and then declined with the decrease of plastic to microwave absorbent ratios (increase of microwave absorbent amounts). The highest oil yields were obtained at plastic to activated carbon/carbon ratios of 10:1 (Prathiba et al. 2018; Rex et al. 2020) and plastic to SiC ratios of 1:2 (Fan et al. 2022). The promoting effect of iron on oil yield from MAP of PS gradually improved with the increasing amount of Fe (Fan et al. 2023b).

Table 16. Oil components from microwave irradiation for pyrolysis of PP at different PP to graphite ratios (Data from (Suriapparao & Vinu, 2015b))

Component (%)	100:1	25:1	5:1
Alkanes	2.27	/	/
Cycloalkanes	7.37	7.56	2.17
Alkenes	29.26	6.31	8.95
Cycloalkenes	/	/	~0.5
Benzene derivates	~0.1	/	~2.5
Condensed ring compounds	~0.3	~1.32	~1
Others	~0.7	~1	~5.26

Oil components from microwave irradiation for pyrolysis of PP at different PP to graphite ratios are listed in Table 16 (Suriapparao & Vinu, 2015b). Oil components from MAP of PS with different amounts of graphites as microwave absorbents primarily consisted of alkenes (6.31-29.26%), cycloalkanes (2.17-7.56%), alkanes (0-2.27%), and benzene derivates (0-2.5%%). Additionally, a small amount of condensed ring compounds (0.3-

1.32%) and cycloalkenes (0-0.5%) were generated. The increased utilization of graphite as a microwave absorber has accelerated the formation of condensed ring compounds, such as benzene derivatives, which increased from 0 to 2.5% at the PP to graphite ratios of 100:1, 25:1, and 5:1, while inhibiting the formation of alkanes, alkenes, and cycloalkanes.

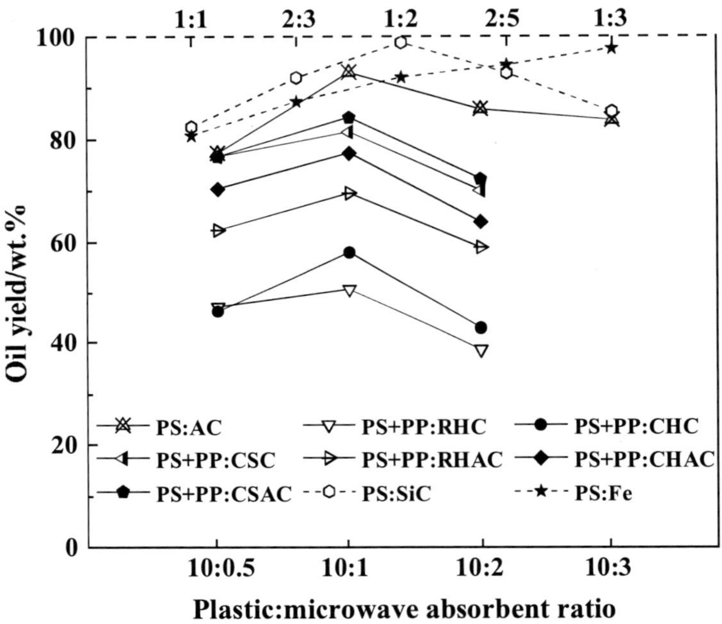

Figure 5. Oil yield from microwave irradiation for pyrolysis of plastics at different plastic to microwave absorbent ratios: PS:AC (Prathiba et al. 2018), PS+PP: RHC /CHC /CSC /RHAC /CHAC /CSAC (Rex et al. 2020), PS: SiC (Fan et al. 2022), and PS: Fe (Fan et al. 2023b).

5. Microwave Irradiation for Pyrolysis of Plastics with Catalyst

5.1. Effect of Catalyst Type

The utilization of catalysts is an effective approach for improving the quality of liquid products during the conversion of plastics to fuel through microwave irradiation for pyrolysis. Table 17 lists oil production from microwave irradiation for pyrolysis of plastic with different catalyst types. The common catalyst is zeolite. Zeolites are hydroaluminosilicate compounds derived from aluminate and silicates (i.e., AlO_4 and SiO_4), exhibiting diverse open cavities and crystal structures that undergo structural enhancement during the chemical reaction process.

Table 17. Oil yield from microwave irradiation for pyrolysis of plastics with different catalyst types

Catalyst	Catalyst type	Oil yield (wt.%)	Reference
Natural zeolite	LDPE	28.12	Juliastuti et al. 2017
HZSM-5	PS	98.3	Suriapparao et al. 2020
HZSM-5	PP	82.1	Suriapparao et al. 2020
HZSM-5	PP+PET	42	Zhang et al. 2024
HZSM-5	LLDPE	41.5	Fan et al. 2021
ZSM-5	HDPE	56.9	Zhou et al. 2021
ZSM-5	HDPE	~58	Dai et al. 2022b
ZSM-5	HDPE	~91	Dai et al. 2022a
ZSM-5	LDPE	32.58	Zhang et al. 2015
ZSM-5	PP+HDPE	48.9	Zhou et al. 2021
ZSM-5+Raney Ni	LDPE	64.41	Zhang et al. 2017
ZSM-5/SiC	HDPE	~48	Zhou et al. 2022
SiC@ZSM-5	LDPE	36	Chen et al. 2022
HY	LDPE	48.97	Ding et al. 2019
NiO+HY	LDPE	56.54	Ding et al. 2019
MS13X	HDPE	~60	Jing et al. 2021
MS4A	HDPE	~78	Jing et al. 2021
MS4A	Plastic bag	~73	Jing et al. 2020
MgO	LDPE	38.5	Fan et al. 2017
Fe/SiC	LDPE	~23	Luo et al. 2023
Ni/SiC	LDPE	~22	Luo et al. 2023
Fe-Ni/SiC	LDPE	~19	Luo et al. 2023
FeAlOx	HDPE	~2.6	Jie et al. 2020
KOH	PS+PP	83.8	Kamireddi et al. 2023

Suriapparao et al. (2020) used HZSM-5 as a catalyst for the conversion of PS through MAP, resulting in an oil yield of 98.3 wt.%. Dai et al. (2021) employed ZSM-5 as a catalyst during the MAP of HDPE and achieved an increase of oil yield to about 91wt.%. In addition, various other zeolite catalysts, including HY (Ding et al. 2019), MS13X (Jing et al. 2021), and MA4A (Jing et al. 2020; Jing et al. 2021), have been found to contribute significantly to the observed positive catalytic effects. Catalytic pyrolysis also frequently employs metal oxides such as CaO, ZnO, MgO, ZrO_2, CeO_2, and TiO_2 to improve the quality of gas products and the stability of liquid products. Metal oxides possess characteristics of catalytic reduction, multivalent state, and acidity, which particularly inhibit oxygenated compounds and reduce the content of aldehydes, acids, and sugars (Bohi et al. 2020). Through MAP of LDPE using MgO as the catalyst, the oil yield was 38.5wt.% (Fan et al. 2017). In addition to studying the catalytic effect of monofunctional catalysts, the application of microwave-coupled bifunctional catalysts has been introduced for plastic pyrolysis. Due to its excellent microwave absorption, SiC can enhance the efficiency of the pyrolysis reaction in the MAP process and significantly contribute to improving the quality of oil (Luque et al. 2012). By utilizing bifunctional catalysts in microwave fields, the advantages of microwaves can be fully harnessed, resulting in increased heating rates and catalytic activity. The surface activation of bifunctional catalysts induced by microwave radiation can enhance their catalytic performance, while the presence of SiC in the reactor can further promote microwave absorption, leading to higher temperatures and faster reaction rates. Chen et al. (2022) and Luo et al. (2023) both employed SiC-supported bifunctional catalysts during the MAP of LDPE, resulting in a pyrolysis oil yield with the range of 19-36 wt.%.

Zeolite catalysts and their modified forms (such as HZSM-5 and HY) have been widely utilized in the catalytic upgrading of fuels owing to their acidity, pore structure, ion exchange characteristics, and shape selectivity. Zeolite catalysts exhibit excellent alkylation and aromatization capabilities, enabling the conversion of plastics into more desirable products such as aromatics, offering a low-energy-consumption approach to improve product quality. The catalytic activity of zeolite catalysts is influenced by the interplay between acid site density, acid strength, and porosity. Additionally, the positive ion mechanism can reduce the temperature required for C-C bond fracture (Selvam et al. 2023). Zeolites possess both Bronsted and Lewis acid sites, with the Bronsted acid site serving as the primary catalytic center. The presence of active sites and diverse pore sizes in molecular sieves contributes to the enhancement of aromatics yield. In the microwave-assisted catalytic pyrolysis of plastics using zeolite catalysts, the primary reaction pathways include the hydrocarbon pool mechanism, cracking, and aromatization. Zeolite is well-suited as a selective catalyst, and the generation of hydrocarbons within the molecular sieve framework depends on the shape selectivity of the pore space and pore size of the molecular sieve. In the research of Ding et al. (2019), NiO catalyst could promote the production of aromatics while inhibiting normal aliphatics, both of which are beneficial for improving the octane number of the oil. What's more, NiO could also promote the extraction of hydrogen atoms and the dehydrogenation of alkanes to olefins. HY catalysts promoted the generation of cycloalkenes and aromatic hydrocarbons from olefins. The conjugated diolefin reacted with the double or triple-bonded hydrocarbon to form cycloalkenes through the Diels-Alder reaction. At the same time, it could also promote the cyclization of olefins and alkanes to produce cycloalkanes. Subsequently, aromatic hydrocarbons were obtained through the hydrogenation of NiO. ZSM-5 catalysts could promote cracking, isomerization, and aromatization, whereas inhibit the production of alkenes, which increase the generation of gasoline range aromatic (18 to 45%), isomerized aliphatic (10.4 to 24.6%) (Zhou et al. 2021).

In the research of Kamireddi et al. (2023), compared with no-catalytic pyrolysis, the heating rate decreased in the microwave-assisted catalytic (KOH catalyst) pyrolysis, and the residence time increased, thereby promoting the secondary cracking of plastics and resulting in an increase in oil production. However, the overall conversion rate of plastics decreased. Table 18 lists oil components from microwave irradiation for pyrolysis of LDPE with different catalyst types by Luo et al. (2023). In this research, SiC powders were loaded with Fe and Ni to synthesize Fe/SiC, Ni/SiC, Fe-Ni/SiC catalysts. The main components include alkanes, alkenes, mononuclear aromatics, and polyaromatic hydrocarbons. The results showed that Ni promoted =C-H bonds to convert to –CH_2- bonds, leading to the generation of saturated alkanes, whereas Fe promoted the fracture of C-C and C-H bonds to form smaller gas molecules. Additionally, when compared to the other two catalysts (Fe/SiC and Ni/SiC), Fe-Ni/SiC had the most significant effect on converting unsaturated aliphatic hydrocarbons (alkenes) to saturated aliphatic hydrocarbons (alkanes), with approximately 36.5 area% alkanes and 32.5 area% alkenes. Among the pyrolysis products, the production of gas and carbon nanotubes was also considerable.

Table 18. Oil components from microwave irradiation for pyrolysis of LDPE with different catalyst types (Data from (Luo et al. 2023))

Component (area%)	Fe/SiC	Ni/SiC	F-Ni/SiC
Alkanes	~30	~31	~36.5
Alkenes	~36	~37	~32.5
Mononuclear aromatics	~20	~17	~9
Polyaromatic hydrocarbons	~9	~8	~16
Others	~5	~7	~6

5.2. Effect of Plastic to Catalyst Ratio

In the microwave irradiation for catalytic pyrolysis of plastic, the catalyst amount is very important. The ratio of plastic to catalyst directly affects the pyrolysis process, and the appropriate ratio can magnify its catalytic promotion effect indefinitely. Furthermore, there may exist a critical value of the plastic-to-catalyst ratio, which impacts both the yield and composition of the pyrolysis oil. Once the ratio exceeds this critical value, the ratio of plastic to catalyst will have no obvious effect on the pyrolysis results. Table 19 lists oil yield from microwave irradiation for pyrolysis of plastic at different plastic to catalyst ratios. During the microwave-assisted catalytic pyrolysis (MACP) of LDPE, the oil yield gradually decreased with the increase in the ratio of plastic to catalyst, suggesting that MgO and HY catalysts restrained the oil yield. As the amount of ZSM-5 increased, the oil yield initially rose and then declined, peaking at a ratio of 2:1 with a maximum value of 32.58 wt.%. This showed that the plastic to catalyst ratio of 2:1 was a critical value, and further increasing the amount of catalyst used would have the opposite effect. For the MACP of LLDPE, the oil yield decreased with the increase in the amount of HZSM-5 catalyst.

Table 19. Oil yield from microwave irradiation for pyrolysis of plastic at different plastic to catalyst ratios

LDPE: MgO	Oil yield (wt.%)	LDPE: ZSM-5	Oil yield (wt.%)	LDPE: HY	Oil yield (wt.%)	LLDPE: HZSM-5	Oil yield (wt.%)
1:0	46.3	4.68:1	22.16	1:0	51.56	1:0	84.1
15:1	~38	4:1	27.39	20:1	~57	20:1	41.5
10:1	30.3	3:1	29.70	15:1	60.01	10:1	~32
5:1	~30	2:1	32.58	10:1	~56.8	20:3	~30
3:1	~29.5	1.32:1	26.57	5:1	~38	5:1	25.4
(Fan et al. 2017)		(Zhang et al. 2015)		(Ding et al. 2019)		(Fan et al. 2021)	

Table 20. Oil components from microwave irradiation for pyrolysis of LDPE at different plastic to MgO ratios (Data from (Fan et al. 2017))

Component (%)	1:0	15:1	10:1	5:1	3:1
Mono-aromatics	10.5	~36	~41	~46	~50
Poly-aromatics	5.3	~4	~5	~5	~5
C5-C12 alkenes	23.3	37.5	~34	~27	~22
C5-C12 alkanes	~7	~11	~8	~9	~8
C12+ alkenes	~38	~7	~1	/	/
C12+ alkanes	~15.9	~4.5	~11	~13	~15

The effect of plastic to catalyst ratio on the oil yield is illustrated by the study of Fan et al. (2017), as listed in Table 20. The inclusion of MgO significantly impacted the production of mono-aromatics, with the content increasing from 10.5% to 50% as the amount of MgO increased. For C5-C12 alkanes and alkenes, the contents showed the tendency of increase first and then decrease. When the ratio of plastic to catalyst was 15:1, the contents of C5-C12 alkanes and alkenes reached the maximum values (37.5 % and 11%), while the production of C12+ alkenes was the lowest at this ratio of plastic to catalyst. At the same time, the addition of MgO inhibited the production of C12+ alkenes compared without catalyst. Table 21 lists oil compositions from microwave irradiation for pyrolysis of LDPE at different plastic to ZSM-5 ratios. With the improvement of ZSM-5 amounts, the selectivity of mono-ring aromatic hydrocarbons gradually decreased, while the selectivity of double-ring aromatic hydrocarbons increased. This suggests that the addition of ZSM-5 can promote the production of poly-ring compounds. Therefore, to achieve the best research results, it is necessary to consider the product yield and target component when selecting the optimal ratio of plastic to catalyst.

Table 21. Oil compositions from microwave irradiation for pyrolysis of LDPE at different plastic to ZSM-5 ratios (Data from (Zhang et al. 2015))

Compositions (%)	4.68:1	3:1	1.32:1
Mono-ring aromatic hydrocarbons	87.16	84.34	77.08
Xylenes	51.29	52.30	43.90
Ethyltoluene	14.67	12.13	11.95
Trimethylbenzene	8.32	7.99	7.45
Indane	1.15	1.16	1.21
Diethylbenzene and isomers	6.38	5.74	5.91
Methylindan and isomers	2.16	2.37	3.63
Dimethylindan and isomers	2.22	1.91	2.03
Pentamethylbenzene	0.16	0.24	0.08
Trimethylindan and isomers	0.81	0.49	0.92
Double-ring aromatic hydrocarbons	12.84	15.66	22.32
Naphthalene	3.14	3.70	4.25
Methylnaphthalene	4.47	5.80	7.87
Dimethylnaphthalene	3.34	3.82	5.94
Ethylnaphthalene	0.83	0.93	1.22
Trimethylnaphthalene and isomers	1.06	1.42	3.04

5.3. Effect of Catalytic Mode

In the microwave irradiation for pyrolysis process, catalysts are generally use in the two modes: in-situ catalytic mode and ex-situ catalytic mode. The catalytic mode can affect the catalytic mechanism and final outcomes of MAP of plastics. The process of simultaneously introducing the catalyst and plastic into the MAP reactor is known as in-situ catalytic pyrolysis. The process in which the catalyst is placed in a separate reactor from the MAP unit and only interacts with the pyrolysis vapors is referred to as ex-situ catalytic pyrolysis (Ruddy et al. 2014). During the in-situ MACP process, the catalyst is mixed with the plastic and actively takes part in the pyrolysis process. This involvement aids in the breakdown of the high polymer structure in the plastic and minimizes the formation of secondary products. However, due to the short residence time of pyrolysis vapors in the in-situ catalytic pyrolysis process, a high ratio of catalyst to plastic is necessary to ensure better-upgrading effects and higher liquid product yields. It is important to consider that the optimal pyrolysis temperature of plastic may not be consistent with the optimal temperature for the catalyst, resulting in suboptimal catalytic effects. For the ex-situ MACP process, the catalyst action and the pyrolysis phases are independent of each other. The catalytic upgrading process can be carried out independently of the MAP process, ensuring that the catalyst performs at the optimal temperature. In the MAP of plastics, the microwave absorbent plays a crucial role by efficiently converting microwave energy into heat energy and transferring it to the feedstock. The combination of microwave absorbent and catalyst can not only satisfy the microwave absorption effect during the MAP, but also effectively enhance the product quality. Therefore, the research of bifunctional catalysts is gradually developed, and the bifunctional catalysts need to be in the in-situ catalytic mode for full function. Table 22 lists oil yield from microwave irradiation for pyrolysis of plastics with different catalytic modes. Zeolite catalysts generally adopt ex-situ catalytic mode, including HZSM-5, ZSM-5, HY, and others. The highest oil yield obtained in the microwave conversion of PS was 98.78 wt.% (Suriapparao et al. 2020), while the lowest oil yield was 32.58 wt.% obtained by MACP of LDPE under the catalytic effect of ZSM-5. Bifunctional catalysts generally used in the in-situ catalytic mode (i.e., SiC@ZSM-5 (Chen et al. 2022), Fe/SiC (Luo et al. 2023), Ni/SiC (Luo et al. 2023), and Fe-Ni/SiC (Luo et al. 2023)). And the oil yields produced from MACP of LDPE were in the range of 19-36 wt.%.

Some studies have found that NiO, CaO, and CuO are predominantly used in in-situ catalytic pyrolysis rather than ex-situ catalytic pyrolysis (Morgan et al. 2017). Oil components from microwave irradiation for pyrolysis of LDPE with different catalytic modes are listed in Table 23. In this research, MACP of LDPE with HY and NiO+HY as catalysts were performed (Ding et al. 2019). HY was used in the ex-situ catalytic mode, whereas NiO was used in the in-situ catalytic mode. Consequently, the NiO+HY catalysts resulted in the initial cracking of LDPE polymer under the catalytic effect of NiO in the microwave reactor. Subsequently, the obtained pyrolysis vapors flowed through the ex-situ catalytic device (HY), producing liquid and gas products. This dual catalytic pyrolysis promoted the Diels-Alder reaction and increased the production of aromatic hydrocarbons. In comparison to the sole HY catalytic effect, the contents of aromatic hydrocarbons exhibited an increasing trend from 35.25% to 38%. In addition, significant production of n-alkenes (18%), iso-alkenes (18%), and iso-alkanes (18%) was observed.

Table 22. Oil yield from microwave irradiation for pyrolysis of plastics with different catalytic modes

Catalyst code	Catalyst	Catalyst type	Oil yield (wt.%)	Reference
Ex-situ	HZSM-5	PS	98.3	Suriapparao et al. 2020
Ex-situ	HZSM-5	PP	82.1	Suriapparao et al. 2020
Ex-situ	HZSM-5	PP+PET	42	Zhang et al. 2024
Ex-situ	HZSM-5	LLDPE	41.5	Fan et al. 2021
Ex-situ	ZSM-5	HDPE	56.9	Zhou et al. 2021
Ex-situ	ZSM-5	HDPE	~58	Dai et al. 2022b
Ex-situ	ZSM-5	LDPE	32.58	Zhang et al. 2015
Ex-situ	ZSM-5	PP+HDPE	48.9	Zhou et al. 2021
Ex-situ	ZSM-5+Raney Ni	LDPE	64.41	Zhang et al. 2017
Ex-situ	ZSM-5/SiC	HDPE	~48	Zhou et al. 2022
Ex-situ	HY	LDPE	48.97	Ding et al. 2019
Ex-situ	MgO	LDPE	38.5	Fan et al. 2017
In-situ	SiC@ZSM-5	LDPE	36	Chen et al. 2022
In-situ	Fe/SiC	LDPE	~23	Luo et al. 2023
In-situ	Ni/SiC	LDPE	~22	Luo et al. 2023
In-situ	Fe-Ni/SiC	LDPE	~19	Luo et al. 2023
In-situ	FeAlOx	HDPE	~2.6	Jie et al. 2020
In-situ	KOH	PS+PP	83.8	Kamireddi et al. 2023
In situ+ex-situ	NiO+HY	LDPE	56.54	Ding et al. 2019

Table 23. Oil components from microwave irradiation for pyrolysis of LDPE with different catalytic modes (Data from (Ding et al. 2019))

Component (%)	Ex-situ (HY)	In situ (NiO)+ex-situ (HY)
PAHs	~3	~4
MAHs	~32.25	~34
N-alkenes	3.03	~18
Cyclo-alkenes	~4	~7
Iso-alkenes	23.14	~19
N-alkanes	13.67	~9
Cyclo-alkanes	~2	~6
Iso-alkanes	18.91	~12

Conclusion

As an important technology in the field of renewable energy, this chapter focuses on organic wastes and introduces the use of microwave irradiation for pyrolysis process of different biomass and plastics. The effect of different influence factors (i.e., biomass or plastics types, microwave power, pyrolysis temperature, microwave absorbent types, plastic to microwave absorbent ratio, catalyst type, plastic to catalyst ratio, and catalytic mode) are detailed, and the yields and components of oil are evaluated.

The conversion of organic wastes into oil products through microwave irradiation of pyrolysis can effectively address the environmental problems associated with waste plastics, while also producing liquid fuels as substitutes for petroleum-based fuels. Additionally, this process can generate additional values from organic wastes, contributing to the circularity of resources, circular economy, and carbon neutrality. Future research should focus on the following areas: (a) exploring the heating and warming principles of biomass when used as both a raw material and microwave absorbent, (b) investigating the action mechanism of microwave irradiation on different microwave absorbents; (c) screening catalysts with excellent performance to improve the oil quantity and quality of oil, and (d) applying microwave radiation for pyrolysis of organic wastes to pilot scale system to achieve mass production.

Acknowledgments

The following financial supports are acknowledged: Financial support was provided by the National Natural Science Foundation of China (52476005, 52076049), Heilongjiang Province "Double First-class" Discipline Collaborative Innovation Achievement Project (LJGXCG2023-080), Heilongjiang Provincial Key R&D Program (2023ZX02C05), and Heilongjiang Provincial Key R&D Program "Unveiling the Leader" Project (2023ZXJ02C04).

References

Abdelsayed, V., Shekhawat, D., Smith, M. W., Link, D., Stiegman, A. E. (2018). Microwave-assisted pyrolysis of Mississippi coal: A comparative study with conventional pyrolysis. *Fuel*, 217, 656-667.

Ahmad, I., Khan, M. I., Khan, H., Ishaq, M., Tariq, R., Gul, K., Ahmad, W. (2014). Pyrolysis Study of Polypropylene and Polyethylene Into Premium Oil Products. *International Journal of Green Energy*, 12(7), 663-671.

Arjharn, W., Liplap, P., Maithomklang, S., Thammakul, K., Chuepeng, S., Sukjit, E. (2022). Distilled Waste Plastic Oil as Fuel for a Diesel Engine: Fuel Production, Combustion Characteristics, and Exhaust Gas Emissions. *ACS Omega*, 7(11), 9720-9729.

Arshad, H., Sulaiman, S. A., Hussain, Z., Naz, M. Y., Moni, M. N. Z. (2020). Effect of Input Power and Process Time on Conversion of Pure and Mixed Plastics into Fuels Through Microwave-Metal Interaction Pyrolysis. *Waste and Biomass Valorization*, 12(6), 3443-3457.

Beneroso, D., Bermudez, J. M., Arenillas, A., Menendez, J. A. (2013). Microwave pyrolysis of microalgae for high syngas production. *Bioresour Technol*, 144, 240-6.

Bhattacharya, M., Basak, T. (2016). A review on the susceptor assisted microwave processing of materials. *Energy*, 97, 306-338.

Bhoi, P. R., Ouedraogo, A. S., Soloiu, V., Quirino, R. (2020). Recent advances on catalysts for improving hydrocarbon compounds in bio-oil of biomass catalytic pyrolysis. *Renewable and Sustainable Energy Reviews*, 121, 109676.

Chang, S. H. (2023). Plastic waste as pyrolysis feedstock for plastic oil production: A review. *Science of the Total Environment*, 877, 162719.

Chen, C., He, S., Qiu, S., Wan, S., Ling, H., Wei, Y. (2023). Characterization and product analysis of Chlorella vulgaris microalgae and oil seed biomass waste by microwave co-pyrolysis. Journal of the *Energy Institute*, 111, 101413.

Chen, H., Xu, H., Zhu, H., Yan, S., Zhang, S., Zhang, H., Guo, X., Hu, X., Gao, W. (2024). A review on bioslurry fuels derived from bio-oil and biochar: Preparation, fuel properties and application. *Fuel*, 355, 129283.

Chen, Z., Monzavi, M., Latifi, M., Samih, S., Chaouki, J. (2022). Microwave-responsive SiC foam@zeolite core-shell structured catalyst for catalytic pyrolysis of plastics. *Environ Pollut*, 307, 119573.

Cui, Y., Zhang, Y., Cui, L., Liu, Y., Li, B., Liu, W. (2023). Microwave-assisted pyrolysis of polypropylene plastic for liquid oil production. *Journal of Cleaner Production*, 411, 137303.

Dai, L., Wang, Y., Liu, Y., Ruan, R., Yu, Z., Jiang, L. (2019). Comparative study on characteristics of the bio-oil from microwave-assisted pyrolysis of lignocellulose and triacylglycerol. *Sci Total Environ*, 659, 95-100.

Dai, L., Zhou, N., Cobb, K., Chen, P., Wang, Y., Liu, Y., Zou, R., Lei, H., Mohamed, B. A., Cheng, Y., Ruan, R. (2022a). Insights into structure–performance relationship in the catalytic cracking of high density polyethylene. *Applied Catalysis B: Environmental*, 318, 121835.

Dai, L., Zhou, N., Lv, Y., Cobb, K., Chen, P., Wang, Y., Liu, Y., Zou, R., Lei, H., Mohamed, B. A., Ruan, R., Cheng, Y. (2022b). Catalytic reforming of polyethylene pyrolysis vapors to naphtha range hydrocarbons with low aromatic content over a high silica ZSM-5 zeolite. *Sci Total Environ*, 847, 157658.

Dai, M., Xu, H., Yu, Z., Fang, S., Chen, L., Gu, W., Ma, X. (2018). Microwave-assisted fast co-pyrolysis behaviors and products between microalgae and polyvinyl chloride. *Applied Thermal Engineering*, 136, 9-15.

Dai, Q., Jiang, X., Jiang, Y., Jin, Y., Wang, F., Chi, Y., Yan, J., Xu, A. (2014). Temperature Influence and Distribution in Three Phases of PAHs in Wet Sewage Sludge Pyrolysis Using Conventional and Microwave Heating. *Energy & Fuels*, 28(5), 3317-3325.

Ding, K., Liu, S., Huang, Y., Liu, S., Zhou, N., Peng, P., Wang, Y., Chen, P., Ruan, R. (2019). Catalytic microwave-assisted pyrolysis of plastic waste over NiO and HY for gasoline-range hydrocarbons production. *Energy Conversion and Management*, 196, 1316-1325.

Dominguez, A., Menendez, J. A., Inguanzo, M., Bernad, P. L., Pis, J. J. (2003). Gas chromatographic-mass spectrometric study of the oil fractions produced by microwave-assisted pyrolysis of different sewage sludges. *J Chromatogr A*, 1012(2), 193-206.

Du, H., Zhong, Z., Zhang, B., Shi, K., Li, Z. (2020). Ex-situ catalytic upgrading of vapors from microwave-assisted pyrolysis of bamboo with chemical liquid deposition modified HZSM-5 to enhance aromatics production. *Journal of Analytical and Applied Pyrolysis*, 149, 104857.

Du, Z., Li, Y., Wang, X., Wan, Y., Chen, Q., Wang, C., Lin, X., Liu, Y., Chen, P., Ruan, R. (2011). Microwave-assisted pyrolysis of microalgae for biofuel production. *Bioresour Technol*, 102(7), 4890-6.

Fan, L., Su, Z., Wu, J., Xiao, Z., Huang, P., Liu, L., Jiang, H., Zhou, W., Liu, S., Ruan, R. (2021). Integrating continuous-stirred microwave pyrolysis with ex-situ catalytic upgrading for linear low-density polyethylene conversion: Effects of parameter conditions. *Journal of Analytical and Applied Pyrolysis*, 157, 105213.

Fan, L., Zhang, Y., Liu, S., Zhou, N., Chen, P., Liu, Y., Wang, Y., Peng, P., Cheng, Y., Addy, M., Lei, H., Ruan, R. (2017). Ex-situ catalytic upgrading of vapors from microwave-assisted pyrolysis of low-density polyethylene with MgO. *Energy Conversion and Management*, 149, 432-441.

Fan, S., Zhang, Y., Cui, L., Maqsood, T., Nižetić, S. (2023a). Cleaner production of aviation oil from microwave-assisted pyrolysis of plastic wastes. *Journal of Cleaner Production*, 390: 136102.

Fan, S., Zhang, Y., Cui, L., Xiong, Q., Maqsood, T. (2023b). Conversion of Polystyrene Plastic into Aviation Fuel through Microwave-Assisted Pyrolysis as Affected by Iron-Based Microwave Absorbents. *ACS Sustainable Chemistry & Engineering*, 11(3), 1054-1066.

Fan, S., Zhang, Y., Liu, T., Fu, W., Li, B. (2022). Microwave-assisted pyrolysis of polystyrene for aviation oil production. Journal of Analytical *and Applied Pyrolysis*, 162, 136102.

Farag, S., Mudraboyina, B. P., Jessop, P. G., Chaouki, J. (2016). Impact of the heating mechanism on the yield and composition of bio-oil from pyrolysis of kraft lignin. *Biomass and Bioenergy*, 95, 344-353.

Ferrera-Lorenzo, N., Fuente, E., Bermudez, J. M., Suarez-Ruiz, I., Ruiz, B. (2014). Conventional and microwave pyrolysis of a macroalgae waste from the Agar-Agar industry. Prospects for bio-fuel production. *Bioresour Technol*, 151, 199-206.

Francis Prashanth, P., Midhun Kumar, M., Vinu, R. (2020). Analytical and microwave pyrolysis of empty oil palm fruit bunch: Kinetics and product characterization. *Bioresour Technol*, 310, 123394.

Fukuda, S. (2015). Pyrolysis investigation for bio-oil production from various biomass feedstocks in Thailand. *International Journal of Green Energy*, 12(3), 215-224.

Gasner, L. L., Denloye, A. O., Regan, T. M. (2007). Microwave and conventional pyrolysis of a bituminous coal. *Chemical Engineering Communications*, 48(4-6), 349-354.

Geyer, R., Jambeck, J. R., Law, K. L. (2017). Production, use, and fate of all plastics ever made. *Science Advances*, 3, e1700782.

Günther, B., Gebauer, K., Barkowski, R., Rosenthal, M., Bues, C.-T. (2012_. Calorific value of selected wood species and wood products. *European Journal of Wood and Wood Products*, 70(5), 755-757.

Guo, M., Song, W., Buhain, J. (2015). Bioenergy and biofuels: History, status, and perspective. *Renewable and Sustainable Energy Reviews*, 42, 712-725.

Halim, S. A., Swithenbank, J. (2016). Characterisation of Malaysian wood pellets and rubberwood using slow pyrolysis and microwave technology. *Journal of Analytical and Applied Pyrolysis*, 122, 64-75.

Huang, W., Gong, F., Fan, M., Zhai, Q., Hong, C., Li, Q. (2012). Production of light olefins by catalytic conversion of lignocellulosic biomass with HZSM-5 zeolite impregnated with 6wt.% lanthanum. *Bioresour Technol*, 121, 248-55.

Huber, G. W., Iborra, S., Corma, A. (2006). Synthesis of transportation fuels from biomass chemistry catalysts and engineering. 106, 4044-4098.

Hussain, Z., Khan, K. M., Basheer, N., Hussain, K. (2011). Co-liquefaction of Makarwal coal and waste polystyrene by microwave–metal interaction pyrolysis in copper coil reactor. *Journal of Analytical and Applied Pyrolysis*, 90(1), 53-55.

Hussain, Z., Khan, K. M., Hussain, K. (2010). Microwave–metal interaction pyrolysis of polystyrene. *Journal of Analytical and Applied Pyrolysis*, 89(1), 39-43.

Jie, X., Li, W., Slocombe, D., Gao, Y., Banerjee, I., Gonzalez-Cortes, S., Yao, B., AlMegren, H., Alshihri, S., Dilworth, J., Thomas, J., Xiao, T., Edwards, P. (2020). Microwave-initiated catalytic deconstruction of plastic waste into hydrogen and high-value carbons. *Nature Catalysis*, 3(11), 902-912.

Jing, X., Dong, J., Huang, H., Deng, Y., Wen, H., Xu, Z., Ceylan, S. (2021). Interaction between feedstocks, absorbers and catalysts in the microwave pyrolysis process of waste plastics. *Journal of Cleaner Production*, 291, 125857.

Jing, X., Wen, H., Gong, X., Xu, Z., Kajetanowicz, A. (2020). Recycling waste plastics packaging to value-added products by two-step microwave cracking with different heating strategies. *Fuel Processing Technology*, 201, 106346.

Juliastuti, S. R., Hendrianie, N., Ramadhan, P. J., Satria, D. H. (2017). Microwave pyrolysis of multilayer plastic waste (LDPE) using zeolite catalyst. in: *AIP Conference Proceedings*, 1840, 110001-1–110001-9.

Kamireddi, D., Terapalli, A., Sridevi, V., Bai, M. T., Surya, D. V., Rao, C. S., Jeeru, L. R. (2023). Microwave-assisted In-situ catalytic co-pyrolysis of polypropylene and polystyrene mixtures: Response surface methodology analysis using machine learning. *Journal of Analytical and Applied Pyrolysis*, 172, 105984.

Li, L., Ma, X., Xu, Q., Hu, Z. (2013). Influence of microwave power, metal oxides and metal salts on the pyrolysis of algae. *Bioresour Technol*, 142, 469-74.

Liu, H., Ma, X., Li, L., Hu, Z., Guo, P., Jiang, Y. (2014). The catalytic pyrolysis of food waste by microwave heating. *Bioresour Technol*, 166, 45-50.

Liu, Y., Chen, T., Gao, B., Meng, R., Zhou, P., Chen, G., Zhan, Y., Lu, W., Wang, H. (2021). Comparison between hydrogen-rich biogas production from conventional pyrolysis and microwave pyrolysis of sewage sludge: is microwave pyrolysis always better in the whole temperature range? *International Journal of Hydrogen Energy*, 46(45), 23322-23333.

Liu, Y., Ran, C., Siddiqui, A. R., Mao, X., Kang, Q., Fu, J., Deng, Z., Song, Y., Jiang, Z., Zhang, T., Dai, J. (2018). Pyrolysis of textile dyeing sludge in fluidized bed and microwave-assisted auger reactor: Comparison and characterization of pyrolysis products. *J Hazard Mater*, 359, 454-464.

Lu, Q., Li, W.-Z., Zhu, X.-F. (2009). Overview of fuel properties of biomass fast pyrolysis oils. *Energy Conversion and Management*, 50(5), 1376-1383.

Ludlow-Palafox, C., A. Chase, H. (2001). Microwave-Induced Pyrolysis of Plastic Wastes. *Industrial & Engineering Chemistry Research* 40, 4749-4756.

Lujaji, F. C., Boateng, A. A., Schaffer, M. A., Mullen, C. A., Mkilaha, I. S. N., Mtui, P. L. (2016). Pyrolysis Oil Combustion in a Horizontal Box Furnace with an Externally Mixed Nozzle. *Energy & Fuels*, 30(5), 4126-4136.

Luo, J., Gong, G., Ma, R., Sun, S., Cui, C., Cui, H., Sun, J., Ma, N. (2023). Study on high-value products of waste plastics from microwave catalytic pyrolysis: Construction and performance evaluation of advanced microwave absorption-catalytic bifunctional catalysts. *Fuel*, 346, 128296.

Luque, R., Menéndez, J.A., Arenillas, A., Cot, J. (2012). Microwave-assisted pyrolysis of biomass feedstocks: the way forward? *Energy Environ. Sci.*, 5(2), 5481-5488.

Miandad, R., Rehan, M., Barakat, M. A., Aburiazaiza, A. S., Khan, H., Ismail, I. M. I., Dhavamani, J., Gardy, J., Hassanpour, A., Nizami, A.-S. (2019). Catalytic Pyrolysis of Plastic Waste: Moving Toward Pyrolysis Based Biorefineries. *Frontiers in Energy Research*, 7, 00027.

Mohan, D., Jr., C. U. P., Steele, P. H. (2006). Pyrolysis of wood biomass for bio oil: a critical review. *Energy & Fuels*, 20, 848-889.

Morgan, H. M., Jr., Bu, Q., Liang, J., Liu, Y., Mao, H., Shi, A., Lei, H., Ruan, R. (2017). A review of catalytic microwave pyrolysis of lignocellulosic biomass for value-added fuel and chemicals. *Bioresour Technol*, 230, 112-121.

Mostafazadeh, A. K., Solomatnikova, O., Drogui, P., Tyagi, R.D. (2018). A review of recent research and developments in fast pyrolysis and bio-oil upgrading. *Biomass Conversion and Biorefinery*, 8(3), 739-773.

Moya, D., Aldás, C., López, G., Kaparaju, P. (2017). Municipal solid waste as a valuable renewable energy resource: a worldwide opportunity of energy recovery by using Waste-To-Energy Technologies. *Energy Procedia*, 134, 286-295.

Mushtaq, F., Abdullah, T. A., Mat, R., Ani, F. N. (2015). Optimization and characterization of bio-oil produced by microwave assisted pyrolysis of oil palm shell waste biomass with microwave absorber. *Bioresour Technol*, 190, 442-50.

Okolie, J. A., Epelle, E. I., Nanda, S., Castello, D., Dalai, A. K., Kozinski, J. A. (2021). Modeling and process optimization of hydrothermal gasification for hydrogen production: A comprehensive review. *The Journal of Supercritical Fluids*, 173, 105199.

Prathiba, R., Shruthi, M., Miranda, L. R. (2018). Pyrolysis of polystyrene waste in the presence of activated carbon in conventional and microwave heating using modified thermocouple. *Waste Manag*, 76, 528-536.

Rex, P., Masilamani, I. P., Miranda, L. R. (2020). Microwave pyrolysis of polystyrene and polypropylene mixtures using different activated carbon from biomass. *Journal of the Energy Institute*, 93(5), 1819-1832.

Ruddy, D. A., Schaidle, J. A., Ferrell Iii, J. R., Wang, J., Moens, L., Hensley, J. E. (2014). Recent advances in heterogeneous catalysts for bio-oil upgrading via "ex situ catalytic fast pyrolysis": catalyst development through the study of model compounds. *Green Chem.*, 16(2), 454-490.

Sakulkit, P., Palamanit, A., Dejchanchaiwong, R., Reubroycharoen, P. (2020). Characteristics of pyrolysis products from pyrolysis and co-pyrolysis of rubber wood and oil palm trunk biomass for biofuel and value-added applications. *Journal of Environmental Chemical Engineering*, 8(6), 104561.

Salema, A. A., Ani, F. N. (2012). Microwave-assisted pyrolysis of oil palm shell biomass using an overhead stirrer. *Journal of Analytical and Applied Pyrolysis*, 96, 162-172.

Schüppel, M., Gräbner, M. (2024). Pyrolysis of heavy fuel oil (HFO) – A review on physicochemical properties and pyrolytic decomposition characteristics for application in novel, industrial-scale HFO pyrolysis technology. *Journal of Analytical and Applied Pyrolysis*, 179, 106432.

Selvam, E., Kots, P. A., Hernandez, B., Malhotra, A., Chen, W., Catala-Civera, J. M., Santamaria, J., Ierapetritou, M., Vlachos, D. G. (2023). Plastic waste upgrade to olefins via mild slurry microwave pyrolysis over solid acids. *Chemical Engineering Journal*, 454, 140332.

Sharuddin, S. D. A., Abnisa, F., Wan Daud, W. M. A., Aroua, M. K. (2016). A review on pyrolysis of plastic wastes. *Energy Conversion and Management*, 115, 308-326.

Suriapparao, D. V., Boruah, B., Raja, D., Vinu, R. (2018). Microwave assisted co-pyrolysis of biomasses with polypropylene and polystyrene for high quality bio-oil production. *Fuel Processing Technology*, 175, 64-75.

Suriapparao, D. V., Pradeep, N., Vinu, R. (2015). Bio-Oil Production from Prosopis juliflora via Microwave Pyrolysis. *Energy & Fuels*, 29(4), 2571-2581.

Suriapparao, D. V., Vinu, R. (2015a). Bio-oil production via catalytic microwave pyrolysis of model municipal solid waste component mixtures. *RSC Advances*, 5(71), 57619-57631.

Suriapparao, D. V., Vinu, R. (2015b). Resource recovery from synthetic polymers via microwave pyrolysis using different susceptors. *Journal of Analytical and Applied Pyrolysis*, 113, 701-712.

Suriapparao, D. V., Vinu, R., Shukla, A., Haldar, S. (2020). Effective deoxygenation for the production of liquid biofuels via microwave assisted co-pyrolysis of agro residues and waste plastics combined with catalytic upgradation. *Bioresour Technol*, 302, 122775.

Tian, Y., Zuo, W., Ren, Z., Chen, D. (2011). Estimation of a novel method to produce bio-oil from sewage sludge by microwave pyrolysis with the consideration of efficiency and safety. *Bioresour Technol*, 102(2), 2053-61.

Undri, A., Meini, S., Rosi, L., Frediani, M., Frediani, P. 2013. Microwave pyrolysis of polymeric materials: Waste tires treatment and characterization of the value-added products. *Journal of Analytical and Applied Pyrolysis*, 103, 149-158.

Undri, A., Rosi, L., Frediani, M., Frediani, P. (2014a). Microwave assisted pyrolysis of corn derived plastic bags. *Journal of Analytical and Applied Pyrolysis*, 108, 86-97.

Undri, A., Rosi, L., Frediani, M., Frediani, P. (2014b). Upgraded fuel from microwave assisted pyrolysis of waste tire. *Fuel*, 115, 600-608.

Wang, D., Li, D., Liu, Y., Lv, D., Ye, Y., Zhu, S., Zhang, B. (2014). Study of a new complex method for extraction of phenolic compounds from bio-oils. *Separation and Purification Technology*, 134, 132-138.

Wang, H., Gross, A., Liu, J. (2022). Influence of methanol addition on bio-oil thermal stability and corrosivity. Chemical Engineering Journal, 433, 133692.

Wei, Y., Lei, H., Wang, L., Zhu, L., Zhang, X., Liu, Y., Chen, S., Ahring, B. (2014). Liquid–liquid extraction of biomass pyrolysis bio-oil. *Energy & Fuels*, 28(2), 1207-1212.

Yang, Z., Kumar, A., Huhnke, R. L. (2015). Review of recent developments to improve storage and transportation stability of bio-oil. *Renewable and Sustainable Energy Reviews*, 50, 859-870.

Yu, F., Ruan, R., Steele, P. (2009). Microwave pyrolysis of corn stover. *Transactions of the ASABE*, 52(5), 1595-1601.

Zevenhoven, R., Karlsson, M., Hupa, M., Frankenhaeuser, M. (1997). Combustion and gasification properties of plastics particles. *J Air Waste Manag Assoc*, 47(8), 861-870.

Zhang, H., Yang, K., Tao, Y., Yang, Q., Xu, L., Liu, C., Ma, L., Xiao, R. (2023). Biomass directional pyrolysis based on element economy to produce high-quality fuels, chemicals, carbon materials - A review. *Biotechnol Adv*, 69, 108262.

Zhang, L., Wu, Q., Fan, L., Liao, R., Zhang, J., Zou, R., Cobb, K., Ruan, R., Wang, Y. (2024). Monocyclic aromatic hydrocarbons production from NaOH pretreatment metallized food plastic packaging waste through microwave pyrolysis coupled with ex-situ catalytic reforming. *Chemical Engineering Journal*, 484, 149777.

Zhang, Q., Chang, J., Wang, T., Xu, Y. (2007). Review of biomass pyrolysis oil properties and upgrading research. *Energy Conversion and Management*, 48(1), 87-92.

Zhang, X., Lei, H., Yadavalli, G., Zhu, L., Wei, Y., Liu, Y. (2015). Gasoline-range hydrocarbons produced from microwave-induced pyrolysis of low-density polyethylene over ZSM-5. *Fuel*, 144, 33-42.

Zhang, X., Lei, H., Zhu, L., Qian, M., Yadavalli, G., Wu, J., Chen, S. (2017). From plastics to jet fuel range alkanes via combined catalytic conversions. *Fuel*, 188, 28-38.

Zhang, Y., Liu, S., Fan, L., Zhou, N., Mubashar Omar, M., Peng, P., Anderson, E., Addy, M., Cheng, Y., Liu, Y., Li, B., Snyder, J., Chen, P., Ruan, R. (2018). Oil production from microwave-assisted pyrolysis of a low rank American brown coal. *Energy Conversion and Management*, 159, 76-84.

Zhou, N., Dai, L., Lv, Y., Li, H., Deng, W., Guo, F., Chen, P., Lei, H., Ruan, R. (2021). Catalytic pyrolysis of plastic wastes in a continuous microwave assisted pyrolysis system for fuel production. *Chemical Engineering Journal*, 418, 129412.

Zhou, N., Dai, L., Lyu, Y., Wang, Y., Li, H., Cobb, K., Chen, P., Lei, H., Ruan, R. (2022). A structured catalyst of ZSM-5/SiC foam for chemical recycling of waste plastics via catalytic pyrolysis. *Chemical Engineering Journal*, 440, 135836.

Chapter 6

Microwave Irradiation for Pyrolysis of Organic Wastes for Char Production

Tianhao Qiu
Chaoyue Liu
Yaning Zhang*
and Wenke Zhao
School of Energy Science and Engineering, Harbin Institute of Technology, Harbin, China

Abstract

Biochar is widely used in various fields due to its good physical and chemical properties. The biochar prepared by microwave-assisted pyrolysis is affected by reaction parameters. This study reviewed biochar preparation by microwave-assisted pyrolysis of organic wastes. The effects of pyrolysis temperature, microwave power and residence time on the properties and yield of biochar produced from microwave-assisted pyrolysis of organic wastes were covered. And the applications of biochar in different fields were also detailed. It was concluded that with the increase of pyrolysis temperature, microwave power, and residence time, the yield of biochar produced from microwave-assisted pyrolysis mainly shows decreasing trends. Both higher pyrolysis temperature and higher microwave power improve the specific surface area and pore volume of biochar. The increase of residence time reduces the yield of biochar whereas increases the high heating value. The activation methods of biochar were also introduced. The preparation of biochar by microwave-assisted pyrolysis has become increasingly popular, and

* Corresponding Author's Email: ynzhang@hit.edu.cn.

In: Characteristics and Applications of Microwave Irradiation
Editors: Yaning Zhang and Wenke Zhao
ISBN: 979-8-89530-170-8

further development is needed to achieve excellent properties and high yield to make the applications of biochar more widespread and popular.

Keywords: biochar, microwave-assisted pyrolysis, high heating value, specific surface area, activation

1. Introduction of Char

1.1. Definition of Biochar

Biochar was first used by Bapat et al. to distinguish activated carbon materials from biomass and fossil fuels in 1999 (Bapat et al. 1999). Lehman et al. (2006) enriched the meaning of the word "biochar" and defined "biochar" as biomass-derived black carbon (Figure 1), that is a material similar in concept to charcoal that can be used to sequester soil carbon pools, improve soil physical and biological properties, and improve soil for plant growth. With the deepening of biochar research, the concept of biochar is gradually becoming clearer. In 2013, the International Biochar Initiative (IBI) defined biochar as a solid material obtained by thermochemical conversion of biomass under oxygen-restricted environmental conditions. With the continuous improvement and innovation of carbonization process and preparation technology, biochars obtained by different principles, technologies and methods vary greatly, which makes it significantly different in application objects, objectives, scope and efficacy, and also makes the concept and definition of biochar show a trend of diversification and complexity. Based on the source and application of biochar, from the perspective of recycling waste biomass resources, it is generally believed that biochar comes from waste biomass such as agriculture and forestry, and is a carbon-rich solid product formed by pyrolysis under a certain carbonization temperature (<700 °C) and under oxygen restriction or hypoxia conditions (Chen et al. 2011). Masebinu et al. (2019) provided a definition that biochar is the carbonaceous residue derived from the thermal conversion of biomass regardless of type, including non-plant biomass like animal and sludge, in an oxygen-starved environment. Zhang et al. (2019 c) defined biochar as s a carbon-rich solid material from the thermo-chemical conversion of biomass (e.g., wood, manure, and leaves, etc.) when temperatures below 900 °C in an oxygen-limited environment. At present, although there are still some different views on the understanding and cognition of the concept and definition of biochar, with the innovation and development of biochar technology, especially the continuous emergence of multi-disciplinary and advanced technologies, the research on the structures and physical and chemical properties of biochar will continue to be deepened. At that time, the concept and definition of biochar will be gradually unified, improved and developed.

Figure 1. Biochar.

1.2. Properties of Biochar

Biochar is a product formed by the thermal cracking reaction process of biomass (Azargohar et al. 2013). In the process of pyrolysis, biomass will form biochar, bio-oil, gas mixture and other materials by aromatic polycyclic structure through intramolecular and intermolecular rearrangement reaction (McGrath et al. 2003; Roy and Dias, 2017). The carbonization processes of different kinds of biomass are different. In general, raw materials for carbon production can be divided into lignocellulose and non-lignocellulosic biomass (Li and Jiang, 2017 a; Rangabhashiyam and Balasubramanian, 2019). Among them, lignocellulosic biomass is mainly derived from plant wastes (Vaibhav and Thallada, 2017), and its main components are cellulose, hemicellulose, lignin, etc. (Yu et al. 2017). Non-lignocellulosic biomass mainly includes animal feces, some plants and their derivatives, and its main components are proteins, lipids, sugars, inorganic substances, and some lignin and cellulose (Toor et al. 2011). Under normal circumstances, lignocellulosic biomass begins to undergo pyrolysis reaction at 200-260 °C, and the branched polymers and short side chains of hemicellulose are first decomposed into oligosaccharides, and then rearranged to form 1, 4-ether D-xylose. 1, 4-ether D-xylose is formed into carbon after dehydration, decarboxylation, aromatization and intramolecular condensation, or decomposed into low molecular weight bio-oil and gas mixture (Liu et al. 2015 b). At around 300 °C, cellulose begins to be decomposed into oligosaccharides, and then the glucoside bond breaks to produce D-glucopyranose, which is rearranged intramolecular to form L-glucosaccharides. Some of the L-glucosans are formed into carbon after decarboxylation, aromatization and intramolecular condensation. The other part is converted into hydroxymethylfurfural after the process of rearrangement and dehydration, and further converted into bio-oil and gas mixture (Liu et al. 2015 b). At 400 °C, lignin began to be decomposed into free radicals in large quantities. Biochar, bio-oil and gas mixture were formed through the processes of free radical substitution, addition, and carbon-carbon coupling (Cao et al. 2017). When the temperature exceeds 500 °C, the polymers and volatile compounds in the pores of biochar will re-polymerize to form secondary carbon or recombine to produce tar with higher molecular weight (Zhang et al. 2014). Non-lignocellulosic biomass, on the other hand, begins pyrolytic reaction at around 200 °C, and the low bond energy bonds (such as hydrogen bonds and calcium hydroxide bonds) of proteins, lipids, carbohydrates and other organic compounds break (Li and Jiang, 2017 a). Bond breakage increases with increasing temperature, and the main pyrolysis process occurs at 300~600 °C. The final pyrolysis forms biochar, bio-oil and gas mixture. In general, the pyrolysis behavior of non-lignocellulosic biomass is more complicated and variable due to the large amount of nitrogen, phosphorus and oxygen contained in proteins, lipids and nucleic acids. In the process of biomass pyrolysis and carbonization, the material form transformation "from biomass to carbon" is completed, and extremely rich multi-microporous carbon frame structures, different kinds of surface functional groups, small organic molecules and mineral salts are formed (Shaheen et al. 2019), which lays an important foundation for its use as functional material such as adsorption and carrier. It can be widely used in agriculture, environment, energy and other fields (Chen et al. 2019). The following is a brief introduction to the physicochemical properties of biochar and the relationship between biochar pyrolysis and preparation conditions.

1.2.1. Acid-Basicity

Biochar is generally alkaline, but it is also acidic. The pH of biochar ranges from 3 to 13 (Zhang et al. 2013). In general, the more ash content in the raw material, the higher the pH of biochar produced (Lehmann, 2007). In terms of material, since the biochar prepared from non-cellulosic biomass has a higher ash content, the pH of non-cellulosic biochar is generally higher than that of cellulosic biochar (Spokas et al. 2011). Under the same carbonization process conditions, the pH values of biochar from different materials were as follows: livestock feces > herbs > woody plants (Lehmann, 2007; Yuan and Xu, 2011 a). It is generally believed that the pH of biochar increases with the increase of carbonization temperature, as shown in Figure 2. The increase of pH value of biochar is mainly due to the fact that the decomposition of acidic functional groups, such as -COOH and -OH, will be accelerated under higher carbonization temperature, thus increasing the pH of biochar (Li et al. 2017 b; Xu et al. 2017), and gradually forming basic functional groups (Tomczyk et al. 2020). It was found that when the temperature increased from 200 °C to 800 °C, the acid functional group of biochar decreased from 4.17 mm·μg^{-1} to 0.22 mm·μg^{-1}, and the basic functional group increased from 0.15 mm·μg^{-1} to 3.55 mm·μg^{-1}. The pH increased from 7.37 to 12.40 (Al-Wabel et al. 2013). During pyrolysis, as the temperature increases, the C≡C, C=C, and C=O bonds break, forming aromatic rings that contain more hydroxyl groups, thereby increasing the pH of biochar (Hassan et al. 2020). Separation of alkali salts from organic matter at higher pyrolysis temperatures also increases the pH of biochar (Pariyar et al. 2020). In other words, higher carbonization temperature is conducive to promoting the formation of alkali metal (Na, K) or alkaline earth metal (Ca, Mg) plasma compounds in ash, such as KOH, NaOH, $MgCO_3$, $CaCO_3$, etc., thus increasing the pH of biochar (Xu et al. 2017). In addition, the carbonization residence time can also affect the pH of biochar to a certain extent, showing that the longer the carbonization residence time, the higher the pH (Conti et al. 2014). But the heating rate has little effect on the pH of biochar (Angin, 2013).

The acid-alkalinity of biochar is one of the important indexes that affect its properties and functions. From the current research, due to the different raw materials, carbonization processes and preparation technologies, the pH range of biochar is wide, basically covering the different states of "acidic, neutral and alkaline". In the actual production and application process, biochar materials with a certain acid-base range can be obtained through carbonization process control, modification and other technical methods, and scientifically applied according to the purpose and target of use.

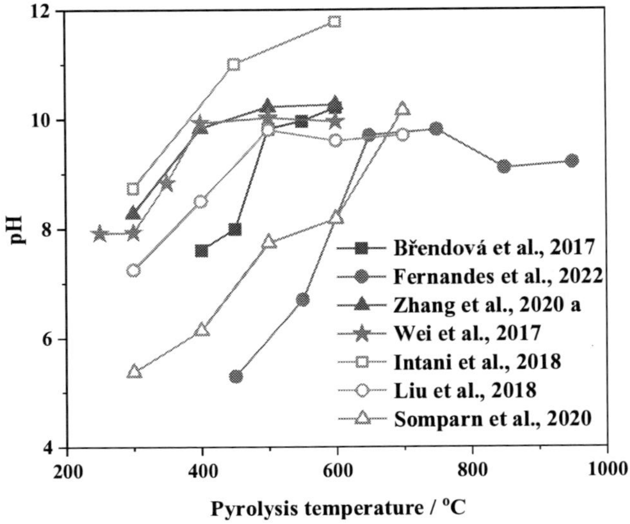

Figure 2. pHs of biochars at different pyrolysis temperatures.

1.2.2. Surface Specific Area

Biochar has a large specific surface area due to its abundant multi-microporous structure (Tan et al. 2017; Zhou et al. 2017). The specific surface area of biochar is related to its raw material and carbonization temperature. In general, cellulosic biomass has abundant internal pores, and the biochar formed after carbonization will retain the fine pore structure of the original biomass and greatly increase the specific surface area, while the pore structure of non-cellulosic biomass is relatively small, so the specific surface area of carbonized biochar is smaller than that of cellulosic biochar.

Carbonization temperature is one of the important factors affecting the specific surface area of biochar (Figure 3). When the carbonization temperature is below 400 °C, the porosity formed in the carbonization process of biomass decreases, which may be related to the pyrolysis separation and incomplete volatilization of volatile substances in biomass (Oliveira et al. 2017; Chen et al. 2017 b). With the increase of carbonization temperature, biomass separates and releases more volatile substances, with more pores and larger specific surface area (Zhao et al. 2018 a). However, when the temperature is higher than 800 °C, the porous structure in biochar will partially collapse, thus blocking the pores and making the pore size smaller (Burhenne et al. 2013). Ash content will also affect the specific surface area of biochar. In general, with the increase of carbonization temperature, the ash content of biochar will increase, which will block some biochar pores and reduce its specific surface area. Lignin-rich biomass is usually decomposed at high temperatures (Leng et al. 2021). The increase in the specific surface area of biochar at higher pyrolysis temperatures may be due to the decomposition of lignin and cellulose, thus forming channel structures during pyrolysis. These channel structures increase the specific surface area and pore volume. Similarly, the increase in pore volume of wood biochar at high pyrolysis temperature is mainly due to the increase in micropore volume caused by the release of volatile compounds. Biochar after pyrolysis can maintain most of the original biomass structure, thus affecting the structure of biochar (Brewer et al. 2014). At lower temperatures, biochar generally retains the pore structure of thin cell walls and channels, and the pore volume is larger. Higher pyrolysis temperatures further damage these structures, which negatively affects the pore volume.

The large specific surface area is one of the important basic characteristics of biochar as a carrier and adsorption functional material. In the process of biochar preparation, the size of specific surface area can be controlled by modification and process control to some extent.

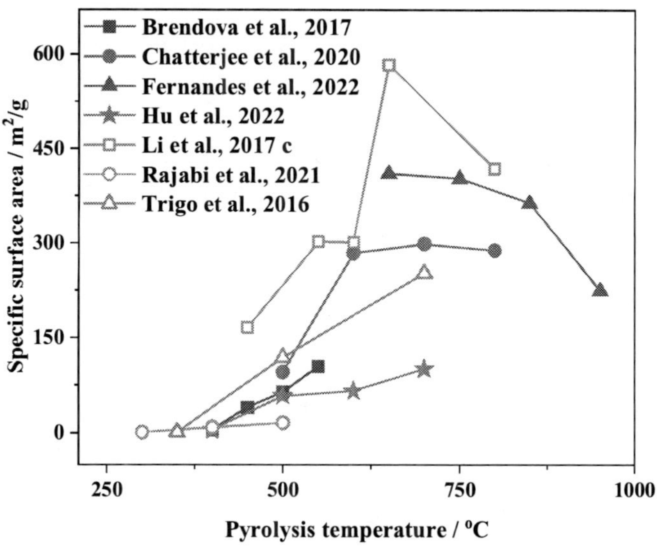

Figure 3. Specific surface areas of biochars at different pyrolysis temperatures.

1.2.3. Element Composition

Biochar is mainly composed of C, O, H, N, K, Ca, Na, Mg and other elements (Zhang et al. 2015 a). Among them, the C element is mainly fixed carbon in the form of aromatic ring, while the alkali metals (K, Na, etc.) and alkaline earth metals (Ca, Mg) exist in the ash in the form of carbonate, phosphate or oxide (Xu et al. 2017). The types and contents of elements in biochar are mainly related to raw materials (Chen et al. 2014). According to the study, the C content of cellulose biochar is higher than that of non-cellulose biochar, and the C content of wood and bamboo biochar is higher (Zornoza et al. 2016). Among other elements, the content of K in cellulosic biochar is relatively high (Subedi et al. 2016), while the content of Ca, Mg, N, P and other elements in non-cellulosic biochar is higher than that in cellulosic biochar. Carbonization temperature is also an important factor affecting the contents and compositions of biochar. In general, the contents of C, alkali metals and minerals in biochar increase with the increase of carbonization temperature, while the contents of N, H, O and other elements decrease with the increase of carbonization temperature (Zhang and Wang, 2016).

The C in biochar is mainly aromatic carbon, which exists in the form of aromatic ring and irregular stacking, so that biochar has high stability and strong decomposition resistance (Ameloot et al. 2013). Biochar also contains certain organic carbon, which is mainly incomplete carbonized biomass, fatty acids, alcohols, phenols, esters, and other material components such as fulvic acid and humic acid. The contents of organic carbon are generally relatively high in newly prepared biochar, non-cellulosic biochar, and biochar formed by low-temperature pyrolysis. N, P, K and other inorganic salt ions in biochar can be used as one of the nutrient sources of plants, microorganisms and other living organisms (Gul et al. 2015), but the nutrient element contents and availabilities of different biochar are different. Generally, biochar prepared with nutritionally rich raw materials contains relatively high nutrient elements, and vice versa (Gul and Whalen, 2016). The availability of elements in biochar is related to freshness, pH, carbonization temperature and other conditions of biochar. In general, the newly prepared biochar can release more nutrient elements such as N, P and K, while the aged biochar is relatively weak in terms of nutrient release amount and speed (Ding et al. 2016). The pH of biochar also affects its elemental availability. When the pH is between 2 and 7, the release of PO_4^{3-} and NH_4^+ from biochar decreases with the increase of pH, while the content of K^+ remains relatively stable (Qian et al. 2015). When the pH drops from 8.9 to 4.5, the release of PO_4^{3-} and NH_4^+ from biochar decreases with the increase of pH. The release of Ca and Mg elements in biochar will increase. In addition, carbonization temperature also affects the state of elements in biochar. With the increase of carbonization temperature, mineral components such as alkali metals in biochar tend to crystallize, reducing its water solubility (Yuan et al. 2011 b).

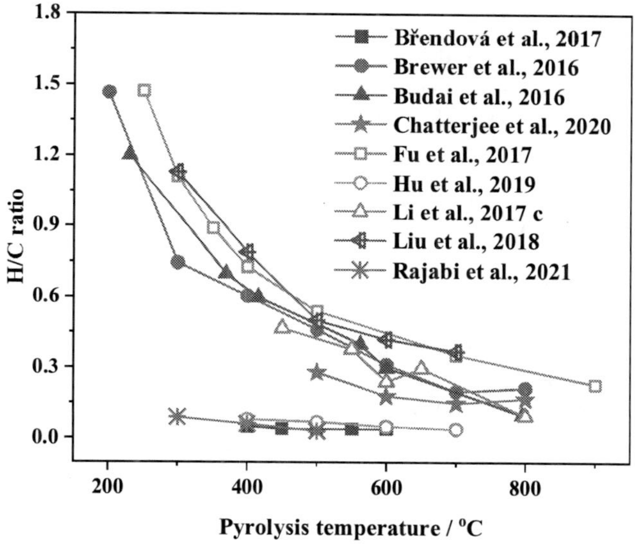

Figure 4. H/C ratios of biochars at different pyrolysis temperatures.

Biochar contains higher C and has a stable aromatic carbon structure, which makes it highly stable (Wiedemeier et al. 2015). The degree of aromatic ring condensation in biochar structure affects the stability and durability of biochar function, while the non-aromatic structure/functional group is conducive to the formation of cluster structure, which also has a certain impact on the stability of biochar. The stability of biochar is usually characterized by H/C (Figure 4) and O/C (Manyà et al. 2014). Low H/C and O/C ratios indicate that biochar has a higher molten aromatic ring structure and strong stability, and vice versa. Raw material type and carbonization temperature are the main factors affecting the stability of biochar. In the preparation of raw materials, lignin is the most stable component, followed by cellulose and hemicellulose (Li et al. 2017 d). The higher the lignin content in raw materials, the higher the aromatic C content formed after carbonization, and the stronger the stability of biochar (Ameloot et al. 2013). Therefore, lignocellulosic biochar is more stable and more resistant to biodegradation than non-lignocellulosic biochar. Metals, metal compounds or minerals in raw materials will reduce the stability of biochar to a certain extent, but some alkali metals, alkaline earth metals and silicon will enhance the stability of biochar. For example, amorphous silicon in biochar interacts with C to form stable Si-C, thereby preventing C from being oxidized (Han et al. 2018). In addition, the size of raw materials will also affect the stability of biochar to a certain extent. Generally, the stability of biochar prepared with large particles/size raw materials is higher than that prepared with small particles/size raw materials (Chen et al. 2017 a), which is mainly due to the longer contact time between gas phase and solid phase materials during pyrolysis carbonization of large particles/size raw materials. More carbon is formed by polymerization, thus increasing the fixed carbon content of biochar and enhancing its stability (Manyà et al. 2014). The aromatics and aromatic cohesion of biochar are determined by pyrolysis carbonization temperature. Most of the unstable components in biochar gradually disappear with the increase of carbonization temperature, the structure of molten aromatic rings increases, and the structure, size and quantity of unstable non-aromatic rings decrease. In addition, the heating rate will also have a certain impact on the stability of biochar, and a slow heating rate is conducive to enhancing the stability of biochar (Crombie and Masek, 2015).

The element composition, content and form of biochar are the "source" of the important characteristics and efficacy of biochar. It is mainly reflected in: 1) releasing large amounts of N, P, K, Ca, Mg and other medium and trace elements, providing certain exogenous nutrients for soil and promoting plant growth (Chen et al. 2013); 2) Ash, an important component of biochar, in addition to containing certain alkali metals, can play a key role in the process of adsorption of metal elements.

1.2.4. Surface Functional Groups

Cellulose, hemicellulose, protein and fat in biomass form a large number of carboxyl, carbonyl, lactone, hydroxyl, ketone and other types of functional groups on the surface and inside of biochar after pyrolysis and carbonization, most of which are oxygen-containing functional groups or alkaline functional groups, so that biochar has good adsorption, hydrophilic/hydrophobic, as well as buffer acid-base, promote ion exchange, and other properties (Qambrani et al. 2017; Antón-Herrero et al. 2018). The surface functional groups of biochar are closely related to the raw materials and carbonization temperature conditions (Li et al. 2017 b). Among different raw materials, non-cellulosic biochar contains more N and S functional groups than cellulosic biochar. Under different carbonization temperatures, the number and density of functional groups of biochar decreased with the increase of carbonization temperature. At 185-200 °C, the types of functional groups on the surface of biochar do not change significantly (Li et al. 2017 b), while when the temperature reaches 300 °C, the contents of carboxyl and carbonyl groups rapidly rise to the highest point, and then decrease with the increase of carbonization temperature. At 400-550 °C, the aliphatic functional groups of biochar gradually disappear with the increase of temperature (Chen et al. 2017 b), and at 600 °C, the alkyl carbon functional groups disappear.

The special structures and physical and chemical properties of biochar make it have certain hydrophobic/water holding capacity. Biochar is hydrophobic, which is related to the reduction of surface oxygen-containing functional groups in the carbonization process (Masebinu et al. 2019). Biochar has water retention, mainly due to its abundant multi-microporous structure, which increases its adsorption power to water (Gray et al. 2014). The water retention performance of biochar is mainly determined by the counteracting, balancing or superimposing effect of its hydrophobic/water retention or water adsorption (Weber and Quicker, 2018). The hydrophobicity of biochar is an important basis for its use as a non-soluble adsorbent. Its properties can be characterized by measuring the number of O and N functional groups in biochar. The lower the number of O and N functional groups, the stronger the hydrophobicity (Masebinu et al. 2019). The hydrophobicity of biochar is closely related to the carbonization temperature. The higher the carbonization temperature, the stronger the hydrophobicity, is related to the decrease of polar functional groups on the surface of biochar and the increase of aromatics (Fang et al. 2014). The water holding capacity of biochar is related to its porous structure and pore connectivity at different carbonization temperatures (Weber and Quicker, 2018). Under low temperature carbonization conditions, biochar has smaller pore size and lower interconnection, and the tar components generated are easy to block the pores, resulting in decreased water holding capacity (Das and Sarmah, 2015), while under high temperature carbonization conditions, biochar has lower water holding capacity. The increase of multi-micropore structure and quantity, enhanced pore connectivity, and enhanced physical water absorption and adsorption capacity of biochar improve its water retention performance (Weber and Quicker, 2018).

1.2.5. Heating Value

As the fourth largest energy source besides coal, natural gas and oil, biomass energy is the only renewable carbon-containing energy at present and has great development potential (He et al. 2015). As a large agricultural country, China is rich in straw resources, and its heat can be released to generate electricity (Peng et al. 2014). Research shows that the theoretical resources of crop straw in China is about 800 million tons, of which about 205 million tons of straw, 150 million tons of wheat straw, 265 million tons of corn straw and 25.84 million tons of cotton straw, accounting for 25%, 18.3%, 32.3% and 3.2% of the total amount of agricultural straw, respectively (Tian et al. 2011). The disadvantages of biomass directly used as fuel are high water content, difficult to crush, low energy density, high oxygen content, and not suitable for storage (Abdullan et al. 2010). Thermal chemical treatment of biomass can improve its energy yield and fuel characteristics and increase its energy density, and the fuel characteristics of treated biomass are better than those of highly volatile coal (Du et al. 2014). Heating value is the most important characteristic of biomass fuel, which determines the value of the fuel, and is an indispensable parameter for the calculation of thermal balance, thermal efficiency and consumption of combustion and other transformations. Heating value represents the heat released when fuel is completely burned per unit mass (or volume), which are important parameters to represent fuel quality. The greater the heat released, the greater the heating value of biochar. At present, the more accurate measurement of heating value depends on the oxygen bomb calorimeter. However, this method is time-consuming and costly (Matin and Chelgani, 2016). The near infrared spectrum in infrared spectroscopy has a wide band, and the important absorption peak in the spectrum reflects the comprehensive information of a variety of compounds, while the middle infrared spectrum determines the characteristic absorption peak of a certain compound. Therefore, mid-infrared spectroscopy technology has obvious advantages in accuracy and stability, and is widely used in preventive medicine, medicine, petroleum, biology, agriculture, food, tobacco and other fields due to its simple, rapid and specific fingerprint characteristics. Calamari et al. (2016) used Fourier transform mid-infrared spectroscopy to determine the titration acidity of milk. Schalk et al. (2017) used mid-infrared technology to realize real-time monitoring of yeast fermentation. Fagan et al. (2011) used near infrared spectroscopy technology combined with least square method to determine the calorific value of biomass raw materials, and its mid-calorific value pre-measurement modulus determination coefficient reached 0.99.

Pyrolysis temperature significantly affects the high heating value of biochar. The trends of heating value of biochar at different pyrolysis temperatures are shown in Figure 5. With the increase in the pyrolysis temperature, the heating value of biochar mainly showed an upward trend.

Nzediegwu et al. (2021) reported that when the reaction temperature was increased from 300 °C to 500 °C, the heating values of biochar were increased in the ranges of 22.3-24.5, 25.9-32.3, 25.6-26.9, and 4.1-4.7 MJ/kg for canola straw, sawdust, wheat straw, and manure pellet, respectively. Parvez et al. (2020) reported that when the reaction temperature was increased from 400 °C to 600 °C, the heating values of biochar were increased in the ranges of 18.3-22.1, 23.2-27.3, and 18.6-21.7 MJ/kg for corn stalk, pinewood, and algae, respectively. On the whole, the heating values of biochar are positively proportional to the pyrolysis temperature, which is mainly related to the increase of C and fixed carbon contents in biochar.

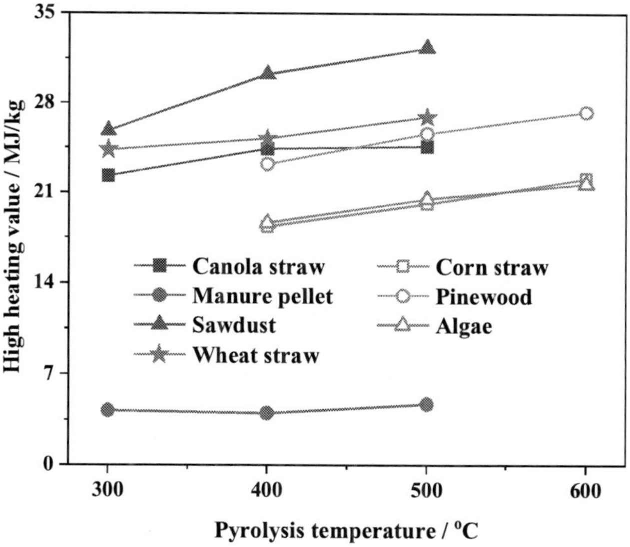

Figure 5. Heating values of biochars at different pyrolysis temperatures.

1.3. Applications of Biochar

Biochar has great application value due to its properties: (a) rich organic carbon, (b) certain mineral components (i.e., P, K, Ca, and Mg), (c) high porosity, (d) large specific surface area (BET), (e) abundant functional groups, (f) strong alkalinity, and (g) high electrical conductivity. Biochar was only applied in agriculture initially. With the development of technology, biochar has been used in various fields, i.e., wastewater treatment, soil remediation, electrochemical conductivity, building insulation, catalyst support, etc.

1.3.1. Removal of Heavy Metals from Wastewater Treatment

Adsorption technology is one of the main methods to treat gaseous pollutants and organic pollutants in water by using porous solid materials as pollutant adsorbents. Biochar usually has large specific surface area, uniform pore structure, and a large proportion of the microporous structure, which are contributed to adsorption.

Chu et al. (2017) confirmed that the peanut shell-biochar showed great adsorption ability to carbamazepine and bisphenol A. The research of Liew et al. (2018 b) concluded that the specific surface area of oil palm waste biochar reached 210 m^2/g, and the adsorption capacity of the biochar to methylene blue was 48 mg/g. When Yek et al. (2018 a) analyzed the pyrolysis of orange peel, it was obtained that the orange peel biochar presented more micropores in the CO_2 pyrolysis atmosphere and more mesopores in the steam atmosphere. The effect of orange peel biochar prepared in the CO_2 atmosphere on the removal of hazardous Congo red dye wastewater from the textile industry was better than that in the steam atmosphere. In addition to organic pollutants, biochar also had a great removal effect on heavy metal pollutants in wastewater. Antunes et al. (2018) found biosolid biochar can effectively remove silver and phosphorus in water, and Yap et al. (2017) also found agricultural biomass biochar had the effect on the removal of lead and cadmium from wastewater. Generally speaking, biochar should be activated to improve the adsorption ability. The hydroxides (KOH and NaOH) were used as chemical activators by Yek et al. (2020 b) to obtain biochar with high specific surface area, pore volume, and fixed carbon contents and the removal rate of malachite green dye in industrial wastewater reached 90%. Fan et al. (2020) also used KOH to activate corn stem-pith biochar. The specific surface area of activated biochar reached 581.85 m^2/g, and the absorbed amount and efficiency on lead in water were 99.95 mg/g and 50.35%, respectively. Nair and Vinu (2016) adopted peroxide-assisted microwave activation of pyrolysis char from Prosopis juliflora biomass and found the biochar showed good adsorption ability to dyes from wastewater.

Besides, there are plenty of studies for the applications of biochar in wastewater treatment, as shown in Table 1. Singh et al. (2021) summarized several mechanisms of removal of heavy metals from wastewater by biochar, as shown in Figure 6. Removal of heavy metal ions from wastewater by biochar can be achieved through various mechanisms like ion exchange, precipitation, complexation, electrostatic interaction, and sorption. Biochar has a uniformly distributed network of pores including micropores (with diameter < 2 nm), mesopores (2-50 nm), and macropores (>50 nm), and this distinguishing feature of the char helps heavy metals to be physically adsorbed onto them. Metal pollutants can be removed through various processes like electrostatic attractions in which negative charges on surfaces of biochar bind positively charged metal ions, and complexation in which ligands and functional groups present on biochar surface form complexes by interacting with various metals.

In summary, some aspects should be paid attention to in wastewater treatment. The surface charge of biochar is negatively charged in general. Therefore, the adsorption of cations is easier, and the adsorption of anions needs to be improved. The original biochar tends to have little advantage in specific surface area and porosity, so it is necessary to consider the activation of biochar. What's more, the solid-liquid separation of biochar and wastewater is also a problem to be considered in water adsorption, which will affect the large-scale promotion of biochar in industrial production.

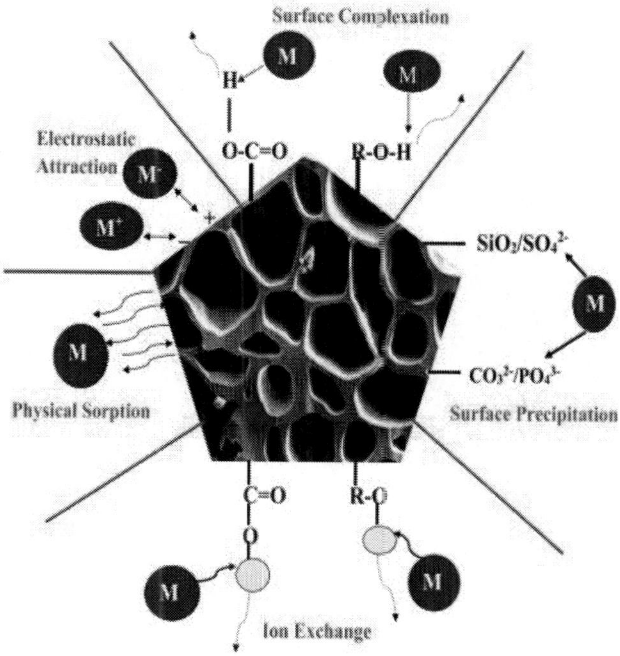

Figure 6. Possible mechanisms for the removal of heavy metals from aqueous solution using biochar as an adsorbent (Singh et al. 2021).

Table 1. Removal of various heavy metals from wastewater biochar

Biomass	Preparation condition	Experimental condition	Adsorption efficiency	Reference
Discarded mushroom-stick	300~800℃	Adsorbent dose: 4 g/L, Metal concen-tration: 0.1 g/L, room temperature, 150 rpm	Adsorption capacity for Pb, Cu, Cd, and Ni was 21.0, 18.8, 11.2, and 9.8 mg/g ,respectively	Wang et al. 2019
Cherry wood and cattle manure	400~600℃	Adsorbent dose: 4 g/L, Metal concentration: 100 mg/L, room temperature, 150 rpm	Adsorption capacity for Pb, Cd, and Ni was 40.8, 24.2, and 25.1 mg/g for cattle manure biochar, respectively; Adsorption capacity for Pb, Cd, and Ni was 29.7, 21.1, and 14.7 mg/g for cattle manure biochar, respectively	Ye et al. 2022
Cotton stalk	400℃	As: 50-200 μg/L, adsorbent dose: 1-3 g/L, pH: 6.0, Temperature: 25℃, 150 rpm	the maximum sorption capacity of CSB was 90 μg /g; 45%-88% for As removal rate	Ahamd et al. 2022
Banana straw	430℃ for 4 h	Adsorbent dose: 1 g/L, Cd(II), Pb(II) and Cu(II): 1 g/L, Temperature: 30℃, 150 rpm	The maximum theoretical saturation adsorption capacities for Cd(II), Cu(II) and Pb(II) were 333.33, 370.37, and 302.58 mg/g	Li et al. 2020
Rice straw	500℃ for 1 h	Adsorbent dose: 2 g/L, Temperature: 22℃, 120 rpm	The equilibrium sorption capacity is 28.49 mg/g for As	Nguyen et al. 2019
Date palm waste-derived biochar	400~600℃	Adsorbent: 0.1 g; Mn: 10 mg/L, 200 rpm	High removal efficiencies and adsorption capacities (73.20%, qe: 3.57 mg/g for manganese and 94.94%, qe: 4.18 mg/g for nitrate)	Fseha et al. 2022 a
Date palm waste -derived biochar	500℃	Metal concentration: 20 mg/L	Adsorption capacity (mg/ g): Cu (26.21) > Fe (3.58) > K (3.23) > Zn (3.01) > Mn (0.78) > Ni (0.09)	Fseha et al. 2022 b
Barley grass	350℃ for 4 h	Contaminated soil with Cr(VI) 6622 mg/kg	Cr(VI) removal: 71%	Chen et al. 2021 b
Wheat straw	500℃ for 2 h	Adsorbent dose: 5 g/L, Cr(VI): 100 mg/L, Temperature: 25℃, 150 rpm	Cr(VI) removal: 99.78%	Jiang et al. 2022 b

Biomass	Preparation condition	Experimental condition	Adsorption efficiency	Reference
Corn straw	700°C	Adsorbent dose: 4 g/L, T: 25°C, 200 rpm, Vanadium (V): 0.78-9.66 mg/L	vanadium(V) adsorption capacities is 41.07 mg/g	Meng et al. 2018 b
Oak wood	500°C	Arsenic: 27-144 μg/L, pH: 7.31-8.07	Arsenic removal: 92-100%	Niazi et al. 2018
Colocasia esculenta	600°C	Adsorbent dose: 0.2-1.0 g/L, pH: 2-7, Temperature: 15-40°C, metal concen-tration: 5-90 mg/L, 100-180 rpm	Removal efficiencies for Fe(II), Cu(II) and As(V) were 97.34%, 94.89% and 84.09, respectively. The maximum sorption capacity was 6.19, 2.31, and 2.2 mg/g for Fe(II), Cu(II) and As(V)	Banerjee et al. 2016
Muskmelon peel	873 K for 1 h	Adsorbent dose: 0.8 g/L, Cu(II) and Zn (II): 50 mg/L, 303 K, 200 rpm	Removal efficiencies for Cu(II): 78.74 mg/g, and Zn(II): 72.99 mg/g	Khan et al. 2017
Hemp biomass	400~700°C for 30 min	Adsorbent dose: 0.1 mg/L, Cr(VI): 100-10000 μg/L, pH: 7.0, Temperature : 25°C	Cr(VI) adsorption capacity: 21.6 mg/g	Asimakidou et al. 2022

1.3.2. Soil Remediation

In recent years, biochar has been paid more attention as the soil remediation agents. Biochar can retain the nutrients in soil, fix the rainwater in the soil surface to weaken the acidity, and improve the pH value of soil. The production of biochar via microwave vacuum pyrolysis conversion of cassava stem presented abundant pores on the surface in the research of Foong et al. (2020 a). The results indicated that the cassava stem biochar was proper to be used as adsorbent and catalyst support, showing good advantages in wastewater treatment and soil nutrient retention. Liew et al. (2018 b), Nam et al. (2018), and Mahari et al. (2020) studied the preparation of palm kernel shell engineered biochar as the additive for agricultural applications. It could be concluded that the mushroom substrate with porous palm kernel shell biochar could provide sufficient water for the growth of oyster mushrooms (up to 99%). The available adsorption sites on the biochar could retain more nutrients (i.e., rice bran and calcium carbonate) and provide attachment sites for many living microorganisms (i.e., mycelium or plant-growth-promoting bacteria), thereby reducing the mortality of young mushrooms due to lack of water and nutrients. In the process of mycelium colonization, biochar also acted as the neutralizer to maintain the optimal pH level (i.e., 6.8-7.2) of the mushroom substrate. The research of Mohamed et al. (2016) found that the biochar produced from pyrolysis of switch grass could promote soil fertility, water holding capacity, and cation exchange capacity. The capacity of water holding and cation exchange improved because of the increase of micropore area. Although biochar provides convenience for soil remediation, the dust, secondary pollution, and other problems caused by biochar in the soil also need to be considered and resolved.

The higher pH and cation exchange capacity produced under microwave treatment enhances the metal precipitation and simultaneous reduction of its solubility. It enhances the exchange of cations and form co-precipitates or get complex with humic matters. Immobilization of heavy metals on biochar surface is regulated more importantly by ash composition of biochar. The higher heating rate of microwave produced biochar with higher ash content containing phosphates (PO_4^{3-}), carbonates (CO_3^{2-}) that increase the precipitation as metal phosphates and carbonates and also provide requisite nutrients (K, Ca, Mg, Al, Fe, Zn and P) to soil. The presence of functional groups like -NH, -OH, C-C, C-O, Si-O-Si and -PO facilitates the performance of metal immobilization process. The mechanism has been illustrated schematically in Figure 7 (Selvam and Paramasivan, 2022).

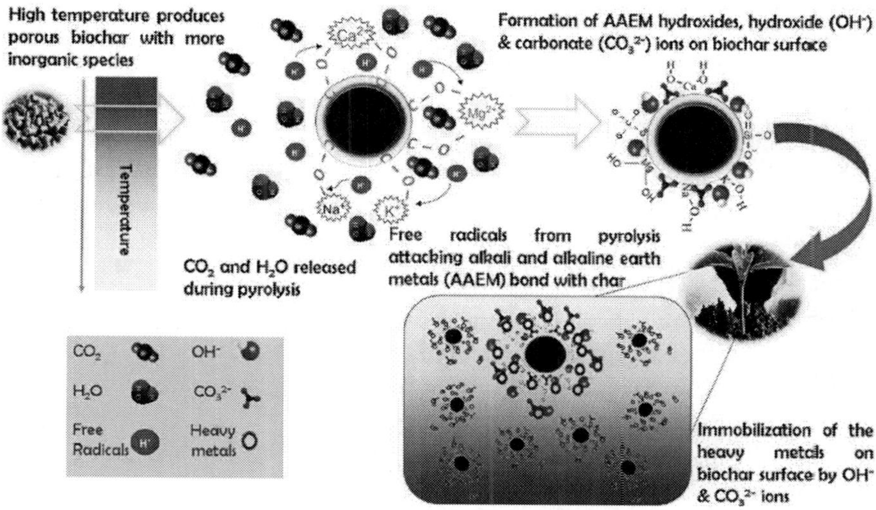

Figure 7. Mechanism of microwave produced biochar in soil heavy metal immobilization (Selvam and Paramasivan, 2022).

Catalyst assisted microwave pyrolysis of sewage sludge biochar increased heavy metals immobilization on the pores of biochar and thus reduced its bioavailability, which could otherwise deteriorate water and soil fertility. The heavy metal ions react with oxygen-containing groups at biochar surface and precipitate as metal hydroxides which then decomposes in presence of alkaline catalyst assisted microwave pyrolysis (Hou et al. 2020). The sewage sludge was granulated and treated with additives by microwave slow pyrolysis. It was found that microwave slow pyrolysis showed potential of immobilizing heavy metals on the structure of biochar presumably inhibiting its fixation, bioavailability for plants increasing the productivity. The oxygen-containing groups at surface of biochar are influenced by higher pH which deprotonates and enhances the adsorption efficiency. The microwave produced biochar at low temperature (300°C) reduced phytotoxicity and bioavailability of heavy metals and also increased rate of plant growth up to 145 % (Mohamed et al. 2021). Microwave irradiated biochar produced from slow pyrolysis beneficially reduced ecotoxicity and bioluminescence in soil and the microwave treated biochar is relatively more effective than other pre-treatment methods and non-activated biochar. Many literatures have stated that biochar is a suitable material with excellent biodegradation property for sustaining prominent environmental and agricultural benefits over fertilizers comprising long term carbon sink, enriching soil nutrient availability, microbial activity, fertility in turn leads to crop productivity (Lyu et al. 2020).

1.3.3. Building Materials

Biochar application in building materials (both concrete and asphalt) is recently being explored. It can sequester carbon, enhance the properties of the building materials and also reduce the embodied energy of such construction materials. Gupta and Kua (2016) explored the use of biochar as concrete admixture that reduces the greenhouse gas emissions when used in building constructions. A study by Consentino et al. (2018) demonstrated the use of biochar as nanoparticles in cement matrix composites (as filler). This improved the mechanical properties of the composites and sooner has wider opportunity for creating new green building materials from it. Biochar is also explored as a replacement for any product that utilizes carbon black in its manufacture (such as tires, plastic, ink, etc.).

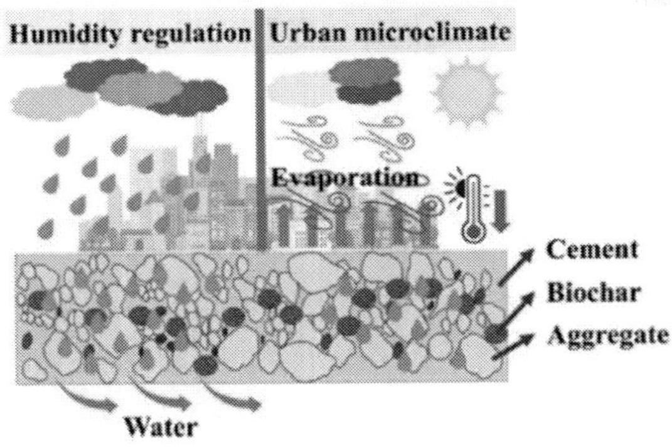

Figure 8. Environmental and technical advantages of biochar composites: humidity regulation and urban microclimate (Zhang et al. 2022 b).

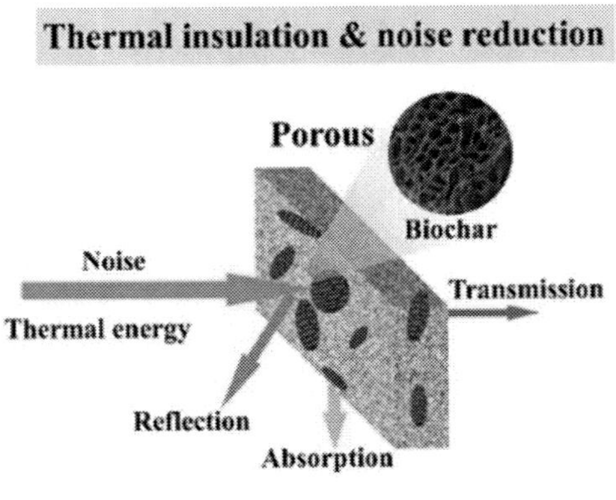

Figure 9. Environmental and technical advantages of biochar composites: thermal insulation and noise reduction (Zhang et al. 2022 b).

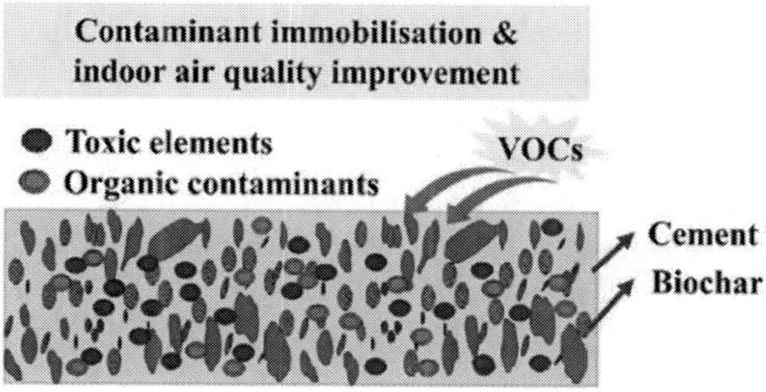

Figure 10. Environmental and technical advantages of biochar composites: contaminant immobilization and indoor air quality improvement (Zhang et al. 2022 b).

Zhang et al. (2022 b) summarized the environmental and technical advantages of biochar-enhanced construction materials. Incorporating biochar into the pervious concrete could reduce albedo and increase water adsorption capacity, mitigating urban heat island issues. The biochar-pervious concrete could absorb more solar radiation than plain pervious concrete, keeping cool through water evaporation and creating an urban microclimate. These findings were validated by Tan et al. (2021) by adopting a low-speed straight flow climatic wind tunnel to investigate the temperature regulation capacity of biochar pervious concrete. Their results demonstrated that the maximum temperature decreased by 6°C for 18 h. Therefore, biochar is considered a promising hygroscopic filler for manufacturing humidity- and temperature-regulating construction materials, enhancing hydrothermal performance and alleviating the urban heat island issues (Park et al. 2021). Figure 8 illustrates the mechanisms of humidity regulation and urban microclimate of biochar pervious concrete. The hydrothermal properties of biochar-pervious concrete should be optimized to enhance the capabilities for regulating humidity and temperature.

Customized biochar possesses 3D porous and 2D flake-like structures, which contribute to the formation of additional pathways for heat transfer. When porous biochar is uniformly distributed throughout the construction materials, it can induce scattered heat propagation, render the heat propagation routes multi-directional, and hinder the effect of unidirectional heat propagation (Jiang et al. 2022 a). This phenomenon effectively slows the expected propagation of heat flow and inhibits the heat transfer through solids, enhancing the thermal insulation potential of biochar-cement composites (Figure 9). The introduction of biochar can increase the porosity of cement-biochar composites. The pores in biochar would break the thermal bridging within the biochar-cement composites, which is responsible for the low thermal conductivity and thermal insulation improvement. Therefore, biochar with high porosity and 3D pore structure is more favored for enhancing the thermal insulation and noise reduction of construction materials.

Biochar has been widely applied for water purification and soil remediation due to its peculiar and tune-able properties, such as high porosity, good stability, and high cation exchange capacity. Many studies have focused on incorporating biochar into construction materials to impart the functional properties (e.g., contaminant immobilization and indoor air quality improvement) to the composites (Figure 10).

Biochar as a cost-effective material was proposed for improving the electromagnetic interference shielding efficacy of cement-based composites (Figure 11). A maximum increase of shielding effectiveness was up to 353% at 1.56 GHz frequency by incorporating 0.5 wt. % biochar compared to plain cement. The electromagnetic shielding performance of a sustainable lightweight biochar-cement-gypsum composite was facilitated by increasing biochar content, which became more pronounced at frequencies above 4 GHz (Natalio et al. 2020). The biochar composite exhibited high shielding efficacy in the microwave range; the mechanisms behind it were not interpreted clearly but attributed to some conventional reasons, such as natural alignment of carbon ultrastructure (e.g., lignin), dissipation of surface currents and polarization in the electric field.

Carbon-negative biochar was also incorporated to improve the thixotropy of 3D concrete printing, which can provide a new direction for versatile 3D concrete printing (Figure 12). The addition of biochar can reduce the density of 3D printing of lightweight concrete and enhance extrudability and build-ability. Developing appropriate mixture compositions for high-performance 3D concrete printing has become an opportunity yet a grand challenge. Incorporating biochar into 3D concrete printing has a promising perspective regarding property enhancement and carbon neutrality.

Biochar, as a low-cost carbon-negative material, has recently been adopted as a supporting scaffold for enhancing the performance of PCMs (Figure 13) (Kim et al. 2021). The biochar-PCMs displayed negligible leakage, good thermal insulation capacity, high chemical compatibility, exudation stability, and shape stability whilst reducing building energy consumption of the referenced building model.

Biochar has been investigated as a potential alternative to graphene nano-platelets in self-sensing cement composites (Figure 14). The addition of biochar at a dosage of 1% (v/v) to cement significantly reduced the electrical resistivity (-42%) whilst maintaining good mechanical properties (Mobili et al. 2021).

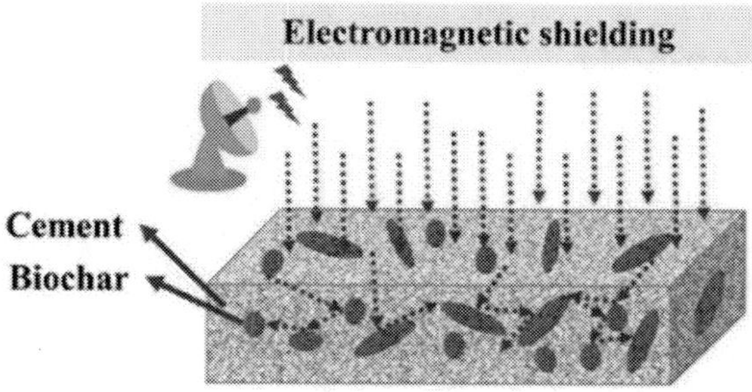

Figure 11. Environmental and technical advantages of biochar composites: electromagnetic shielding (Zhang et al. 2022 b).

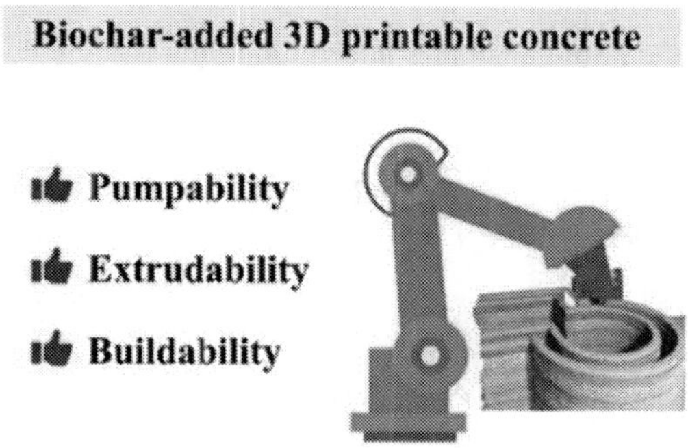

Figure 12. Environmental and technical advantages of biochar composites: biochar-added 3D printable concrete (Zhang et al. 2022 b).

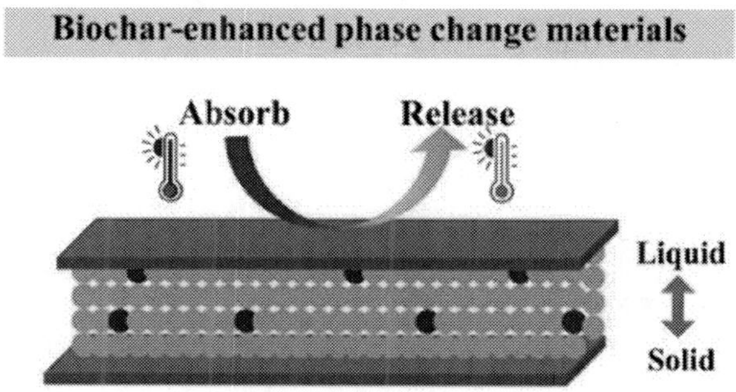

Figure 13. Environmental and technical advantages of biochar composites: biochar-enhanced phase change materials (Zhang et al. 2022 b).

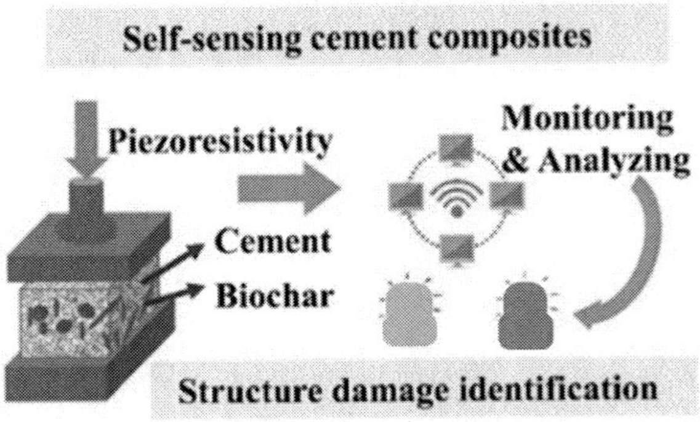

Figure 14. Environmental and technical advantages of biochar composites: self-sensing cement composites (Zhang et al. 2022 b).

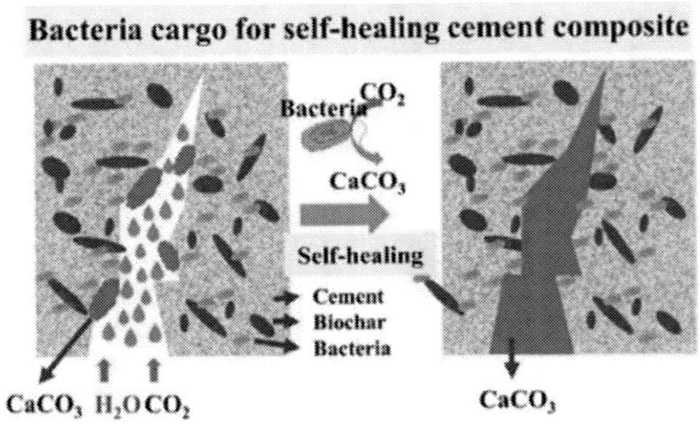

Figure 15. Environmental and technical advantages of biochar composites: bacteria cargo for self-healing cement composites (Zhang et al. 2022 b).

Biochar as bacteria cargo for self-healing composites can boost the sealing of cracks (Figure 15). This provides a relatively low-cost and sustainable solution for solving the crack issue in concrete, which is also beneficial for expanding the life span and durability of concrete. Further study is required to improve the survival rate of bacteria exposed to alkaline conditions of concrete.

1.3.4. Electrochemical Conductivity

The rich functional groups, good cation exchange performance, and excellent conductivity of biochar make it be used as the electrochemical material, i.e., fuel cell, super-capacitor, etc. However, due to some electrochemical properties (i.e., high intra-matrix diffusion rates of ions, a small number of storage sites, etc.), a series of modified technologies will be adopted to improve the quality of biochar to meet the need of application. The study of Kuo et al. (2021) indicated the cellulosic-derived biochar obtained by microwave NH_3 activation contained Fe-N groups in the carbon core and N-doped graphite in the carbon shell. There were many active sites formed in the cellulosic-derived biochar, which exhibited strong electro-catalytic performance and oxygen reduction reaction properties with the four-electron transfer reaction pathway. Yadav et al. (2016) used mushroom biochar to make electrodes in paper-based flexible super-capacitors. The mushroom biochar was activated by KOH solution, and the slurry of the active material was stuck on the surface of polyethylene terephthalate (PET) with incium tin oxide as the collector. The super-capacitor showed better performance. Table 2 shows the preparation methods of biochar, electrolyte and application in electrochemical sensors and super-capacitors.

2. Microwave Irradiation for Pyrolysis of Biomass for Value-Add Biochar

2.1. Introduction

With the goal of "Carbon emission peak and carbon neutrality" being put forward, the energy industry is under enormous pressure to control carbon emissions in the future, accompanied by opportunities and challenges (Liu et al. 2022; Wu et al. 2022). It is now important about the development of low-carbon and even zero-carbon technologies using renewable and clean energy. Among them, biomass accounts for the largest proportion in clean energy, which is twice that of hydropower and 3.5 times that of wind power. Due to its ubiquity and availability, biomass can provide up to 25 % of global energy demand (Rawat et al. 2022). Biomass resources are mainly divided into the following three categories: agricultural and forestry waste, solid waste (domestic waste), and animal waste (Bildirici and Özaksoy, 2013). Agricultural and forestry waste accounts for a large proportion in biomass resources. The annual output agricultural straw is 740 Mt in China, which is equal to 317 Mt of standard coal in calories. And about 460 Mt of standard coal in calories are reached added with various fuelwood and forestry wastes (Haarlemmer et al. 2016). Forestry and agricultural residues are ideal raw materials to process using numerous techniques. They belong to lignocellulosic biomass, consisting of aromatics and carbohydrates with some organic extracts and inorganic minerals (Byambajav et al. 2018; Zhang et al. 2021). Compared with coal, oil, natural gas, and other fossil fuels, the technological processing using agricultural and forestry wastes to transform generally have the higher energy conversion efficiency and lower cost.

There are many ways to convert agricultural and forestry waste into value-added products, depending on the biomass type, biomass output, utilization method, environmental requirements, local economic conditions, and special project requirements, etc. (McGrath, 2002; Zhang et al. 2022 a). Currently, three conversion methods are most commonly used: physical conversion, thermochemical conversion, and biochemical conversion. Among them, thermochemical conversion technology is the hotspot of biomass conversion technology, mainly including drying, torrefaction, carbonization, pyrolysis, liquefaction, and gasification (Azizi and Mowla, 2016; Wang et al. 2020 a). In recent years, many scholars have tried to introduce microwave into the thermochemical conversion process of biomass, and many breakthroughs have been made. Microwaves is a high-frequency electromagnetic wave. Different from infrared, radio wave, and visible light, the frequency of microwave belongs to 300 MHz-300 GHz, and the wavelength ranges between 1 mm and 1 m. Specifically, microwave heating is caused by the absorbing medium in the microwave field which absorbs microwave energy through microwave irradiation, and converts it into internal energy. The temperature gradually rises from inside to outside in materials, and the temperature gradient is the same as the direction of volatile emission.

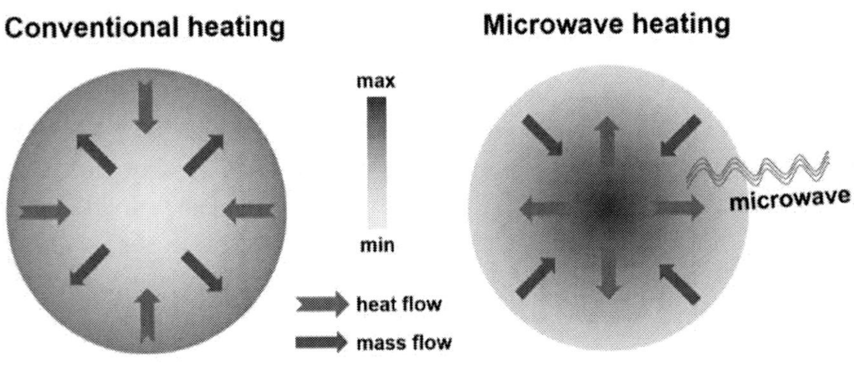

Figure 16. Difference between microwave pyrolysis and conventional pyrolysis.

Table 2. Preparation methods of biochar, electrolyte and application in electrochemical sensors and super-capacitors

Biomass	Preparation condition	Electrode	Characterization	Performance	Reference
Spent coffee grounds	300~700 °C, 60 min, 5 °C/min	Carbon paste electrode	Differential pulse adsorptive anodic stripping voltammetry	Under optimized conditions, it was observed a linear dynamic range from 0.128 to 2.44 μmol/L, limits of detection and quantification of 4.5 nmol/L and 15 nmol/L, respectively	Oliveira et al. 2021
Magnolia grandiflora fruit	800 °C, 2 h, 5 °C/min	Multilayer activated biochar/ Zr-metal–organic frameworks	Differential pulse adsorptive anodic stripping voltammetry	Both the Pb^{2+} and Hg^{2+} can be detected with the detection limit down to 1.0 ng/L	Zou et al. 2021
Kelp	900 °C, 3 h, 5 °C /min	Glassy carbon electrode	Square wave anodic stripping voltammetry	A linear dynamic range of 0~1.2 μM limits of detection of 10.3 nmol were obtained	Guan et al. 2020
Microalgae, rice straw and sorghum straw	500 °C, 30 min	Carbon dots	Fluorescence emission	Sorghum straw based biochar and microalgae based biochar can be used for the detection of Cu (II) and Zn (II) ions. Limits of detection for Cu (II) is 0.5 μm and 9μm for Zn (II)	Plácido et al. 2019
Withered rose	750~900 °C, 1.5h	The working electrode was fabricated by thoroughly mixing biochar, acetylene black and polyvinylidene fluoride with a mass ratio of 8:1:1 in N-methyl2-pyrrolidone to form a slurry.	Cyclic voltammetry, galvano-static charge-discharge and cycling stability	Specific capacity was 208 F/g and density of super-capacitor was 0.5 A/g	Zhao et al. 2018 b

Table 2. (Continued)

Biomass	Preparation condition	Electrode	Characterization	Performance	Reference
Peanut shell	600~800 °C, 1 h, 5°C/min	A carbon film comprising of the as-prepared samples, acetylene black and PTFE binder with a 8:1:1 mass ratio was pressed on a Ti mesh	Cyclic voltammetry, galvano-static charge-discharge and cycling stability	Specific capacity was 447 F/g and density of super-capacitor was 0.2 A/g	Lei et al. 2021
Rice straw	800~900 °C, 1 h	The electrode slurry was prepared by biochar, conductive material and polytetrafluoroethylene binder homogeneously in ethanol with a weight ratio of 8:1:1.	Cyclic voltammetry, galvano-static charge-discharge	Specific capacity was 324 F/g and density of super-capacitor was 0.5 A/g	Charoensook et al. 2021
Sword bean shell	600~900 °C, 2 h	The working electrode was fabricated using synthesized samples as active materials, acetylene black, and polyvinyli-dene fluoride at 8:1:1 wt ratio.	Cyclic voltammetry, galvano-static charge-discharge	Specific capacity was 264 F/g and density of super-capacitor was 1.0 A/g	Chen et al. 2021 a

Figure 16 shows the difference between microwave pyrolysis and conventional electric pyrolysis. Above all, microwave heating has the following advantages: (a) uniform heating (Ke et al. 2019), (b) fast heating (Chen et al. 2022 a; Zhang et al. 2020 b), (c) selective heating and (d) energy-saving and efficient heating (Zhang et al. 2017). Microwave has the penetrability. Regardless of the shape and size of the materials, as long as microwave radiation occurs, the polar molecules inside the material will collide with each other. Electromagnetic energy acts on medium molecules and is directly converted into heat energy to transfer heat to the outside uniformly (Zhang et al. 2019 b). The temperature of the core in the materials is much higher than that in the outer surface, so that the materials are heat up rapidly (Wang et al. 2020 b). What's more, the capability of the materials to absorb microwave determines the capability to heat, which directly affects the thermal efficiency of microwave heating (Zhang et al. 2017). When some materials with weak capability to absorb microwave, they can be mixed with the microwave absorbents to improve the heat efficiency. At the same time, a controlled heating requirement for special materials can also be achieved.

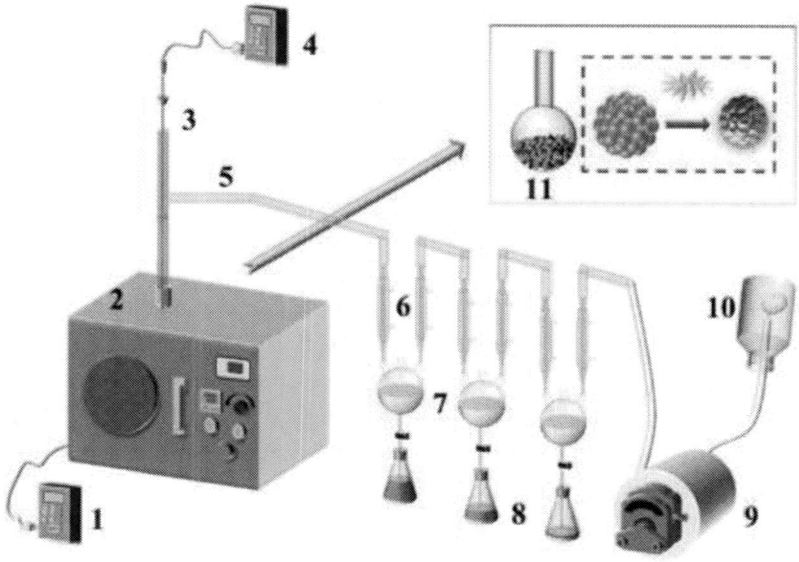

Figure 17. Diagram of experimental setup for microwave pyrolysis: (1) power meter, (2) microwave oven, (3) K-type thermocouple, (4) temperature reader, (5) connecting pipe, (6) condensing pipe; (7) three-mouth flask; (8) oil collecting bottle; (9) vacuum pump, (10) gas collecting bottle, and (11) quartz reactor.

With the popular utilization of agricultural and forestry waste, the preparation of added-value products through microwave pyrolysis (MP) of agricultural and forestry wastes has been widely investigated. Figure 17 shows the schematic diagram of a typical microwave-assisted pyrolysis experimental system. The experimental setup mainly includes heating module, condensing module, and other auxiliary elements. Microwave oven (2) and quartz reactor (11) make up the heating module. The condensing module includes condensing pipe (6), three-mouth flask (7), and oil collecting bottle (8). The other auxiliary elements include power meter (1), K-type thermocouple (3), temperature reader (4), connecting pipe (5), vacuum pump (9), and gas collecting bottle (10). Firstly, the condensing module was weighed. Then the microwave absorbents and raw materials were evenly distributed in the quartz reactor to weigh the quality. The experimental setup was connected as shown in Figure 17, and the quartz reactor was wrapped with the thermal insulation foam to fill the inner cavity of the microwave oven. The vacuum pump was kept on for about 10 min to create an inert environment. The pyrolysis reaction started when the microwave oven was turned on and the parameters were set. When the temperature of the mixtures of biomass and absorbents reached the specified pyrolysis temperature, the time started to be recorded. The pyrolysis temperature was controlled by turning the microwave oven on and off. The experiment came to an end when the microwave oven was turned off after the designated reaction time. After the experiment, the condensing module and quartz reactor was reweighed. The biochar production was the mass difference between the quartz reactor before and after experiment. The oil production was the mass difference of condensing module before and after experiment. The other substances left, except biochar and oil, were gas production.

In the process of performing experiments, some systematic errors and random errors will inevitably occur. Systematic errors include the tightness of the experimental system related to the connecting pipeline, the vacuum degree of the experimental system determined by the operation of the vacuum pump, the insulation property of the experimental system supported by thermal insulation foam, etc. Under the same conditions, when the same physical quantity is measured, some random errors may also occur to the measured values due to various random factors. Therefore, in order to minimize experimental errors, each experiment was repeated three times. The biochar yield, oil yield, and gas yield were calculated by Eq. (1), Eq. (2), and Eq. (3). The energy values of biomass and biochar were calculated by Eq. (4) and Eq. (5), respectively, and the energy recovery efficiency was calculated by Eq. (6).

$$Y_{biochar} = \frac{m_{biochar}}{m_{biomass}} \times 100 \tag{1}$$

$$Y_{oil} = \frac{m_{oil}}{m_{biomass}} \times 100 \tag{2}$$

$$Y_{gas} = 100 - Y_{biochar} - Y_{oil} \tag{3}$$

$$Q_{biomass} = HHV_{biomass} \times m_{biomass} \tag{4}$$

$$Q_{biochar} = HHV_{biochar} \times m_{biochar} \tag{5}$$

$$\eta_r = \frac{Q_{biochar}}{Q_{biomass}} \times 100 \tag{6}$$

where

$Y_{biochar}$ denotes the biochar yield (wt. %),
$m_{biochar}$ denotes the quantity of biochar (g),
$m_{biomass}$ denotes the quantity of biomass (g),
Y_{oil} denotes the oil yield (wt. %),
m_{oil} denotes the quantity of oil (g),
Y_{gas} denotes the gas yield (wt. %),
$Q_{biomass}$ denotes the energy of biomass (kJ),
$HHV_{biomass}$ denotes the higher heating value of biomass (MJ/kg),
$HHV_{biochar}$ denotes the higher heating value of biochar (MJ/kg),
$Q_{biochar}$ denotes the energy of biochar (kJ),
η_r denotes the energy recovery efficiency (%).

Nzediegwu et al. (2021), Quillope et al. (2021), and Liu et al. (2021) studied the yield of product (i.e., biochar) caused by different reaction parameters (microwave power, exposure time, and reaction temperature). Xiang et al. (2022) and Nzediegwu et al. (2021) explored the characteristics (i.e., pore volume, surface area, micropore area, −OH functional group, pH value, higher heating value) of biochar through microwave pyrolysis as affected by raw material types (i.e., corn straw, rapeseed straw, sawdust, wheat straw, and fecal particles). The results confirmed that non-lignocellulosic biomass-derived biochar was better suited for soil amendment, and lignocellulosic biomass-derived biochar was better suited for solid fuel. Chu et al. (2017) compared the yield (35.5-87 % vs. 29.8-87.9 %) and specific surface area (4.93-587 m^2/g vs. 1.17-406 m^2/g) of biochar via microwave pyrolysis and conventional pyrolysis (CP) of peanut shells (PS). Fodah et al. (2021 b) screened different microwave absorbents (SiC and activated carbon) for maize straw biochar production with more pores, high HHV, low sulfur content, low nitrogen content, and no carbon impurities. Mamaeva et al. (2016) and Liu et al. (2021) confirmed that the catalyst contributed to the yield of biochar and the synthesis of carbon nanotubes, and catalyst types and biomass/catalyst ratio also had different effect on these. When rice straw was used for microwave heating, Zhang et al. (2015 c) detailed the influence of pretreatment methods (washing, baking, and washing-baking combined) during the process. Yek et al. (2020 a) adopted the method about microwave wet roasting combining microwave radiation and steam to study the influence of different microwave heating methods (wet roasting without steam and with steam) on the pyrolysis of palm shell, especially the catalytic effect on biochar. Li et al. (2022) presented a novel method of microwave pyrolysis combined with conventional preheating treatment of straw for biochar production, which can accelerate the heating rate and improve the biochar yield. Lin et al. (2021) constructed a coupling model of multiple physical fields (temperature, structural mechanics, and electromagnetic) to visualize the conversion among kinds of microscopic energy fields (material properties, temperature field, and electromagnetic field) during microwave pyrolysis of cellulose. Georgin et al. (2016), Xiang et al. (2022), and Huang et al. (2008) explored the adsorbed performance about biochar through microwave pyrolysis of peanut shell, corn straw, and straw for DB38 dyes, RR141 dyes, benzene, o-xylene, and Cu^{2+}, which confirmed the application of biochar as adsorbents.

2.2. Effect of Pyrolysis Temperature

Biochar yields obtained from different pyrolysis temperatures are shown in Figure 18. Generally speaking, for the microwave-assisted pyrolysis of organic wastes, the biochar yield decreased monotonously when the reaction temperature was increased. It is common that the yield of biochar obtained from microwave-assisted pyrolysis of biomass decreases with the increase of pyrolysis temperature. This tendency was reported by most of scholars, i.e., the yield of biochar produced from microwave-assisted pyrolysis of horse manure was reduced from 78.98 wt. % to 12 wt. % in the temperature range of 350–550°C (Mong et al. 2020). The others include switch grass (Mohamed et al. 2016), corn stover (Zhu et al. 2015), furfural residue (Mao et al. 2019), wood pellet (Zhou et al. 2020), peanut shell (Chu et al. 2017), oil palm empty fruit bunch (Idris et al. 2021), dried grain with soluble (Lei et al. 2011), sludge (Lin et al. 2020), sewage sludge (Zhou et al. 2018), municipal sewage sludge (Fang et al. 2021), alkali lignin (Fan et al. 2019), bio-solid (Antunes et al. 2017 b), rice straw and camellia oleifera shell (Wang et al. 2018 b), oil sand and sawdust (Li et al. 2019), textile dyeing sludge and red wood waste (Zhou et al. 2021), chromolaena odorata and soybean soap stock (Wang et al. 2018 a), corn stalk, pinewood, and algae (Parvez et al. 2020), canola straw, sawdust, wheat straw, and manure pellet (Nzediegwu et al. 2021), pig manure, rice straw, and their mixtures (Meng et al. 2018 a), prairie cordgrass (Zhou et al. 2013), etc. Being similar to conventional pyrolysis, the biochar yield shows a decreasing trend with the increase of pyrolysis temperature during microwave-assisted pyrolysis of biomass. This may be because high pyrolysis temperature prolongs the reaction time, and the further increase of pyrolysis temperature promotes the secondary decomposition of volatile compounds to form small molecular hydrocarbons, resulting in the reduction of biochar yield.

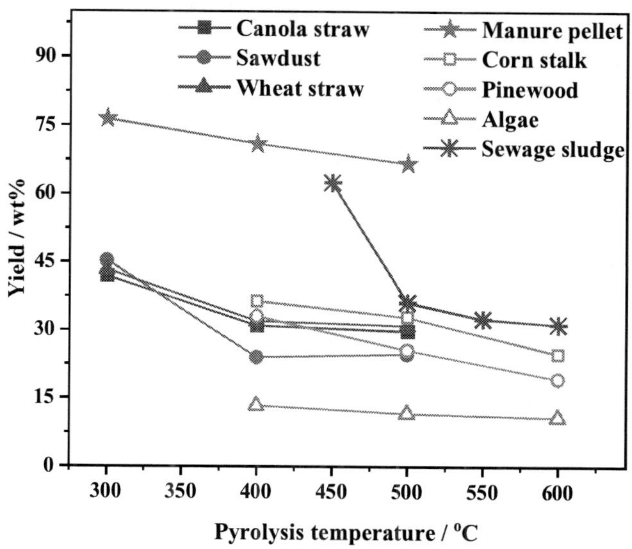

Figure 18. Biochar yields at different pyrolysis temperatures.

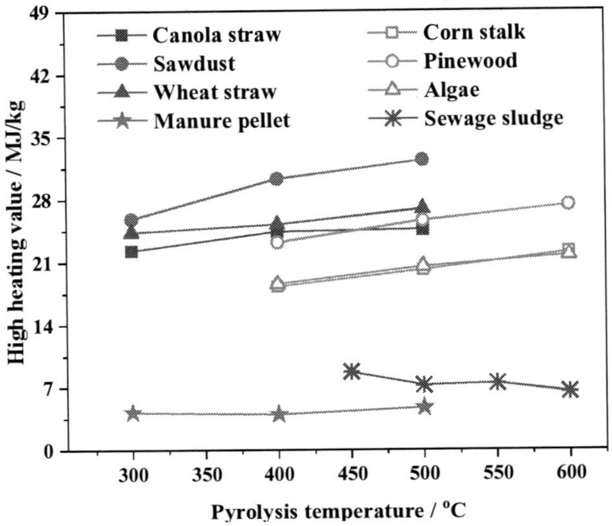

Figure 19. HHVs of biochars at different pyrolysis temperatures.

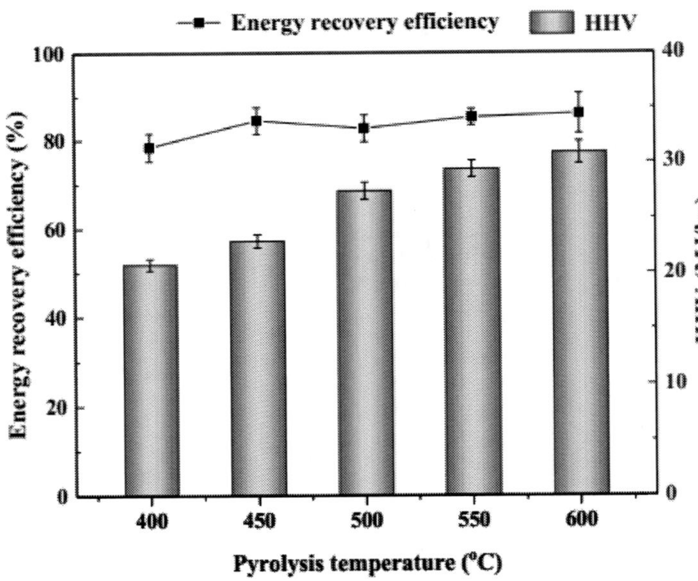

Figure 20. Biochar HHVs and energy recovery efficiencies at different pyrolysis temperatures.

Table 3. Yields and HHVs of biochars obtained from different pyrolysis temperatures

Feedstocks	Temperature (°C)	Yield (wt. %)	HHV (MJ/kg)	Reference
Sewage sludge	450-600	32.98-62.26	6.48-8.70	Zhou et al. 2018
Corn stalk	400-600	24.8-36.4	18.3-22.1	Parvez et al. 2020
Pinewood	400-600	19.3-33.1	23.2-27.3	Parvez et al. 2020
Algae	400-600	10.8-13.4	18.6-21.7	Parvez et al. 2020
Canola straw	300-500	29.8-41.9	22.3-24.5	Nzediegwu et al. 2021
Sawdust	300-500	24.7-45.3	25.9-32.3	Nzediegwu et al. 2021
Wheat straw	300-500	31.0-43.3	25.6-26.9	Nzediegwu et al. 2021
Manure pellet	300-500	66.7-76.4	4.1-4.7	Nzediegwu et al. 2021
Horse manure	350-550	12-78.98		Mong et al. 2020
Switchgrass	300-400	44.23-46.71		Mohamed et al. 2016
Corn stover	550-650	25.1-25.7		Zhu et al. 2015
Furfural residue	450-750	43.62-59.00		Mao et al. 2019
Wood pellets	630-800	12.5-19.8		Zhou et al. 2020
Peanut shell	200-600	35.5-87.0		Chu et al. 2017
Oil palm	300-600	30.51-47.47		Idris et al. 2021
Dried grain	516-684	22.62-65.76		Lei et al. 2011
Sludge	400-800	42-65		Lin et al. 2020
Sewage sludge	300-700	49.94-65.39		Fang et al. 2021
Alkali lignin	500-700	38-48		Fan et al. 2019
Biosolids	300-800	77-91		Antunes et al. 2017 b
Rice straw	400-600	32.44-41.96		Wang et al. 2018 b
Camellia oleifera shell	400-600	24.85-35.47		Wang et al. 2018 b
Oil sands and sawdust	500-1100	35-59		Li et al. 2019
Red wood waste	450-750	48.75-64.75		Zhou et al. 2021
Soybean soapstock	200-400	21.58-23.12		Wang et al. 2018 a
Pig manure	300-700	37-62		Meng et al. 2018 a
Rice straw	300-700	26-53		Meng et al. 2018 a
Prairie cordgrass	529.3-670.7	13.8-47.7		Zhou et al. 2013

Pyrolysis temperature significantly affects the properties of biochar produced from microwave-assisted pyrolysis. The variation trends of high heating value (HHV) of biochar at different pyrolysis temperatures are shown in Figure 19 and Table 3. The HHV of biochar increased gradually with the increase in the pyrolysis temperature. Nzediegwu et al. (2021) reported that when the pyrolysis temperature was increased from 300 °C to 500 °C, the HHVs of biochar were increased in the ranges of 22.3-24.5, 25.9-32.3, 25.6-26.9, and 4.1-4.7 MJ/kg for canola straw, sawdust, wheat straw, and manure pellet, respectively. Parvez et al. (2020) reported that when the pyrolysis temperature was increased from 400 °C to 600 °C, the HHVs of biochar were increased in the ranges of 18.3-22.1, 23.2-27.3, and 18.6-21.7 MJ/kg for corn stalk, pinewood, and algae, respectively. On the whole, the HHVs of biochar obtained from microwave-assisted pyrolysis are positively proportional to the pyrolysis temperature, which is mainly related to the increase of C and fixed carbon contents in biochar. It can be estimated that HHV of biochar is closely related to the pyrolysis temperature. For different kinds of biomass, pyrolysis temperature ranges, and the other reaction conditions, the changing trends of biochar properties are not consistent. But in general, the HHV increase with the improvement of pyrolysis temperature. This is because the pyrolysis reaction of biomass advances with the enhancement of pyrolysis temperature. The H and O in feedstocks are converted into small molecular gas products, while C is fixed in biochar (Mong et al. 2020; Masek et al. 2013).

Fan et al. (2023) studied the production of value-added biochar through microwave pyrolysis of peanut shell and evaluated the energy recovery efficiency during the whole process, as shown in Figure 20. The HHV of peanut shell is 17.69 MJ/kg. The biochar HHV was monotonously improved when the pyrolysis temperature rose, and this was because high pyrolysis temperature resulted in the growing amount of fixed carbon, carbon, and hydrogen contents in biochar. The maximum biochar HHV was 30.86 MJ/kg obtained at 600 °C. A key index for assessing the conversion of peanut shells into biochar via microwave pyrolysis is the energy recovery efficiency. The HHV of peanut shell measured was 17.69 MJ/kg, and the obtained peanut shell energy was 265.35 kJ. According to the HHVs of biochar, the output biochar energy were 207.99-228.36 kJ. The energy recovery efficiency was improved from 78.38% to 86.06% as the enhancement in pyrolysis temperature. The high HHV of biochar resulted in the high energy recovery efficiency at 600 °C due to little difference in biochar yields at different pyrolysis temperatures.

2.3. Effect of Microwave Power

During microwave-assisted pyrolysis, the influence of microwave power on biochar yield is listed in Table 4 and Figure 21. The yield of biochar changes as: (a) decreases monotonously, (b) increases first and then decreases, (c) decreases first and then increases, and (d) fluctuates. In general, the biochar yield of microwave-assisted pyrolysis of biomass is on the decline with the improvement of microwave power. The results were observed by Fang et al. (2021) for municipal sewage sludge, Fodah et al. (2021 a) for corn stover, Liew et al. (2018 b), and Nam et al. (2018) for palm kernel shell, Nhuchhen et al. (2018) for wood pellet, Salema et al. (2017) for corn stalk, Liu et al. (2021)

for apple wood, Foong et al. (2020 a) for cassava stem, Mubarak et al. (2016) for palm oil empty fruit bunch, and Yap et al. (2017) for coconut shell. When the microwave power increased, the biochar yield increased first and then decreased, and this was produced from microwave-assisted pyrolysis of medium density fiberboard (Haeldermans et al. 2019) and dried grain with soluble (Lei et al. 2011). Ellison et al. (2020) used pine wood to conduct microwave-assisted pyrolysis in the microwave power range of 600-1500 W, the biochar yield showed the tendency of decreasing first and then increasing. With the increase of microwave power from 300 W to 1000 W, the yield of biochar obtained from microwave-assisted pyrolysis of pinus pinaster fluctuated, ranging from 16 wt. % to 26 wt. % (Martín et al. 2017). It is concluded that microwave power is a key factor affecting the yield of biochar in microwave-assisted pyrolysis of biomass. In most situations, the yield of biochar decreased with the increase of microwave power. This is because high microwave power accelerates the heating rate. However, high heating rate can shorten the time required for pyrolysis reaction, resulting in incomplete pyrolysis. Therefore, an optimal microwave power is needed for the preparation of biochar obtained from microwave-assisted pyrolysis of organic wastes.

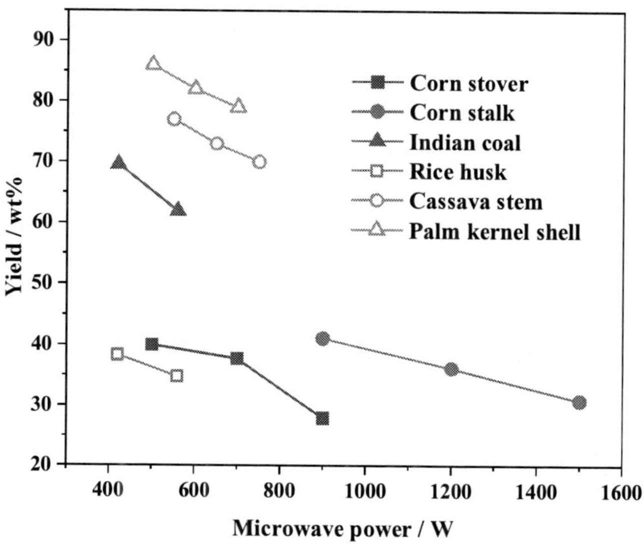

Figure 21. Biochar yields at different microwave powers.

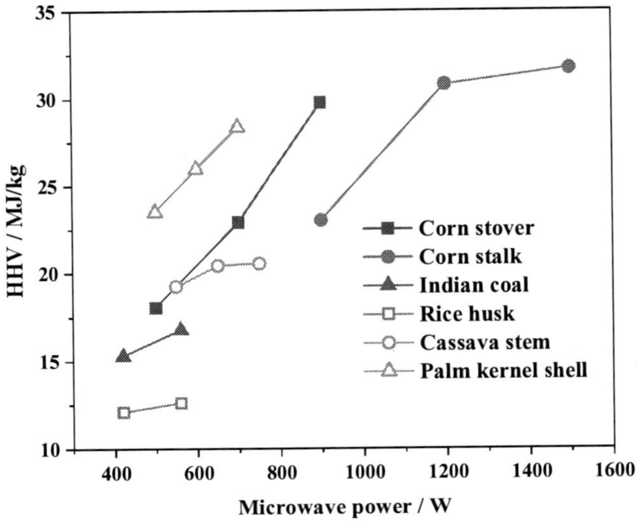

Figure 22. HHVs of biochars at different microwave powers.

Table 4. Yields and HHVs of biochars obtained from different microwave powers

Feedstocks	Microwave power (W)	Yield (wt. %)	HHV (MJ/kg)	Reference
Corn stover	500-900	28-40	18.05-29.77	Fodah et al. 2021 a
Palm kernel shell	500-700	33-38	23-26	Liew et al. 2018 b
Wood pellets	2000-3000	26.2-32.4	30.55-31.81	Nhuchhen et al. 2018
Corn stalk	900-1500	30.9-41.1	23-31.7	Salema et al. 2017
Cassava stem	550-750	70-77	19.24-20.55	Foong et al. 2020 a
Palm kernel shell	500-700	79-86	23.49-28.36	Ge et al. 2020
Indian coal	420-560	61.9-69.6	15.3-16.8	Rajasekhar and Vinu, 2018
Rice husk	420-560	34.8-38.3	12.1-12.6	Rajasekhar and Vinu, 2018
Indian coal and rice husk	420-560	45.7-61.2	12.0-14.6	Rajasekhar and Vinu, 2018
Sewage sludge	1000-1800	51.26-59.5		Fang et al. 2021
Palm kernel shell	550-750	28-36		Nam et al. 2018
Apple wood	500-700	32.2-54		Liu et al. 2021
Palm oil	600-1200	38-94		Mabarak et al. 2016
Coconut shell	600-1200	55-92		Yap et al. 2017
Fiberboard	750-1250	43.8-44.5		Haeldermans et al. 2019
Dried grain	600-1000	22.62-65.76		Lei et al. 2011
Pinewood	600-1500	10-65		Ellison et al. 2020
Pinus pinaster	300-1000	16-26		Martín et al. 2017
Palm kernel shell	550-750	28.1-36		Mahari et al. 2020

Haeldermans et al. (2019) reported the optimal microwave power of 1000 W for medium density fiberboard, and Lei et al. (2011) reported the optimal microwave power of 700 W for dried grain with soluble. Different heating rates lead to different pyrolysis advancement of biomass. At the same time, different microwave powers cause uneven distribution of heat energy inside the materials and local hotspots, all of which may cause other changing trends in the yield of biochar (Lin et al. 2021; Kumar et al. 2020).

During the microwave-assisted pyrolysis, microwave power is an important factor. Microwave power affects the heating rate and determines the extent that biomass, microwave absorbents, and other substances absorb microwave energy and convert it into heat energy. High microwave power can enhance the microwave density and specific energy input in the microwave cavity (microwave density and specific energy input are physical quantities that measure microwave power and microwave energy supplied by a unit mass of matter, respectively) (Motasemi and Afzal, 2013; Arpia et al. 2021 a). High microwave power can enhance the interaction between substances and microwave field, leading to the rotation, collision, torsion, and friction of the molecules inside the substances, and improving the potential to convert the microwave absorbed into heat energy (Asomaning et al. 2018; Bundhoo, 2018). When the microwave power is low, the interaction between substances and microwave field is too weak to cause molecular cleavage. The increase of microwave power will increase the intensity of the microwave field, and the molecular vibration will be violent. So, the pyrolysis reaction of biomass will be more violent. The influence of microwave power on the HHVs of biochar obtained from microwave-assisted pyrolysis of biomass is listed in Table 4 and Figure 22. In most situations, microwave power has a positive effect on the HHV of biochar, and the increases in microwave power increased the HHV of biochar, i.e., the HHV of biochar from corn stover increased from 18.05 MJ/kg to 29.77 MJ/kg when the microwave power was increased from 500 W to 900 W (Fodah et al. 2021 a). The other results include 19.24-20.55 MJ/kg vs 550-750 W for cassava stem (Foong et al. 2020 b), 23-26 MJ/kg vs 500-700 W for palm kernel shell (Liew et al. 2018 b), 23.49-28.36 MJ/kg vs 500-700 W for palm kernel shell (Ge et al. 2020), 15.3-16.8 MJ/kg vs 420-560 W for Indian coal (Rajasekhar and Vinu, 2018), and 23-31.7 MJ/kg vs 900-1500 W for corn stalk (Salema et al. 2017). When the microwave power was in the range of 2000-3000 W, Nhuchhen et al. (2018) found that the HHV of the biochar obtained from microwave-assisted pyrolysis of wood pellet first increased and then decreased, and the maximum HHV was 31.81 MJ/kg at the microwave power of 2500 W. The strong microwave field caused by high power makes the reactant molecules vibrate violently during microwave-assisted pyrolysis of biomass. Because of the instability of magnetic field strength caused by microwave power variation, the hotspot phenomenon often occurs in the microwave field. When the hotspots are concentrated in a particular area, there will be the uneven distribution of heat energy in the materials (Lin et al. 2021; Kumar et al. 2020). Therefore, the HHVs of biochar of various biomass also have different trends within the range of different microwave power.

Fan et al. (2023) studied the production of value-added biochar through microwave pyrolysis of peanut shell and evaluated the energy recovery efficiency during the whole process, as shown in Figure 23. The obtained peanut shell energy was 265.35 kJ, and the biochar energy were in the range of 219.08-259.94 kJ at different microwave powers. The energy recovery efficiency steadily declined as microwave power increased. And it reached a highest value of 97.96% at 350 W, which was determined by the highest biochar yield (86 wt. %).

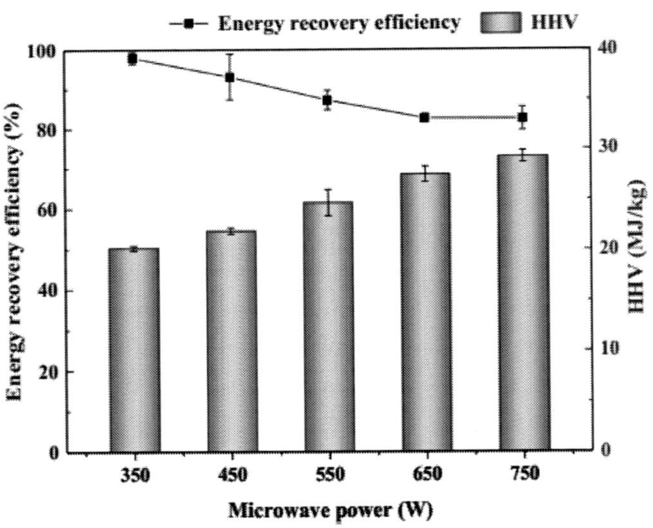

Figure 23. Biochar HHVs and energy recovery efficiencies at different microwave powers.

2.4. Effect of Residence Time

The effect of residence time on the biochar yield of microwave-assisted pyrolysis mainly presents two changes: (a) increases monotonously and (b) decreases monotonously, as shown in Figure 24 and Table 5. The first phenomenon was observed by Bartolil et al. (2016) for Arundo donax, leave and rhizome. The second phenomenon was reported by Fang et al. (2021) for municipal sewage sludge, Ge et al. (2020) for palm kernel shell, Lei et al. (2011) for dried grain with soluble, Bartoli et al. (2016) for Arundo donax stem, Yap et al. (2017) for coconut shell, and Zhou et al. (2013) for prairie cordgrass. Generally speaking, the prolongation of residence time can promote the increase of C content in biochar, and inhibit the contents of H, O, and N. In the research of Bartoli et al. (2016), all the C contents of biochar (70.4-77.4 wt% of Arundo donax leave biochar, 69.4-77.4 wt. % of Arundo donax stem biochar, and 85.5-87.5 wt. % Arundo donax rhizome biochar) prepared by microwave-assisted pyrolysis of the three raw materials increased with increases in the residence time (18-40 min). The HHVs of biochar showed similar changes (i.e., 27.17-28.69 MJ/kg for 5-15 min, plam kernel shell (Ge et al. 2020)), as shown in Figure 25.

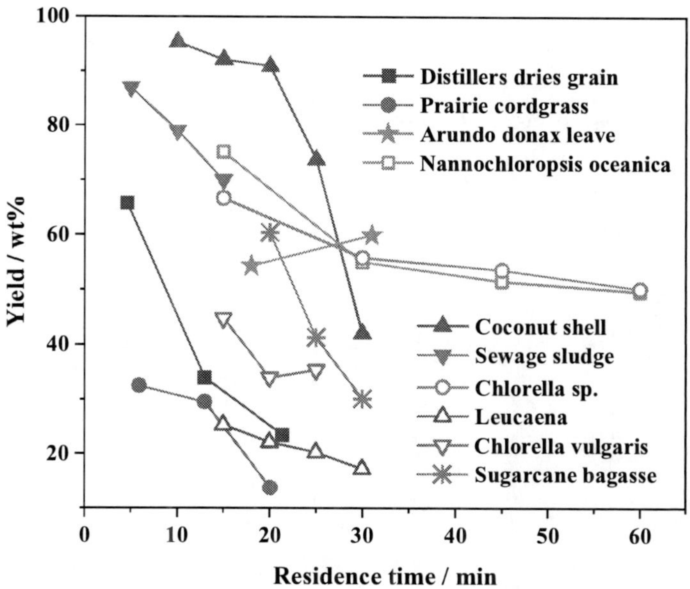

Figure 24. Biochar yields at different residence times.

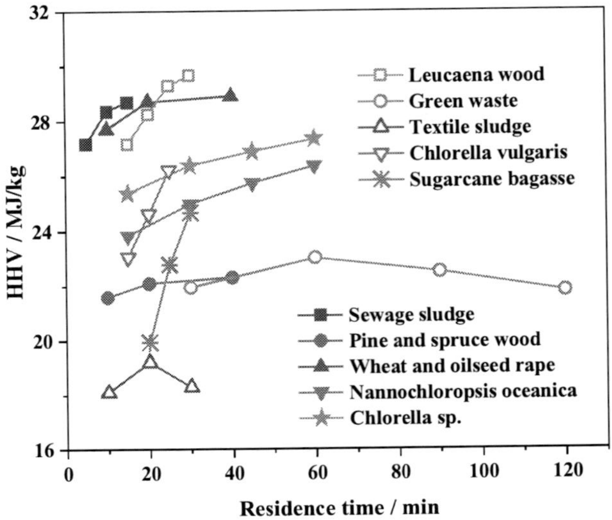

Figure 25. HHVs of biochars at different residence times.

Fan et al. (2023) studied the production of value-added biochar through microwave pyrolysis of peanut shell and evaluated the energy recovery efficiency during the whole process, as shown in Figure 26.

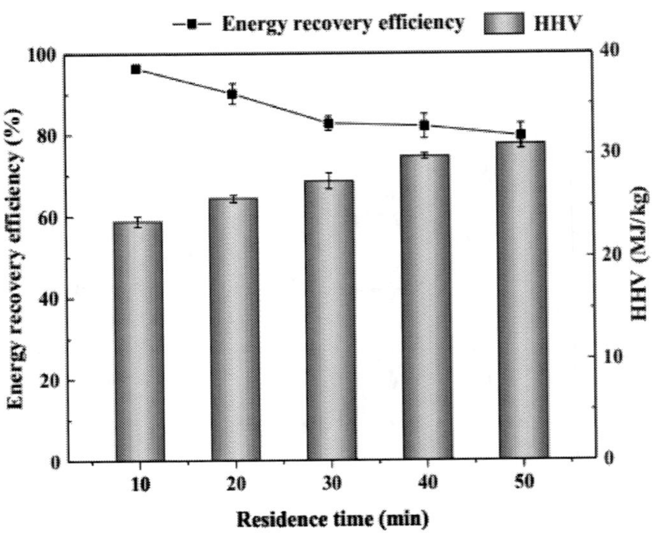

Figure 26. Biochar HHVs and energy recovery efficiencies at different residence times.

Table 5. Yields and HHVs of biochars obtained from different residence times

Feedstocks	Residence time (min)	Yield (wt. %)	HHV (MJ/kg)	Reference
Distillers dried grain with solubles	4.6-21.4	23.53-65.76		Lei et al. 2011
Prairie cordgrass	5.9-20.1	13.8-32.5		Zhou et al. 2013
Coconut shell	10-30	42.1-95.2		Yap et al. 2017
Sewage sludge	5-15	70-87	27.17-28.36	Ge et al. 2020
Arundo donax leave	18-31	54.3-59.9		Bartolil et al. 2016
Pine and spruce wood	10-40		27.7-28.9	Crombie and Masek, 2015
Wheat and oilseed rape	10-40		21.6-22.3	Crombie and Masek, 2015
Nannochloropsis oceanica	15-60	49.59-75.13	23.824-26.359	Zhang et al. 2019a
Chlorella sp.	15-60	50.12-66.65	25.385-27.336	Zhang et al. 2019a
Leucaena wood	15-30	17.25-25.28	27.16-29.65	Huang et al. 2017
Green waste	30-120		21.8-23.01	Shao et al. 2013
Textile sludge	10-30		18.1-19.2	Zheng et al. 2020
Chlorella vulgaris	5-10		23.82-25.52	Yu et al. 2020
Chlorella sp.	5-10		22.05-23.15	Yu et al. 2020
Chlorella vulgaris	15-25	35.4-44.8	23.05-26.22	Chen et al. 2022b
Sugarcane bagasse	20-30	30.1-60.4	19.93-24.66	Arpia et al. 2021b

As the residence time gradually increased, the biochar HHVs were in the range of 23.48-31.02 MJ/kg, showing a monotonically increasing trend. The C/H ratio severely affects the biochar characteristics, which is proportional to the HHV, and it can also represent the aromatic degree and structural stability of biochar. The microwave pyrolysis of peanut shells was more sufficient with time going by, and small molecular compounds containing O and H elements were gradually precipitated. Therefore, the C/H ratio gradually improved, and the HHV of biochar increased. When the input energy of peanut shell is constant, the energy recovery efficiency is affected by biochar energy. Due to the large differences in biochar yields at different residence times, the output biochar energy decreased from 255.93 to 210.94 kJ. The energy recovery efficiency was proportional to the output biochar energy, and also showed a monotonically decreasing trend. The energy recovery efficiency was between 79.49 and 96.45%, and the maximum energy recovery efficiency (96.45%) was occurred at the residence time of 10 min.

2.5. Summary

This section reviewed the influence of the reaction parameters (mainly including pyrolysis temperature, microwave power, and residence time) on the yield and HHV of biochar prepared by the microwave-assisted pyrolysis. It was concluded that with the increase of pyrolysis temperature, microwave power, and residence time, the yield of biochar produced from microwave-assisted pyrolysis mainly shows decreasing trends, while the high heating value increases.

3. Microwave Irradiation for Pyrolysis of Biomass for Porous Biochar

3.1. Introduction

Due to the characteristics of rich porous structures, biochar has great potentials in applications such as soil fertility improvement, carbon sequestration, adsorbent, and recycling of agricultural by-product. This has created a multidisciplinary field of interest in science and engineering for the application of biochar (Ahmad et al. 2014; Gan et al. 2018). Although, the specific surface area of the biochar from the microwave-assisted pyrolysis of different agro-residues was found to be lower than activated carbon (600-1200 m^2/g), the microwave-assisted pyrolysis is considered as a preferred method to produce biochar with high specific surface area. The surface area and pores volume of the biochar produced from the microwave-assisted pyrolysis is also reported to be higher than most of the biochar produced during the conventional pyrolysis process.

The pore structure properties of biochar are strongly controlled by the type of feedstock and the pyrolysis conditions, such as pyrolysis temperature, microwave power, and residence time. Many scholars have studied the microwave-assisted pyrolysis process of different biomass such as wood (Ren et al. 2012), rice (Du et al. 2010), oil palm (Salema and Ani, 2011), corn straw (Zhao et al. 2010), microalgae (Hu et al. 2012) and sewage sludge (Lin et al. 2012). The pyrolysis process temperature plays important role in the production distribution and control on the biochar nature from the given quantity of the biomass feedstock. The increased surface area of the biochar at the higher production temperature was because of volatile substances release from biomass leaving channel structures. Such structure enhances the biochar surface area and pore structure (Li et al. 2013). The parameter of microwave power plays vital role in biomass pyrolysis since the thermal change rate affects the characteristics of the product to some extent. At the lower heating rate, the likelihood of secondary pyrolysis reactions and thermal biomass cracking are reduced and consequent to the high yield of biochar. Nevertheless, pyrolysis with high heating rate enhances the yield of liquid, gaseous product and restricts the chance of biochar formation. The heating rate determines the porosity of biochar and surface area through heat and mass transfer resistance (Bouchelta et al. 2012). The slow or conventional pyrolysis process performed with a residence time duration from few minutes to days at moderate temperatures, slower heating rates mostly favored the production of biochar (Chen et al. 2003). The longer residence time of vapor in the pyrolysis reactor along with intermediate temperature and lower condensation rate, the secondary reactions also leads to the generation of high molecular weight compound like biochar (Bridgwater, 2012). The low pyrolysis temperatures and long residence time favors the biochar formation. The residence time has great impact on the finishing point of carbonization process, biochar features of specific surface area and pores volumes. The longer prolonged residence time may affect the biochar surface area due to the structural melting, elemental realignment and shrinkage (Yavari et al. 2017). The residence time determines over the composition of pyrolysis product, the yield of biochar reduced with the conditions of increased residence time and uniform pyrolysis temperature. The residence time controls over the surface area, micro and macro-pores distributions on biochar. The increased residence time up to 2 h with the pyrolysis temperature of 500-900°C showed enhanced surface area and pore area, whereas had negative effect when the residence time go beyond 2 h (Zhang et al. 2015 b). The effect of residence time on the pyrolysis process is always determined only along with the parameters like feedstock biomass, temperature, heating rate and other parameters (Fassinou et al. 2009).

In general, the pore structure characteristics of biochar can be determined by adsorption experiments, the most common being nitrogen adsorption experiments at 77 K. The adsorption and desorption isotherms of biochar under different relative pressures can be obtained through adsorption experiments, as shown in Figure 27. Further analysis of the adsorption and desorption isotherms using Brunauer-Emmett-Teller (BET), Barrett-Joiner-Halenda (BJH), t-plot, Density Functional Theory (DFT) and other methods can obtain data such as specific surface area, average pore size, total pore volume, micropore volume and pore size distribution.

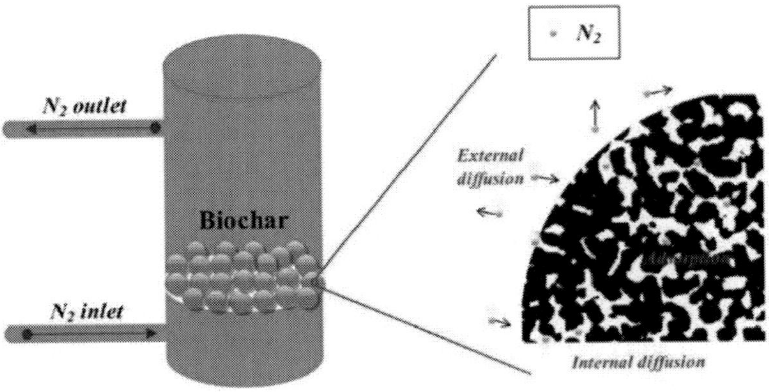

Figure 27. Schematic diagram of nitrogen adsorption experiment.

3.2. Effect of Pyrolysis Temperature

As shown in Figure 28 and Table 6, the specific surface area and pore volume of the biochar obtained from microwave-assisted pyrolysis also increased with the increase of pyrolysis temperature (Fang et al. 2021; Huang et al. 2015; Mong et al. 2020; Antunes et al. 2017b). The maximum specific surface area (SSA) and pore volume (PV) of biochar were 698.4 m^2/g and 0.302 cm^3/g prepared from horse manure at 550°C, respectively (Mong et al. 2020). This may be because as the microwave-assisted pyrolysis process goes on, the reactions of feedstocks become more sufficient and the pores of the biochar expand due to the flow of more and more gas products were generated. Therefore, in order to make the pore structure (i.e., specific surface area and pore volume) of biochar more ideal, the pyrolysis temperature during microwave-assisted pyrolysis can be appropriately increased. It can be estimated that the specific surface area and pore volume of biochar are closely related to the pyrolysis temperature. For different kinds of biomass, pyrolysis temperature ranges, and the other reaction conditions, the changing trends of biochar properties are not consistent. But in general, the specific surface areas and pore volumes increase with the improvement of pyrolysis temperature. This is because the pyrolysis reaction of biomass advances with the enhancement of pyrolysis temperature. The advancement of reaction also enlarged the pore structure and increased the specific surface area of biochar.

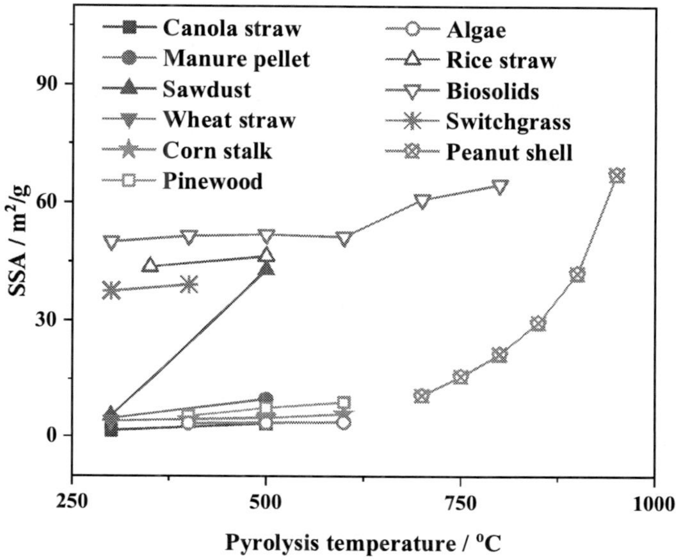

Figure 28. Specific surface areas of biochars at different pyrolysis temperatures.

Table 6. SSAs and PVs of biochars obtained from different pyrolysis temperatures

Feedstocks	Temperature (°C)	SSA (m^2/g)	PV (cm^3/g)	Reference
Manure pellet	300-500	4.5-9.7		Nzediegwu et al. 2021
Canola straw	300-500	1.5-3.2		Nzediegwu et al. 2021
Sawdust	300-500	5.2-43		Nzediegwu et al. 2021
Wheat straw	300-500	3.9-4.8		Nzediegwu et al. 2021
Peanut shell	200-600	4.93-587		Chu et al. 2017
Corn stalk	400-600	4.2-5.8		Parvez et al. 2020
Pinewood	400-600	5.2-8.9		Parvez et al. 2020
Algae	400-600	3.3-3.6		Parvez et al. 2020
Sludge	300-700	35.17-73.97		Fang et al. 2021
Rice straw	350-500	43.7-46.4		Huang et al. 2015
Horse manure	350-550	48.3-698.4	0.047-0.302	Mong et al. 2020
Biosolids	300-800	50.06-64.67		Antunes et al. 2017b
Switchgrass	300-400	37.54-39.21	0.115-0.1299	Mohamed et al. 2016
Biosolids	400-800	54.6-148.71	0.119-0.15	Antunes et al. 2017a
Peanut shell	700-950	10.76-67.29	0.011-0.084	Qiu et al. 2023

3.3. Effect of Microwave Power

As shown in Figure 29 and Table 7, as the improvement of microwave power, the specific surface area and pore volume of the biochar obtained from microwave-assisted pyrolysis mainly showed an increase trend (i.e., palm kernel shell (Liew et al. 2018b; Nam et al. 2018; Ge et al. 2020); prosopis juliflora (Nair and Vinu, 2016)).

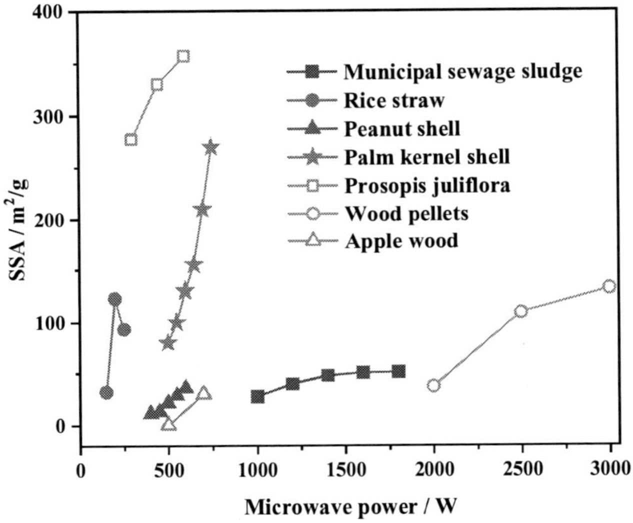

Figure 29. Specific surface areas of biochars at different microwave powers.

Table 7. SSAs and PVs of biochars obtained from different microwave powers

Feedstocks	Microwave power (W)	SSA (m^2/g)	PV (cm^3/g)	Reference
Municipal sewage sludge	1000-1800	27.27-51.05		Fang et al. 2021
Rice straw	150-250	31.6-122.2	0.034-0.083	Huang et al. 2015
Peanut shell	400-600	11.37-35.66	0.011-0.041	Qiu et al. 2023
Palm kernel shell	500-700	80-210	0.03-0.10	Liew et al. 2018b
Palm kernel shell	550-750	100-270	0.03-0.11	Nam et al. 2018
Palm kernel shell	500-700	890-1320	0.48-0.7	Ge et al. 2020
Prosopis juliflora	300-600	277.5-357	0.07-0.13	Nair and Vinu, 2016
Wood pellets	2000-3000	36.65-131.77	0.021-0.067	Nhuchhen et al. 2018
Apple wood	500-700	0.44-29.77	1.1-9.7	Liu et al. 2021

However, in the research of Huang et al. (2015), the specific surface area and pore volume of the biochar obtained from rice straw were increased first and then decreased. The maximums were occurred at the microwave power of 200 W (122.2 m^2/g for specific surface area and 0.083 cm^3/g for pore volume). Therefore, a microwave power of 200 W was the best choice to produce the biochar with the highest specific surface area in this research.

3.4. Effect of Residence Time

As shown in Figure 30 and Table 8, generally speaking, the specific surface area showed mainly three trends with the increase of residence time: (a) increases monotonously (i.e., 9.1-18.3 m^2/g vs 10-20 min for corn stover (Zhu et al. 2015), 44.12-51.19 m^2/g vs 90-150 min for municipal sewage sludge (Fang et al. 2021)), (b) increases first and then decreases, (i.e., 47.99-51.7 m^2/g vs 90-150 min for municipal sewage sludge (Fang et al. 2021), 940-1320 m^2/g vs 5-15 min for plam kernel shell (Ge et al. 2020)), and (c) decreases first and then increases (i.e., 38.06-150.85 m^2/g vs 10-30 min for bio-solid (Antunes et al. 2017 a)). The pore volume shows two tendencies: (a) increases first and then decreases (i.e., 0.47-0.70 cm^3/g vs 5-15 min for palm kernel shell (Ge et al. 2020)) and (b) decreases first and then increases (i.e., 0.096-0.182 cm^3/g vs 10-30 min for bio-solid (Antunes et al. 2017 a)). The results showed that the optimal residence times in the researches of Ge et al. (2020) and Fang et al. (2021) were 10 min and 120 min, respectively.

Table 8. SSAs and PVs of biochars obtained from different residence times

Feedstocks	Residence time (min)	SSA (m^2/g)	PV (cm^3/g)	Reference
Municipal sewage sludge	90-150	47.99-51.7		Fang et al. 2021
Biosolids	10-30	39.67-148.81	0.102-0.169	Antunes et al. 2017a
Peanut shell	60-180	4.682-26.703	0.008-0.030	Qiu et al. 2023
Palm kernel shell	5-15	940-1110	0.2-0.24	Ge et al. 2020
Corn stover	10-20	9.1-18.3	0.009-0.021	Zhu et al. 2015
Oil palm shell	20-60	228.1-258.62		Tripathi et al. 2020

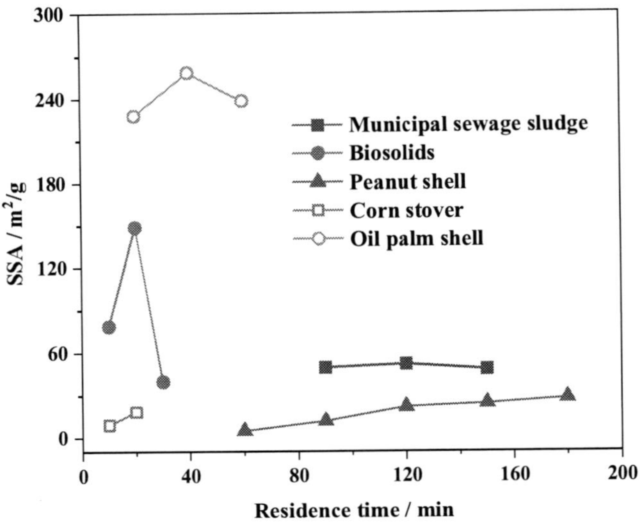

Figure 30. Specific surface areas of biochar at different residence times.

3.5. Summary

This section reviewed the influence of the reaction parameters (mainly including pyrolysis temperature, microwave power, and residence time) on the SSA of biochar prepared by the microwave-assisted pyrolysis. It was concluded that with the increase of pyrolysis temperature, microwave power, and residence time, the SSA of biochar produced from microwave-assisted pyrolysis mainly shows increasing trends. Biochar prepared from microwave-assisted pyrolysis has broad application prospects because of the large specific surface area and rich pore structure. It can be used in carbon dioxide capture, hydrogen storage, and catalyst support. The relationship between SSA and experimental conditions summarized here can guide the preparation of biochar with different pore structures. These biochars can be used to absorb gas particles with a specific diameter, such as carbon dioxide, hydrogen and so on. In addition, biochar can also be used as catalyst carrier to prepare catalysts by adsorbing some catalyst particles with specific sizes. The larger the pore volume and specific surface area, the larger the adsorption capacity of small molecules by biochar will be.

4. Activation of Biochar

4.1. Introduction

In recent times, agriculture and forestry industries have been transforming bio-waste into biochar for the management of the bio-waste generated on a daily basis. This biochar is then used as a soil conditioner and energy generation (Hu et al. 2015). Biochar contains higher oxygenated surface functional groups (phenolic, carboxylic, carbonyl) and fixed carbon than biomass and hence, has been used as catalyst, adsorbent and carbon electrode in catalyst process and water adsorption, electrochemical capacitor and super-capacitor. It was observed that the biochar obtained through biomass pyrolysis has a relatively low specific surface area, deprived porosity and lean surface functional group (Liu et al. 2015 b; Liu et al. 2015 a). Some of the applications are limited because of the low specific surface area (<200 m^2/g) and porosimetry. Activation is a technique to increase the physical characteristics and absorption capacity of biochar. The process of increasing specific surface area and pore density is called activation. In general, there are two types of activation techniques, i.e., physical and chemical.

4.2. Physical Activation

The char derived from biomass through the thermochemical process is auto-gasified due to the presence of steam, carbon dioxide, etc. at a temperature of about 700-900°C. This loss of mass in the form of gas and volatiles from the solid biomass makes the solid product, i.e., biochar more porous. The method, where the increase in porosity is initiated due to high temperature in an oxidative environment, is called physical activation or thermal activation (Prauchner and Rodríguez-reinoso, 2012). Most volatile carbon portions of carbonize precursors can be removed by physical activation and the closed pores in the biochar matrix can be opened and diffused with other pores (Liu et al. 2015 b; Liu et al. 2015 a). As a consequence, the specific surface area of physically activated biochar can be substantially enhanced and a robust microporous structure with a low intensity of mesopores can be observed in the physically activated biochar matrix.

Table 9. Characteristics of physical activation method

Feedstock	Activation agent	Temperature (°C)	Holding time (min)	SSA (m^2/g)	Pore volume (cm^3/g)	Reference
Coconut shell	CO_2	750	240	613	0.437	Guo et al. 2009
Coconut shell	CO_2	800	240	860	0.590	Guo et al. 2009
Coconut shell	CO_2	850	240	1305	0.689	Guo et al. 2009
Coconut shell	CO_2	900	240	1391	0.838	Guo et al. 2009
Coconut shell	CO_2	950	240	1323	0.878	Guo et al. 2009
Furfural residue	Pyrolysis gas	800	20	167	0.134	Yin et al. 2018
Furfural residue	Pyrolysis gas	800	40	257	0.186	Yin et al. 2018
Furfural residue	Pyrolysis gas	800	60	384	0.281	Yin et al. 2018
Furfural residue	Pyrolysis gas	800	120	567	0.380	Yin et al. 2018
Furfural residue	Pyrolysis gas	700	60	328	0.202	Yin et al. 2018
Furfural residue	Pyrolysis gas	600	60	263	0.121	Yin et al. 2018
Furfural residue	Pyrolysis gas	900	60	334	0.232	Yin et al. 2018
Burcucumber	Steam	300	45	1.22	0.003	Rajapaksha et al. 2015
Burcucumber	Steam	700	45	7.10	0.038	Rajapaksha et al. 2015
Wheat straw	Steam	800	78	246.2	0.158	Koltowski et al. 2016
Coconut	Steam	800	78	626,8	0.336	Koltowski et al. 2016
Willow	Steam	800	78	840.6	0.576	Koltowski et al. 2016
Wood	Steam	600	0.5-5	389	0.161	Zhu et al. 2018
Wood	Steam	700	0.5-5	586	0.168	Zhu et al. 2018
Wood	Steam	800	0.5-5	672	0.212	Zhu et al. 2018
Oak	CO_2	700	60-120	642-644	0.404-0.411	Zhang et al. 2004
Oak	CO_2	800	60-120	845-985	0.601-0.640	Zhang et al. 2004
Corn hulls	CO_2	700	60-120	902-977	0.328-0.335	Zhang et al. 2004
Corn hulls	CO_2	800	60-120	975-1010	0.680-0.834	Zhang et al. 2004
Corn stover	CO_2	700	60-120	432-660	0.332-0.487	Zhang et al. 2004
Corn stover	CO_2	800	60-120	616-712	0.422-0.549	Zhang et al. 2004
Macro-algae	Steam	700	60	57.9	0.02	Cho et al. 2013

Table 10. Characteristics of chemical activation method

Feedstock	Activation agent	Temperature (°C)	Holding time (min)	SSA (m^2/g)	Pore volume (cm^3/g)	Reference
Rice straw	Char/KOH=0.1	700	2	772.3	0.422	Cha et al. 2010
Sludge char	Char/KOH=0.1	700	2	782.6	0.606	Cha et al. 2010
Coconut shell	Char/H_3PO_4=0.3	500	2		0.652	Prauchner and Rodríguez-reinoso, 2012
Coconut shell	Char/$ZnCl_2$=0.4	500	2		0.625	Prauchner and Rodríguez-reinoso, 2012
Olive stoned	Char/$ZnCl_2$=0.15	500	7	750	0.31	Nakagawa et al. 2007
Olive stoned	Char/H_3PO_4=1.2	450	2	1500	0.63	Nakagawa et al. 2007
Olive stoned	Char/$ZnCl_2$=0.25	600	1	249.3	0.151	Angin, 2013
Olive stoned	Char/$ZnCl_2$=0.25	700	1	491.9	0.249	Angin, 2013
Olive stoned	Char/$ZnCl_2$=0.25	800	1	772	0.358	Angin, 2013
Olive stoned	Char/$ZnCl_2$=0.25	900	1	801.5	0.393	Angin, 2013
Olive stoned	Char/$ZnCl_2$=0.33-1	900	1	619.3-719.8	0.33-0.37	Angin, 2013
Soybean oil cake	K_2CO_3	600	1	643.54	0.336	Tay et al. 2009
Soybean oil cake	K_2CO_3	800	1	1352.6	0.68	Tay et al. 2009
Soybean oil cake	KOH	600	1	600.05	0.299	Tay et al. 2009
Soybean oil cake	KOH	800	1	618.54	0.291	Tay et al. 2009
Organic sewage sludge	H_2SO_4	650	1	408	0.523	Zhang et al. 2005
Organic sewage sludge	H_3PO_4	650	1	289	0.436	Zhang et al. 2005
Organic sewage sludge	$ZnCl_2$	650	1	555	0.752	Zhang et al. 2005

Physical activation is highly dependent on temperature, the degree of activation, type of precursor and activation agent (if any) which is represented in Table 9. In general, the higher the activation temperature and time, the greater the porosity growth. However, it also leads to enlarged pore size distribution. If air is used as an activation agent, then it is a challenge to control the activation process because with air and char (rich in carbon), the reaction will tend to shift toward combustion if not regulated, leading to excessive ash formation and decreased activated carbon yield. However, it is still a favored route for the activation of biochar as oxygen present in the air requires less activation energy compared to CO_2 and steam and hence the overall activation process becomes economical (Dawson et al. 2003). Lima et al. (2010) produced biochar from broiler litter through fast pyrolysis and activated it by steam. Later, the effect specific surface area, porosimetry, and adsorption of heavy metal ion (Cu^{2+}, Cd^{2+}, Zn^{2+}, and Ni^{2+}) were examined. Due to the steam activation (800°C and 45 min), the specific surface area and pore volume increased from 136 to 789 m^2/g and 0.052 to 0.344 cm^3/g, respectively. Besides, it was observed that metal ions' adsorption performance enhanced due to the developed surface area and porosity. Koltowaski (2017) prepared activated biochar in fluidized bed reactor (at a heating rate of 10 °C/min and temperature: 800°C) under steam and CO_2 atmosphere and studied the effect of activated biochar on soil toxicity reduction. The specific surface area and pore volume increase from 11.4 to 512 m^2/g and 0.002-0.169 cm^3/g in CO_2 and 11.4-840 m^2/g and 0.002-0.225 cm^3/g in steam, respectively. It was also observed that the increase in the surface area leads to a decrease in the toxicity of water leachates. On the other hand, Nabais et al. (2008) used coffee endocarp biochar for steam and CO_2 activation. The CO_2-activated biochar showed the higher surface area and pore volume than steam-activated biochar with the same bun-off value. Therefore, it was concluded that this inconsistency is due to the different physical and chemical properties of biomass.

4.3. Chemical Activation

Chemical activation is a heat treatment process in which raw biochar reacts with the chemical activating agent in the temperature range of 450-900 °C. The chemical activation process is still unclear with respect to physical activation. Two types of reactions, dehydration and oxidation, take place inside the biochar and they lead to the microspores formation. The advantages of chemical activation over the physical activation are (1) lower temperature, (2) greater carbon yield, (3) high surface area (~ 3600 m^2/g), (4) qualitative and quantitative micro-porosity (well developed and controlled), and (5) higher efficiency. However, chemical activation does harm equipment due to chemical corrosion. The corrosiveness of this chemical agent can be intense at high temperatures. These chemicals are however suppressing the tar formation, and removing partial carbon molecules from biochar structure and volatile compounds (Azargohar and Dalai, 2008).

Acid, alkali, and oxidation treatment are additionally done on biochar to enhance the physicochemical properties of biochar. The acids such as HCl, $ZnCl_2$, $MgCl_3$, HNO_3, H_2SO_4 and H_3PO_4; an alkali such as KOH, NaOH and K_2CO_3; and oxidants like H_2O_2 and $KMnO_4$ are often used in chemical activation of biochar. The KOH activation of biochar expands the carbon lattice by metallic potassium (K) intercalation which leads to an increase in surface area and porosity. Overall, the chemical activation efficiency is higher than physical activation. However, the efficiency of chemical activation is sensitive to activation temperature, type, and concentration of activation agent, feedstock types, etc. and these in turn influence the physical characteristics of the activated biochar structure.

However, $ZnCl_2$, H_3PO_4, and KOH are commonly used at an industrial scale (Ströbel et al. 2006; Gogotsi et al. 2009). The degree of impregnation (material to agent ratios) is generally in the series of 1:0.5 to 1:3 based on dry content. However, the effect of chemical activation depends on the type of chemical used, the intensity of mixing, temperature, and period of successive activation. It was also observed that pore volume shrank at high chemical concentration; this is because at higher chemical concentration, the carbon structure physically collapsed. After the chemical activation, the activated biochar must be washed to remove the activation agent from the surface, and it can be recycled. Thus, chemical activation at the industrial scale demands substantial effort in the washing process to reduce the environmental impact, which in turn increases activation cost. Dehkhoda et al. (2014) examined the influence of KOH activation temperature (675 and 1000 °C) on the surface area, pores size distribution, porosity, and electro-sorption capacity of woody biochar. It was observed that physical properties such as surface area (1.66-990 m^2/g), pore size (1.28-3.22 nm) and pore volume (0.3-0.9 cm^3/g) increase by KOH activation at 675°C. In contrast, KOH activation at 1000 °C resulted in decrease of the surface area, pore volume, and pore size due to the burn-off effect; collapses the micropores present in the biochar matrix and promotes the expansion of the graphite-like structure in biochar. Moreover, electrosorption capacity of activated biochar at 675 °C is two times higher than that of higher temperatures because of its higher porosimetry and oxygenated functional groups present on the surface. The effect of biochar material, activation agent, and temperature on activated biochar is shown in Table 10.

4.4. Summary

The suitability of biochar for any particular application is highly dependent on its physical and surface properties such as specific surface area, pore volume, surface functional groups, etc. Therefore, biochar is subjected to different types of physical and chemical activation methods to improve its surface properties. The activated microwave carbonization enhances pore volume and surface area of the biochar. There is a significant difference in terms of elemental content, pore size and surface area between microwave biochar and activated biochar. The improvement of pore size and surface area of the activated microwave biochar has increased the biochar capability as an adsorbent. It is concluded that the presence of functional groups, surface area, and pore volume plays a crucial role in enhancing the potential of microwave biochar. Activated biochar also shows a better C/N ratio as compared to the untreated biochar. An excellent C/N ratio in biochar has the potential to be applied as so amender and plant growth enhancer.

Conclusion

This chapter reviewed the microwave-assisted pyrolysis of organic wastes for char production. The influence of the reaction parameters (mainly including pyrolysis temperature, microwave power, and residence time) on the biochar properties (high heating value, specific surface area, and pore volume) and yield during the microwave-assisted pyrolysis has been detailed. It is concluded that with the increase of pyrolysis temperature, microwave power, and residence time, the yield of biochar produced from microwave-assisted pyrolysis mainly shows decreasing trends. Both higher pyrolysis temperature and higher microwave power improve the specific surface area and pore volume of biochar. The increase of residence time reduces the yield of biochar whereas increases the high heating value. The activation methods of biochar were also introduced. The preparation of biochar by microwave-assisted pyrolysis has become increasingly popular, and further development is needed to achieve excellent properties and high yields to make the application of biochar more widespread and popular.

Acknowledgments

The following financial supports are acknowledged: Financial support was provided by the National Natural Science Foundation of China (52476005, 52076049), Heilongjiang Province "Double First-class" Discipline Collaborative Innovation Achievement Project (LJGXCG2023-080), Heilongjiang Provincial Key R&D Program (2023ZX02C05), and Heilongjiang Provincial Key R&D Program "Unveiling the Leader" Project (2023ZXJ02C04).

References

Abdullan, H., Mediaswanti, K., Wu, H. (2010). Biochar as a fuel: 2. significant differences in fuel quality and ash properties of biochars from various biomass componets of mallee trees. *Energy & Fuel*, 24 (3), 1972-1979.

Ahmad, I., Ghaffar, A., Zakir, A., Khan, Z., Saeed, M., Rasool, A., Jamal, A., Mahoub, A., Marzeddu, S., Boni, M. (2022). Activated biochar is an effective technique for arsenic removal from contaminated drinking water in Pakistan. *Sustainability*, 14.

Ahmad, M., Rajapaksha, A., Lim, J., Zhang, M., Bolan, N., Mohan, D., Vithanage, M., Lee, S., Ok, Y. (2014). Biochar as a sorbent for contaminant management in soil and water: a review. *Chemosphere*, 99, 19-33.

Al-Wabel, M., Al-Omran, A., El-Naggar, A., Nadeem, M., Usman, A. (2013). Pyrolysis temperature induced changes in characteristics and chemical composition of biochar produced from conocarpus wastes. *Bioresource Technology*, 131, 374-379.

Ameloot, N., Graber, E., Verheijen, F., De Neve, S. (2013). Interactions between biochar stability and soil organisms: review and research needs. *European Journal of Soil Science,* 64, 379-390.

Angin, D. (2013). Effect of pyrolysis temperature and heating rate on biochar obtained from pyrolysis of safflower seed press cake. *Bioresource Technology*, 128, 593-597.

Antón-Herrero, R., García-Delgado, C., Alonso-Izquierdo, M., García-Rodríguez, G., Cuevas, J., Eymar, E. (2018). Comparative adsorption of tetracyclines on biochars and stevensite: looking for the most effective adsorbent. *Applied Clay Science*, 160, 162-172.

Antunes, E., Jacob, M., Brodie, G., Schneider, P. (2018). Isotherms, kinetics and mechanism analysis of phosphorus recovery from aqueous solution by calcium-rich biochar produced from biosolids via microwave pyrolysis. *Journal of Environmental Chemical Engineering,* 6 (1), 395-403.

Antunes, E., Jacob, M., Brodie, G., Schneider, P. (2017a). Silver removal from aqueous solution by biochar produced from biosolids via microwave pyrolysis. *Journal of Environmental Management*, 203, 264-272.

Antunes, E., Schumann, J., Brodie, G., Jacob, M., Schneider, P. (2017b). Biochar produced from biosolids using a single-mode microwave: characterisation and its potential for phosphorus removal. *Journal of Environmental Management*, 196, 119-126.

Arpia, A., Chen, W., Lam, S., Rousset, P., de Luna, M. (2021a). Sustainable biofuel and bioenergy production from biomass waste residues using microwave-assisted heating: a comprehensive review. *Chemical Engineering Journal*, 403, 126233.

Arpia, A., Chen, W., Ubando, A., Tabatabaei, M., Lam, S., Culaba, A., De Luna, M. (2021b). Catalytic microwave-assisted torrefaction of sugarcane bagasse with calcium oxide optimized via taguchi approach: product characterization and energy analysis. *Fuel*, 305, 121543.

Asimakidou, T., Karfaridis, D., Kalaitzidou, K., Simeonidis, K., Chrissafis, K. (2022). Optimization of tin oxyhydroxide-decorated biochar for improved hexavalent chromium uptake from drinking water. *Journal of Environmental Chemical Engineering*, 10, 108051.

Asomaning, J., Haupt, S., Chae, M., Bressler, D. (2018). Recent developments in microwave-assisted thermal conversion of biomass for fuels and chemicals. *Renewable & Sustainable Energy Reviews*, 92, 642-657.

Azargohar, R., Dalai, A. (2008). Steam and KOH activation of biochar: experimental and modeling studies. *Microporous and Mesoporous Materials*, 110 (2-3), 413-421.

Azargohar, R., Jacobson, K., Powell, E., Dalai, A. (2013). Evaluation of properties of fast pyrolysis products obtained, from Canadian waste biomass. *Journal of Analytical and Applied Pyrolysis*, 104, 330-340.

Azizi, S., Mowla, D. (2016). CFD modeling of algae flash pyrolysis in the batch fluidized bed reactor including heat carrier particles. *International Journal of Chemical Reactor Engineering*, 14, 462-480.

Banerjee, S., Mukherjee, S., LaminKa-ot, A., Joshi, S., Mandal, T., Halder, G. (2016). Biosorptive uptake of Fe^{2+}, Cu^{2+} and As^{5+} by activated biochar derived from colocasia esculenta: isotherm, kinetics, thermodynamics, and cost estimation. *Journal of Advanced Research*, 7, 597-610.

Bapat, H., Manahan, S., Larsen, D. (1999). An activated carbon product prepared from milo (Sorghum Vulgare) grain for use in hazardous waste gasification by chemchar cocurrent flow gasification. *Chemosphere*, 39 (1), 23-32.

Bartoli, M., Rosi, L., Giovannelli, A., Frediani, P., Frediani, M. (2016). Production of bio-oils and bio-char from arundo donax through microwave assisted pyrolysis in a multimode batch reactor. *Journal of Analytical and Applied Pyrolysis*, 122, 479-489.

Bildirici, M., Özaksoy, F. (2013). The relationship between economic growth and biomass energy consumption in some European countries. *Journal of Renewable and Sustainable Energy*, 5, 023141.

Bouchelta, C., Medjram, M., Zoubida, M., Chekkat, F., Ramdane, N., Bellat, J. (2012). Effects of pyrolysis conditions on the porous structure development of date pits activated carbon. *Journal of Analytical and Applied Pyrolysis*, 94, 215-222.

Břendová, K., Száková, J., Lhotka, M., Krulikovská, T., Puncochár, M., Tlustos, P. (2017). Biochar physicochemical parameters as a result of feedstock material and pyrolysis temperature: predictable for the fate of biochar in soil. *Environmental Geochemistry and Health*, 39, 1381-1395.

Brewer, C., Chuang, V., Masiello, C., Gonnermann, H., Gao, X., Dugan, B., Driver, L., Panzacchi, P., Zygourakis, K., Davies, C. (2014). New approaches to measuring biochar density and porosity. *Biomass and Bioenergy*, 66, 176-185.

Brewer, C., Hall, E., Schmidt-Rohr, K., Laird, D., Brown, R., Zygourakis, K. (2016). Temperature and reaction atmosphere effects on the properties of corn stover biochar. *Environmental Progress & Sustainable Energy*, 36, 696-707.

Bridgwater, A. (2012). Review of fast pyrolysis of biomass and product upgrading. *Biomass Bioenergy*, 38, 68-94.

Budai, A., Rasse, D., Lagomarsino, A., Lerch, T., Paruch, L. (2016). Biochar persistence, priming and microbial responses to pyrolysis temperature series. *Biology and Fertility of Soils*, 52, 749-761.

Bundhoo, Z. (2018). Microwave-assisted conversion of biomass and waste materials to biofuels. *Renewable & Sustainable Energy Reviews*, 82, 1149-1177.

Burhenne, L., Damiani, M., Aicher, T. (2013). Effect of feedstock water content and pyrolysis temperature on the structure and reactivity of spruce wood char produced in fixed bed pyrolysis. *Fuel*, 107, 836-847.

Byambajav, E., Paysepar, H., Nazari, L., Xu, C. (2018). Co-pyrolysis of lignin and low rank coal for the production of aromatic oils. *Fuel Processing Technology*, 181, 1-7.

[85] Calamari, L., Gobbi, L., Bany, P. (2016). Improving the prediction ability of FT-MIR spectroscopy to assess titratable acidity in cow's milk. *Food Chemistry*, 192 (2), 477-484.

Cao, X., Sun, S., Sun, R. (2017). Application of biochar-based catalysts in biomass upgrading: a review. *RSC Advances*, 7, 48793-48805.

Cha, J., Choi, J., Ko, J., Park, Y., Park, S., Jeong, K., Kim, S., Jeon, J. (2010). The low-temperature SCR of NO over rice straw and sewage sludge derived char. *Chemical Engineering Journal*, 156 (2), 321-327.

Charoensook, K., Huang, C., Tai, H., Lanjapalli, V., Chiang, L., Hosseini, S., Lin, Y., Li, Y. (2021). Preparation of porous nitrogen-doped activated carbon derived from rice straw for high-performance supercapacitor application. *Journal of the Taiwan Institute of Chemical Engineers*, 120, 246-256.

Chatterjee, R., Sajjadi, B., Chen, W., Mattern, D., Hammer, N., Raman, V., Dorris, A. (2020). Effect of pyrolysis temperature on physicochemical properties and acoustic-based amination of biochar for efficient CO_2 adsorption. *Frontiers in Energy Research*, 8, 1-18.

Chen, C., Fan, D., Ling, H., Huang, X., Yang, G., Cai, D., Zhao, J., Bi, Y. (2022a). Microwave catalytic co-pyrolysis of chlorella vulgaris and high density polyethylene over activated carbon supported monometallic: characteristics and bio-oil analysis. *Bioresource Technology*, 363,127881.

Chen, G., Andries, J., Luo, Z., Spliethoff, H. (2003). Biomass pyrolysis/gasification for product gas production: the overall investigation of parametric effects. *Energy Conversion and Management*, 44, 1875-1884.

Chen, J., Li, S., Liang, C., Xu, Q., Li, Y., Qin, H., Fuhrmann, J. (2017a). Response of microbial community structure and function to short-term biochar amendment in an intensively managed bamboo (phyllostachys praecox) plantation soil: effect of particle size and addition rate. *Science of the Total Environment*, 574, 24-33.

Chen, J., Zhang, D., Wu, M., Wang, P. (2014). Elemental composition and thermal stability of two different biochars. *Environmental Chemistry*, 33, 417-422.

Chen, T., Luo, L., Luo, L., Deng, J., Wu, X., Fan, M., Du, G., Zhao, W. (2021a). High energy density supercapacitors with hierarchical nitrogen-doped porous carbon as active material obtained from bio-waste. *Renewable Energy*, 175, 760-769.

Chen, W., Arpia, A., Chang, J., Kwon, E., Park, Y., Culaba, A. (2022b). Catalytic microwave torrefaction of microalga chlorella vulgaris FSP-E with magnesium oxide optimized via taguchi approach: a thermo-energetic analysis. *Chemosphere*, 290, 133374.

Chen, W., Meng, J., Han, X., Lan, Y., Zhang, W. (2019). Past, present, and future of biochar. *Biochar*, 1, 75-87.

Chen, W., Zhang, W., Meng, J. (2013). Advances and prospects in research of biochar utilization in agriculture. *Scientia Agricultura Sinica*, 46, 3324-3333.

Chen, W., Zhang, W., Meng, J., Xu, Z. (2011). Researches on biochar application technology. *Strategic Study of CAE*, 13 (2), 83-89.

Chen, X., Dai, Y., Fan, J., Xu, X., Cao, X. (2021b). Application of iron-biochar composite in topsoil for simultaneous remediation of chromium-contaminated soil and groundwater: immobilization mechanism and long-term stability. *Journal of Hazardous Materials*, 405, 124226.

Chen, Y., Zhang, X., Chen, W., Yang, H., Chen, H. (2017b). The structure evolution of biochar from biomass pyrolysis and its correlation with gas pollutant adsorption performance. *Bioresource Technology*, 246, 101-109.

Cho, H., Baek, K., Jeon, J., Park, S., Suh, D., Park, Y. (2013). Removal characteristics of copper by marine macro-algae-derived chars. *Chemical Engineering Journal*, 217, 205-211.

Chu, G., Zhao, J., Chen, F., Dong, X., Zhou, D., Liang, N., Wu, M., Pan, B., Steinberg, C. (2017). Physi-chemical and sorption properties of biochars prepared from peanut shell using thermal pyrolysis and microwave irradiation. *Environmental Pollution*, 227, 372-379.

Conti, R., Rombolà, A., Modelli, A., Torri, C., Fabbri, D. (2014). Evaluation of the thermal and environmental stability of switchgrass biochars by Py-GC-MS. *Journal of Analytical and Applied Pyrolysis*, 110, 239-247.

Cosentino, I., Restuccia, L., Ferro, G., Tulliani, J. (2018). Influence of pyrolysis parameters on the efficiency of the biochar as nanoparticles into cement-based composites. *Procedia Structural Integrity*, 13, 2132-2136.

Crombie, K., Masek, O. (2015). Pyrolysis biochar systems, balance between bioenergy and carbon sequestration. *Global Change Biology Bioenergy*, 7 (2), 349-361.

Das, O., Sarmah, A. (2015). The love-hate relationship of pyrolysis biochar and water: a perspective. *Science of the Total Environment*, 512-513, 682-685.

Dawson, E., Parkes, G., Barnes, P., Chinn, M. (2003). An investigation of the porosity of carbons prepared by constant rate activation in air. *Carbon*, 41 (3), 571-578.

Dehkhoda, A., Ellis, N., Gyenge, E. (2014). Electrosorption on activated biochar: effect of thermo-chemical activation treatment on the electric double layer capacitance. *Journal of Applied Electrochemistry*, 44 (1), 141-157.

Ding, Y., Liu, Y., Liu, S., Li, Z., Tan, X., Huang, X., Zeng, G., Zhou, L., Zheng, B. (2016). Biochar to improve soil fertility: a review. *Agronomy for Sustainable Development*, 36, 36.

Du, J., Liu, P., Liu, Z., Sun, D., Tao, C. (2010). Fast pyrolysis of biomass for bio-oil with ionic liquid and microwave irradiation. *Journal of Fuel Chemistry Technology*, 38 (5), 554-559.

Du, S., Chen, W., Lucas, J. (2014). Pretreatment of biomass by torrefaction and carbonization for coal blend used in pulverized coal injection. *Bioresource Technology*, 161 (3), 333-339.

Ellison, C., Hoff, R., Marculescu, C., Boldor, D. (2020). Investigation of microwave-assisted pyrolysis of biomass with char in a rectangular waveguide applicator with built-in phase-shifting. *Applied Energy*, 259, 114217.

Fagan, C., Everard, C., Mcdonnell, K. (2011). Prediction of moisture, calorific value, ash and carbon content of two dedicated bioenergy crops using near-infrared spectroscopy. *Bioresource Technology*, 102 (8), 5200-5206.

Fan, L., Song, H., Lu, Q., Leng, L., Li, K., Liu, Y., Wang, Y., Chen, P., Ruan, R., Zhou, W. (2019). Screening microwave susceptors for microwave-assisted pyrolysis of lignin: comparison of product yield and chemical profile. *Journal of Analytical and Applied Pyrolysis*, 142, 104623.

Fan, S., Cui, L., Li, H., Guang, M., Liu, H., Qiu, T., Zhang, Y. (2023). Value-added biochar production from microwave pyrolysis of peanut shell. *International Journal of Chemical Reactor Engineering.*

Fan, S., Sun, Y., Yang, T., Chen, Y., Yan, B., Li, R., Chen, G. (2020). Biochar derived from corn stalk and polyethylene co-pyrolysis: characterization and Pb(II) removal potential. *RSC Advances*, 10 (11), 6362-6376.

Fang, Q., Chen, B., Lin, Y., Guan, Y. (2014). Aromatic and hydrophobic surfaces of wood-derived biochar enhance perchlorate adsorption via hydrogen bonding to oxygen-containing organic groups. *Environmental Science & Technology*, 48, 279-288.

Fang, Z., Liu, F., Li, Y., Li, B., Yang, T., Li, R. (2021). Influence of microwave-assisted pyrolysis parameters and additives on phosphorus speciation and transformation in phosphorus-enriched biochar derived from municipal sewage sludge. *Journal of Cleaner Production*, 287, 125550.

Fassinou, W., Van de Steene, L., Tourea, S., Volle, G., Girard, P. (2009). Pyrolysis of pinus pinaster in a two-stage gasifier: influence of processing parameters and thermal cracking of tar. *Fuel Processing Technology*, 90, 75-90.

Fernandes, B., Mendes, K., Tornisielo, V., Teófilo, T., Takeshita, V., Chagas, P., Lins, H., Souza, M., Silva, D. (2022). Effect of pyrolysis temperature on eucalyptus wood residues biochar on availability and transport of hexazinone in soil. *International Journal of Environmental Science and Technology*, 19, 499-514.

Fodah, A., Ghosal, M., Behera, D. (2021a). Bio-oil and biochar from microwave-assisted catalytic pyrolysis of corn stover using sodium carbonate catalyst. *Journal of the Energy Institute*, 94, 242-251.

Fodah, A., Ghosal, M., Behera, D. (2021b). Quality assessment of bio-oil and biochar from microwave-assisted pyrolysis of corn stover using different adsorbents. *Journal of the Energy Institute*, 98, 63-76.

Foong, S., Abdul, L. Liew, R., Yek, R., Lam, S. (2020a). Production of biochar for potential catalytic and energy applications via microwave vacuum pyrolysis conversion of cassava stem. *Materials Science for Energy Technologies*, 3, 728-733.

Foong, S., Liew, R., Yang, Y., Cheng, Y., Yek, P., Mahari, W., Lee, X., Han, C., Vo, D., Le, Q., Aghbashlo, M., Tabatabaei, M., Sonne, C., Peng, W., Lam, S. (2020b). Valorization of biomass waste to engineered activated biochar by microwave pyrolysis: progress, challenges, and future directions. *Chemical Engineering Journal*, 389, 124401.

Fseha, Y., Sizirici, B., Yildiz, I. (2022a). Manganese and nitrate removal from groundwater using date palm biochar: application for drinking water. *Environmental Advances*, 8, 100237.

Fseha, Y., Sizirici, B., Yildiz, I. (2022b). Phoenix dactylifera (date palm)-derived biochar application for the adsorptive removal of multiple inorganics from groundwater for drinking water purposes. *Arabian Journal for Science and Engineering.*

Fu, P., Yi, W., Li, Z., Bai, X., Wang, L. (2017). Evolution of char structural features during fast pyrolysis of corn straw with solid heat carriers in a novel V-shaped down tube reactor. *Energy Conversion and Management*, 149, 570-578.

Gan, Y., Ong, H., Show, P., Ling, T., Chen, W., Yu, K., Abdullah, R. (2018). Torrefaction of microalgal biochar as potential coal fuel and application as bio-adsorbent. *Energy Conversion and Management*, 265, 152-162.

Ge, S., Foong, S., Ma, N., Liew, R., Mahari, W., Xia, C., Yek, P., Peng, W., Nam, W., Lim, X., Liew, C., Chong, C., Sonne, C., Lam, S. (2020). Vacuum pyrolysis incorporating microwave heating and base mixture modification: an integrated approach to transform biowaste into eco-friendly bioenergy products. *Renewable & Sustainable Energy Reviews*, 127, 109871.

Georgin, J., Dotto, G., Mazutti, M., Foletto, E. (2016). Preparation of activated carbon from peanut shell by conventional pyrolysis and microwave irradiation-pyrolysis to remove organic dyes from aqueous solutions. *Journal of Environmental Chemical Engineering*, 4, 266-275.

Gogotsi, Y., Portet, C., Osswald, S., Simmons, J., Yidirim, T., Laudisio, G., Fischer, J. (2009). Importance of pore size in highpressure hydrogen storage by porous carbons. *International Journal of Hydrogen Energy*, 34 (15), 6314-6319.

Gray, M., Johnson, M., Dragila, M., Kleber, M. (2014). Water uptake in biochars: the roles of porosity and hydrophobicity. *Biomass & Bioenergy*, 61, 196-205.

Guan, J., Fang, Y., Zhang, T., Wang, L., Zhu, H., Du, M., Zhang, M. (2020). Kelp-derived activated porous carbon for the detection of heavy metal ions via square wave anodic stripping voltammetry. *Electrocatalysis*, 11, 59-67.

Gul, S., Whalen, J. (2016). Biochemical cycling of nitrogen and phosphorus in biochar-amended soils. *Soil Biology & Biochemistry*, 103, 1-15.

Gul, S., Whalen, J., Thomas, B., Sachdeva, V., Deng, H. (2015). Physico-chemical properties and microbial responses in biochar-amended soils: mechanisms and future directions. *Agriculture Ecosystems & Environment*, 206, 46-59.

Guo, S., Peng, J., Li, W., Yang, K., Zhang, L., Zhang, S., Xia, H. (2009). Effects of CO_2 activation on porous structures of coconut shell-based activated carbons. *Applied Surface Science*, 255, 8443-8449.

Gupta, S., Kua, H. (2016). Factors determining the potential of biochar as a carbon capturing and sequestering construction material: critical review. *Journal of Materials in Civil Engineering*, 29 (9), 04017086.

Haarlemmer, G., Guizani, C., Anouti, S., Déniel, M., Roubaud, A., Valin, S. (2016). Analysis and comparison of bio-oils obtained by hydrothermal liquefaction and fast pyrolysis of beech wood. *Fuel*, 174, 180-188.

Haeldermans, T., Claesen, J., Maggen, J., Carleer, R., Yperman, J., Adriaensens, P., Samyn, P., Vandamme, D., Cuypers, A., Vanreppelen, K., Schreurs, S. (2019). Microwave assisted and conventional pyrolysis of MDF-characterization of the produced biochars. *Journal of Analytical and Applied Pyrolysis*, 138, 218-230.

Han, L., Ro, K., Wang, Y., Sun, K., Sun, H., Libra, J., Xing, B. (2018). Oxidation resistance of biochars as a function of feedstock and pyrolysis condition. *Science of the Total Environment*, 616-617, 335-344.

Hassan, M., Liu, Y., Naidu, R., Parikh, S., Du, J., Qi, F., Willett, I. (2020). Influences of feedstock sources and pyrolysis temperature on the properties of biochar and functionality as adsorbents: a meta-analysis. *Science of the Total Environment*, 744, 140714.

He, X., Feng, D., Ao, F., Wang, C. (2015). Research on the characteristic and application of bio-carbon. *Fuel & Chemical Processes*, 46 (4), 1-3, 7.

Hou, D., O'Connor, D., Igalavithana, A., Alessi, D., Luo, J., Tsang, D., Sparks, D., Yamauchi, Y., Rinklebe, J., Ok, Y. (2020). Metal contamination and bioremediation of agricultural soils for food safety and sustainability. *Nature Reviews Earth & Environment*, 1, 366-381.

Hu, E., Shang, S., Wang, N., Nan, X., Zhong, S., Yuan, Z. (2019). Influence of the pyrolytic temperature and feedstock on the characteristics and naphthalene adsorption of crop straw-derived biochars. *BioResources*, 14, 2885-2902.

Hu, E., Tian, Y., Yang, Y., Dai, C., Li, M., Li, C., Shao, S. (2022). Pyrolysis behaviors of corn stover in new two-stage rotary kiln with baffle. *Journal of Analytical and Applied Pyrolysis*, 161.

Hu, Q., Shao, J., Yang, H., Yao, D., Wang, X., Chen, H. (2015). Effects of binders on the properties of bio-char pellets. *Applied Energy*, 157, 508-516.

Hu, Z., Ma, X., Chen, C. (2012). A study on experimental characteristic of microwave-assisted pyrolysis of microalgae. *Bioresource Technology*, 107, 487-493.

Huang, Y., Cheng, P., Chiueh, P., Lo, S. (2017). Leucaena biochar produced by microwave torrefaction: fuel properties and energy efficiency. *Applied Energy*.

Huang, Y., Chiueh, P., Shih, C., Lo, S., Sun, L., Zhong, Y., Qiu, C. (2015). Microwave pyrolysis of rice straw to produce biochar as an adsorbent for CO_2 capture. *Energy*, 84, 75-82.

Huang, Y., Kuan, W., Lo, S., Lin, C. (2008). Total recovery of resources and energy from rice straw using microwave-induced pyrolysis. *Bioresource Technology*, 99, 8252–8258.

Idris, R., Chong, W., Ali, A., Idris, S., Hasan, M., Ani, F., Chong, C. (2021). Phenol-rich bio-oil derivation via microwave-induced fast pyrolysis of oil palm empty fruit bunch with activated carbon. *Environmental Technology & Innovation*, 21, 101291.

Intani, K., Latif, S., Cao, Z., Müller, J. (2018). Characterisation of biochar from maize residues produced in a self-purging pyrolysis reactor. *Bioresource Technology*, 265, 224-235.

Jiang, D., Qin, J., Zhou, X., Li, Q., Yi, D., Wang, B. (2022a). Improvement of thermal insulation and compressive performance of Al_2O_3–SiO_2 aerogel by doping carbon nanotubes. *Ceramics International*, 48, 16290-16299.

Jiang, Y., Dai, M., Yang, F., Ali, I., Naz, I., Peng, C. (2022b). Remediation of chromium (VI) from groundwater by metal-based biochar under anaerobic conditions. *Water*, 14.

Ke, C., Zhang, Y., Gao, Y., Pan, Y., Li, B., Wang, Y., Ruan, R. (2019). Syngas production from microwave-assisted air gasification of biomass: part 1 model development. *Renewable Energy*, 140, 772-778.

Khan, T., Mukhlif, A., Khan, E. (2017). Uptake of Cu^{2+} and Zn^{2+} from simulated wastewater using muskmelon peel biochar: isotherm and kinetic studies. *Egyptian Journal of Basic and Applied Sciences*, 4, 236-248.

Kim, Y., Yun, B., Nam, J., Choi, J., Wi, S., Kim, S. (2021). Evaluation of thermal properties of phase change material-integrated artificial stone according to biochar loading content. *Construction and Building Materials*, 305, 124682.

Koltowski, M., Charmas, B., Skubiszewska-Zieba, J., Oleszczuk, P. (2017). Effect of biochar activation by different methods on toxicity of soil contaminated by industrial activity. *Ecotoxicology and Environmental Safety*, 136, 119-125.

Koltowski, M., Hilber, I., Bucheli, T., Oleszczuk, P. (2016). Effect of steam activated biochar application to industrially contaminated soils on bioavailability of polycyclic aromatic hydrocarbons and ecotoxicity of soils. *Science of the Total Environment*, 567, 1023-1031.

Kumar, N., Grekov, D., Pré, P., Alappat, B. (2020). Microwave mode of heating in the preparation of porous carbon materials for adsorption and energy storage applications: an overview. *Renewable & Sustainable Energy Reviews*, 124, 109743.

Kuo, H., Lin, Y., Chiang, C., Liu, S. (2021). FeN@N-doped graphitic biochars derived from hydrothermal-microwave pyrolysis of cellulose biomass for fuel cell catalysts. *Journal of Analytical and Applied Pyrolysis*, 153, 104991.

Lehmann, J. (2007). Bio-energy in the black. *Frontiers in Ecology and the Environment*, 5, 381-387.

Lehmann, J., Gaunt, J., Rondon, M. (2006). Biochar sequestration in terrestrial ecosystems: a review. *Mitigation and Adaptation Strategies for Global Change*, 11, 403-427.

Lei, H., Ren, S., Wang, L., Bu, Q., Julson, J., Holladay, J., Ruan, R. (2011). Microwave pyrolysis of distillers dried grain with solubles (DDGS) for biofuel production. *Bioresource Technology*, 102, 10, 6208-6213.

Lei, W., Yang, B., Sun, Y., Xiao, L., Tang, D., Chen, K., Sun, J., Ke, J., Zhuang, Y. (2021). Self-sacrificial template synthesis of heteroatom doped porous biochar for enhanced electrochemical energy storage. *Journal of Power Sources*, 488, 229455.

Leng, L., Xiong, Q., Yang, L., Li, H., Zhou, Y., Zhang, W., Jiang, S., Li, H., Huang, H. (2021). An overview on engineering the surface area and porosity of biochar. *Science of the Total Environment*, 763, 144204.

Li, A., Deng, H., Jiang, Y., Ye, C., Yu, B., Zhou, X., Ma, A. (2020). Superefficient removal of heavy metals from wastewater by Mg-loaded biochars: adsorption characteristics and removal mechanisms. *Langmuir*, 36, 9160-9174.

Li, D., Jiang, H. (2017a). The thermochemical conversion of non-lignocellulosic biomass to form biochar: a review on characterizations and mechanism elucidation. *Bioresource Technology*, 246, 57-68.

Li, H., Dong, X., Silva, E., Oliveira, L., Chen, Y., Ma, L. (2017b). Mechanisms of metal sorption by biochars: biochar characteristics and modifications. *Chemosphere*, 178, 466-478.

Li, H., Li, J., Fan, X., Li, X., Gao, X. (2019). Insights into the synergetic effect for co-pyrolysis of oil sands and biomass using microwave irradiation. *Fuel*, 239, 219-229.

Li, M., Liu, Q., Guo, L., Zhang, Y., Lou, Z., Wang, Y., Qian, G. (2013). Cu(II) removal from aqueous solution by spartina alterniflora derived biochar. *Bioresource Technology*, 141, 83-88.

Li, S., Lü, J., Zhang, T., Cao, Y., Li, J. (2017c). Relationship between biochars' porosity and adsorption of three neutral herbicides from water. *Water Science and Technology*, 75, 482-489.

Li, W., Dang, Q., Brown, R., Laird, D., Wright, M. (2017d). The impacts of biomass properties on pyrolysis yields, economic and environmental performance of the pyrolysis-bioenergy-biochar platform to carbon negative energy. *Bioresource Technology*, 241, 959-968.

Li, X., Peng, B., Liu, Q., Zhang, H. (2022). Microwave pyrolysis coupled with conventional pre-pyrolysis of the stalk for syngas and biochar. *Bioresource Technology*, 348, 126745.

Liew, R., Azwar, E., Yek, P., Lim, X., Cheng, C., Ng, J., Jusoh, A., Lam, W., Ibrahim, M., Ma, N., Lam, S. (2018a). Microwave pyrolysis with KOH/NaOH mixture activation: a new approach to produce micromesoporous activated carbon for textile dye adsorption. *Bioresource Technology*, 266, 1-10.

Liew, R., Nam, W., Chong, M., Phang, X., Su, M., Yek, P., Ma, N., Cheng, C., Chang, C., Lam, S. (2018b). Oil palm waste: an abundant and promising feedstock for microwave pyrolysis conversion into good quality biochar with potential multi-applications. *Process Safety and Environmental Protection*, 115, 57-69.

Lima, I., Boateng, A., Klasson, K. (2010). Physicochemical and adsorptive properties of fast-pyrolysis bio-chars and their steam activated counterparts. *Journal of Chemical Technology and Biotechnology*, 85 (11), 1515-1521.

Lin, J., Sun, S., Xu, D., Cui, C., Ma, R., Luo, J., Fang, L., Li, H. (2021). Microwave directional pyrolysis and heat transfer mechanisms based on multiphysics field stimulation: design porous biochar structure via controlling hotspots formation. *Chemical Engineering Journal*, 429, 13219.

Lin, K., Lai, N., Zeng, J., Chiang, H. (2020). Microwave-pyrolysis treatment of biosludge from a chemical industrial wastewater treatment plant for exploring product characteristics and potential energy recovery. *Energy*, 199, 117446.

Lin, Q., Chen, G., Liu, Y. (2012). Scale-up of microwave heating process for the production of bio-oil from sewage sludge. *Journal of Analytical and Applied Pyrolysis*, 94, 114-119.

Liu, C., Liu, X., He, Y., An, X., Fan, D., Wu, Z. (2021). Microwave-assisted catalytic pyrolysis of apple wood to produce biochar: co-pyrolysis behavior, pyrolysis kinetics analysis and evaluation of microbial carriers. *Bioresource Technology*, 320, 124345.

Liu, N., Charrua, A., Weng, C., Yuan, X., Ding, F. (2015a). Characterization of biochars derived from agriculture wastes and their adsorptive removal of atrazine from aqueous solution: a comparative study. *Bioresource Technology*, 198, 55-62.

Liu, W., Jiang, H., Yu, H. (2015b). Development of biochar-based functional materials: toward a sustainable platform carbon material. *Chemical Reviews*, 115, 12251-12285.

Liu, Z., Niu, W., Chu, H., Zhou, T., Niu, Z. (2018). Effect of the carbonization temperature on the properties of biochar produced from the pyrolysis of crop residues. *BioResources*, 13, 3429-3446.

Liu, Z., Sun, T., Yu, Y., Ke, P., Deng, Z., Lu, C., Huo, D., Ding, X. (2022). Near-real-time carbon emission accounting technology toward carbon neutrality. *Engineering*, 14, 44-51.

Lyu, H., Tang, J., Cui, M., Gao, B., Shen, B. (2020). Biochar/iron (BC/Fe) composites for soil and groundwater remediation: synthesis, applications, and mechanisms. *Chemosphere*, 246, 125609.

Mahari, W., Nam, W., Sonne, C., Peng, W., Phang, X., Liew, R., Yek, P., Lee, X., Wen, O., Show, P., Chen, W., Chang, J., Lam, S. (2020). Applying microwave vacuum pyrolysis to design moisture retention and pH neutralizing palm kernel shell biochar for mushroom production. *Bioresource Technology*, 312, 123572.

Mamaeva, A., Tahmasebi, A., Tian, L., Yu, J. (2016). Microwave-assisted catalytic pyrolysis of lignocellulosic biomass for production of phenolic-rich bio-oil. *Bioresource Technology*, 211, 382-389.

Manyà, J., Ortigosa, M., Laguarta, S., Manso, J. (2014). Experimental study on the effect of pyrolysis pressure, peak temperature, and particle size on the potential stability of vine shoots-derived biochar. *Fuel*, 133, 163-172.

Mao, X., Kang, Q., Liu, Y., Siyal, A., Ao, W., Ran, C., Fu, J., Deng, Z., Song, Y., Dai, J. (2019). Microwave-assisted pyrolysis of furfural residue in a continuously operated auger reactor: biochar characterization and analysis. *Energy*, 168, 573-584.

Martín, M., Sanz, A., Nozal, L., Castro, F., Alonso, R., Aguirre, J., González, S., Matía, M., Novella, J., Peinado, M., Vaquero, J. (2017). Microwave-assisted pyrolysis of mediterranean forest biomass waste: bioproduct characterization. *Journal of Analytical and Applied Pyrolysis*, 127, 278-285.

Masebinu, S., Akinlabi, E., Muzenda, E., Aboyade, A. (2019). A review of biochar properties and their roles in mitigating challenges with anaerobic digestion. *Renewable and Sustainable Energy Reviews*, 103, 291-307.

Masek, O., Budarin, V., Gronnow, M., Crombie, K., Brownsort, P., Fitzpatrick, E., Hurst, P. (2013). Microwave and slow pyrolysis biochar: comparison of physical and functional properties. *Journal of Analytical and Applied Pyrolysis*, 100, 41-48.

Matin, S., Chelgani, S. (2016). Estimation of coal gross calorific value based on various analyses by random forest method. *Fuel*, 177 (8), 274-278.

McGrath, T., Chan, W., Hajaligol, M. (2003). Low temperature mechanism for the formation of polycyclic aromatic hydrocarbons from the pyrolysis of cellulose. *Journal of Analytical and Applied Pyrolysis*, 66, 51-70.

McKendry, P. (2002). Energy production from biomass (part 2): conversion technologies. *Bioresource Technology*, 83, 47-54.

Meng, J., Liang, S., Tao, M., Liu, X., Brookes, P., Xu, J. (2018a). Chemical speciation and risk assessment of Cu and Zn in biochars derived from co-pyrolysis of pig manure with rice straw. *Chemosphere*, 200, 344-350.

Meng, R., Chen, T., Zhang, Y., Lu, W., Liu, Y., Lu, T., Liu, Y., Wang, H. (2018b). Development, modification, and application of low-cost and available biochar derived from corn straw for the removal of vanadium (v) from aqueous solution and real contaminated groundwater. *RSC Advances*, 8, 21480-21494.

Mobili, A., Giosuè, C., Bellezze, T., Revel, G., Tittarelli, F. (2021). Gasification char and used foundry sand as alternative fillers to graphene nanoplatelets for electrically conductive mortars with and without virgin/recycled carbon fibres. *Applied Sciences*, 11, 1-16.

Mohamed, B., Ellis, N., Kim, C., Bi, X., Chen, W. (2021). Engineered biochars from catalytic microwave pyrolysis for reducing heavy metals phytotoxicity and increasing plant growth. *Chemosphere*, 271, 129808.

Mohamed, B., Ellis, N., Kim, C., Bi, X., Emam, A. (2016). Engineered biochar from microwave-assisted catalytic pyrolysis of switchgrass for increasing waterholding capacity and fertility of sandy soil. *Science of the Total Environment*, 566-567, 387-397.

Mong, G., Chong, C., Ng, J., Chong, W., Lam, S., Ong, H., Ani, F. (2020). Microwave pyrolysis for valorisation of horse manure biowaste. *Energy Conversion and Management*, 220, 113074.

Motasemi, F., Afzal, M. (2013). A review on the microwave-assisted pyrolysis technique. *Renewable & Sustainable Energy Reviews*, 28, 317-330.

Mubarak, N., Sahu, J., Abdullah, E., Jayakumar, N. (2016). Plam oil empty fruit bunch based magnetic biochar composite comparison for synthesis by microwave-assisted and conventional heating. *Journal of Analytical and Applied Pyrolysis*, 120, 521-528.

Nabais, J., Nunes, P., Carrott, P., Carrott, M., Macías-García, A., Díaz-Díez, M. (2008). Production of activated carbons from coffee endocarp by CO_2 and steam activation. *Fuel Processing Technology*, 89 (3), 262-268.

Nair, V., Vinu, R. (2016). Peroxide-assisted microwave activation of pyrolysis char for adsorption of dyes from wastewater. *Bioresource Technology*, 216, 511-519.

Nakagawa, Y., Molina-Sabio, M., Rodríguez-Reinoso, F. (2007). Modification of the porous structure along the preparation of activated carbon monoliths with H_3PO_4 and $ZnCl_2$. *Microporous and Mesoporous Materials*, 103, 29-34.

Nam, W., Phang, X., Su, M., Liew, R., Ma, N., Bin Rosli, M., Lam, S. (2018). Production of bio-fertilizer from microwave vacuum pyrolysis of palm kernel shell for cultivation of oyster mushroom (pleurotus ostreatus). *Science of the Total Environment*, 624, 9-16.

Natalio, F., Corrales, T., Feldman, Y., Lew, B., Graber, E. (2020). Sustainable lightweight biochar-based composites with electromagnetic shielding properties. *ACS Omega*, 5, 32490-32497.

Nguyen, T., Pham, T., Thi, H., Nguyen, T., Nguyen, M., Dinh, T., Nguyen, M., Do, T., Phuong, T., Hoang, T. (2019). Synthesis of iron-modified biochar derived from rice straw and its application to arsenic removal. *Journal of Chemistry*, 5295610.

Nhuchhen, D., Afzal, M., Dreise, T., Salema, A. (2018). Characteristics of biochar and biooil produced from wood pellets pyrolysis using a bench scale fixed bed, microwave reactor. *Biomass Bioenergy*, 119, 293-303.

Niazi, N., Bibi, I., Shahid, M., Ok, Y., Shaheen, S., Rinklebe, J., Wang, H., Murtaza, B., Islam, E., Nawaz, M., Lüttge, A. (2018). Arsenic removal by Japanese oak wood biochar in aqueous solutions and well water: investigating arsenic fate using integrated spectroscopic and microscopic techniques. *Science of the Total Environment*, 621, 1642-1651.

Nzediegwu, C., Arshad, M., Ulah, A., Naeth, M., Chang, S. (2021). Fuel, thermal and surface properties of microwave-pyrolyzed biochars depend on feedstock type and pyrolysis temperature. *Bioresource Technology*, 320, 124282.

Oliveira, F., Patel, A., Jaisi, D., Adhikari, S., Lu, H., Khanal, S. (2017). Environmental application of biochar: current status and perspectives. *Bioresource Technology*, 246, 110-122.

Oliveira, G., Gevaerd, A., Mangrich, A., Marcolino, L., Bergamini, M. (2021). Biochar obtained from spent coffee grounds: evaluation of adsorption properties and its application in a voltammetric sensor for lead (II) ions. *Microchemical Journal*, 165, 106114.

Pariyar, P., Kumari, K., Jain, M., Jadhao, P. (2020). Evaluation of change in biochar properties derived from different feedstock and pyrolysis temperature for environmental and agricultural application. *Science of the Total Environment*, 713, 136433.

Park, J., Kim, Y., Jeon, J., Yun, B., Kang, Y., Kim, S. (2021). Analysis of biochar-mortar composite as a humidity control material to improve the building energy and hygrothermal performance. *Science of the Total Environment*, 775, 145552.

Parvez, A., Afzal, M., Jiang, P., Wu, T. (2020). Microwave-assisted biomass pyrolysis polygeneration process using a scaled-up reactor: product characterization, thermodynamic assessment and bio-hydrogen production. *Biomass Bioenergy*, 139, 105651.

Peng, C., Luo, H., Kong, J. (2014). Advance in estimation and utilization of crop residues resources in China. *Journal of China Agricultural Resources and Regional Planning*, 35 (3), 14-20.

Plácido, J., Bustamante-López, S., Meissner, K., Kelly, D., Kelly, S. (2019). Microalgae biochar-derived carbon dots and their application in heavy metal sensing in aqueous systems. *Science of the Total Environment*, 656, 531-539.

Prauchner, M., Rodríguez-reinoso, F. (2012). Chemical versus physical activation of coconut shell: a comparative study. *Microporous and Mesoporous Materials*, 152, 163-171.

Qambrani, N., Rahman, M., Wonc, S., Shim, S., Ra, C. (2017). Biochar properties and eco-friendly applications for climate change mitigation, waste management, and wastewater treatment: a review. *Renewable & Sustainable Energy Reviews*, 79, 255-273.

Qian, K., Kumar, A., Zhang, H., Bellmer, D., Huhnke, R. (2015). Recent advances in utilization of biochar. *Renewable & Sustainable Energy Reviews*, 42, 1055-1064.

Qiu, T., Li, C., Guang, M., Zhang, Y. (2023). Porous carbon material production from microwave-assisted pyrolysis of peanut shell. *Carbon Research*, 2, 45.

Quillope, J., Carpio, R., Gatdula, K., Detras, M., Doliente, S. (2021). Optimization of process parameters of self-purging microwave pyrolysis of corn cob for biochar production. *Heliyon*, 7, e08417.

Rajabi, H., Mosleh, M., Mandal, P., Lea-Langton, A., Sedighi, M. (2021). Sorption behaviour of xylene isomers on biochar from a range of feedstock. *Chemosphere*, 268.

Rajapaksha, A., Vithanage, M., Ahmad, M., Seo, D., Cho, J., Lee, S., Lee, S., Ok, Y. (2015). Enhanced sulfamethazine removal by steam-activated invasive plant-derived biochar. *Journal of Hazardous Materials*, 290, 43-50.

Rajasekhar, B., Vinu, R. (2018). Microwave-assisted co-pyrolysis of high ash Indian coal and rice husk: product characterization and evidence of interactions. *Fuel Processing Technology*, 178, 41-52.

Rangabhashiyam, S., Balasubramanian, P. (2019). The potential of lignocellulosic biomass precursors for biochar production: performance, mechanism and wastewater application: a review. *Industrial Crops and Products*, 128, 405-423.

Rawat, I., Kumar, R., Mutanda, T., Bux, F. (2022). Dual role of microalgae: phycoremediation of domestic wastewater and biomass production for sustainable biofuels production. *Applied Energy*, 88, 3411-3424.

Ren, S., Lei, H., Wang, L., Bu, Q., Chen, S., Wu, J., Julson, J., Ruan, R. (2012). Biofuel production and kinetics analysis for microwave pyrolysis of douglas fir sawdust pellet. *Journal of Analytical and Applied Pyrolysis*, 94, 163-169.

Roy, P., Dias, G. (2017). Prospects for pyrolysis technologies in the bioenergy sector: a review. *Renewable & Sustainable Energy Reviews*, 77, 59-69.

Salema, A., Afzal, M., Bennamoun, L. (2017). Pyrolysis of corn stalk biomass briquettes in a scaled-up microwave technology. *Bioresource Technology*, 233, 353-362.

Salema, A., Ani, F. (2011). Microwave induced pyrolysis of oil palm biomass. *Bioresource Technology*, 102, 3388-3395.

Schalk, R., Geoerg, D., Staubach, J., Raedle, M., Methner, F., Beuermann, T. (2017). Evaluation of a newly developed mid-infrared sensor for real-time monitoring of yeast fermentations. *Journal of Bioscience and Bioengineering*, 123 (5), 651-657.

Selvam, S., Paramasivan, B. (2022). Microwave assisted carbonization and activation of biochar for energy-environment nexus: A review. *Chemosphere*, 286, 131631.

Shaheen, S., Niazi, N., Hassan, N., Bibi, I., Wang, H., Tsang, D., Ok, Y., Bolan, N., Rinklebe, J. (2019). Wood-based biochar for the removal of potentially toxic elements in water and wastewater: a critical review. *International Materials Reviews*, 64, 216-247.

Shao, Y., Long, Y., Wang, H., Liu, D., Shen, D., Chen, T. (2018). Hydrochar derived from green waste by microwave hydrothermal carbonization. *Renewable Energy*, 135.

Singh, A., Sharma, R., Pant, D., Malaviya, P. (2021). Engineered algal biochar for contaminant remediation and electrochemical applications. *Science of the Total Environment*, 774, 145676.

Somparn, W., Panyoyai, N., Khamdaeng, T., Tippayawong, N., Tantikul, S., Wongsiriamnuay, T. (2020). Effect of process conditions on properties of biochar from agricultural residues. *IOP Conference Series: Earth and Environmental Science*, 463.

Spokas, K., Novak, J., Stewart, C., Cantrell, K., Uchimiya, M., Dusaire, M., Ro, K. (2011). Qualitative analysis of volatile organic compounds on biochar. *Chemosphere*, 85, 869-882.

Ströbel, R., Garche, J., Moseley, P., Jörissen, L., Wolf, G. (2006). Hydrogen storage by carbon materials. *Journal of Power Sources*, 159 (2), 781-801.

Subedi, R., Taupe, N., Pelissetti, S., Petruzzelli, L., Bertora, C., Leahy, J., Grignani, C. (2016). Greenhouse gas emissions and soil properties following amendment with manure-derived biochars: influence of pyrolysis temperature and feedstock type. *Journal of Environmental Management*, 166, 73-83.

Tan, K., Qin, Y., Du, T., Li, L., Zhang, L., Wang, J. (2021). Biochar from waste biomass as hygroscopic filler for pervious concrete to improve evaporative cooling performance. *Construction and Building Materials*, 287:123078.

Tan, Z., Lin, C., Ji, X., Rainey, T. (2017). Returning biochar to fields: a review. *Applied Soil Ecology*, 116, 1-11.

Tay, T., Ucar, S., Karagöz, S. (2009). Preparation and characterization of activated carbon from waste biomass. *Journal of Hazardous Materials*, 165 (1-3), 481-485.

Tian, Y., Zhao, L., Sun, L., Meng, H. (2011). Analysis and evalution on agricultural biomass resources. *Strategic Study of CAE*, 13 (2), 24-28.

Tomczyk, A., Sokołowska, Z., Boguta, P. (2020). Biochar physicochemical properties: pyrolysis temperature and feedstock kind effects. *Reviews in Environmental Science and Bio-Technology*, 19, 191-215.

Toor, S., Rosendahl, L., Rudolf, A. (2011). Hydrothermal liquefaction of biomass: a review of subcritical water technologies. *Energy*, 36, 2328-2342.

Trigo, C., Cox, L., Spokas, K. (2016). Influence of pyrolysis temperature and hardwood species on resulting biochar properties and their effect on azimsulfuron sorption as compared to other sorbents. *Science of the Total Environment*, 566-567, 1454-1464.

Tripathi, M., Bhatnagar, A., Mubarak, N., Sahu, J., Ganesan, P. (2020). RSM optimization of microwave pyrolysis parameters to produce OPS char with high yield and large BET surface area. *Fuel*, 277, 118184.

Vaibhav, D., Thallada, B. (2017). A comprehensive review on the pyrolysis of lignocellulosic biomass. *Renewable Energy*, 129, 695-716.

Wang, X., Li, X., Liu, G., He, Y., Chen, C., Liu, X., Li, G., Gu, Y., Zhao, Y. (2019). Mixed heavy metal removal from wastewater by using discarded mushroom-stick biochar: adsorption properties and mechanisms. *Environmental Science-Processes & Impacts*, 21, 584-592.

Wang, X., Zhong, Z., Jin, B. (2020a). Experimental evaluation of biomass medium-temperature gasification with rice straw as the fuel in a bubbling fluidized bed gasifier. *International Journal of Chemical Reactor Engineering*, 18, 20190147.

Wang, Y., Ke, L., Peng, Y., Yang, Q., Liu, Y., Wu, Q., Tang, Y., Zhu, H., Dai, L., Zeng, Z., Jiang, L., Ruan, R. (2020b). Ex-situ catalytic fast pyrolysis of soapstock for aromatic oil over microwave-driven HZSM-5@SiC ceramic foam. *Chemical Engineering Journal*, 402, 126239.

Wang, Y., Wu, Q., Duan, D., Ruan, R., Liu, Y., Dai, L., Zhou, Y., Zhao, Y., Zhang, S., Zeng, Z., Jiang, L., Yu, Z. (2018a). Ex-situ catalytic upgrading of vapors from fast microwave-assisted co-pyrolysis of chromolaena odorata and soybean soapstock. *Bioresource Technology*, 261, 306-312.

Wang, Y., Zeng, Z., Tian, X., Dai, L., Jiang, L., Zhang, S., Wu, Q., Wen, P., Fu, G., Liu, Y., Ruan, R. (2018b). Production of bio-oil from agricultural waste by using a continuous fast microwave pyrolysis system. *Bioresource Technology*, 269, 162-168.

Weber, K., Quicker, P. (2018). Properties of biochar. *Fuel*, 217, 240-261.

Wei, S., Zhu, M., Song, J., Peng, P. (2017). Comprehensive characterization of biochars produced from three major crop straws of China. *BioResources*, 12, 3316-3330.

Wiedemeier, D., Abiven, S., Hockaday, W., Keiluweit, M., Kleber, M., Masiello, C., McBeath, A., Nico, P., Pyle, L., Schneider, M., Smernik, R., Wiesenberg, G., Schmidt, M. (2015). Aromaticity and degree of aromatic condensation of char. *Organic Geochemistry*, 78, 135-143.

Wu, Z., Huang, X., Chen, R., Mao, X., Qi, X. (2022). The United States and China on the paths and policies to carbon neutrality. *Journal of Environmental Management*, 320, 115785.

Xiang, W., Zhang, X., Cao, C., Quan, G., Wang, M., Zimmerman, A., Gao, B. (2022). Microwave-assisted pyrolysis derived biochar for volatile organic compounds treatment: characteristics and adsorption performance. *Bioresource Technology*, 355, 127274.

Xu, X., Zhao, Y., Sima, J., Zhao, L., Masek, O., Cao, X. (2017). In-dispensable role of biochar-inherent mineral constituents in its environmental applications: a review. *Bioresource Technology*, 101, 887-899.

Yadav, P., Basu, A., Suryawanshi, A., Game, O., Ogale, S. (2016). Highly stable laser-scribed flexible planar microsupercapacitor using mushroom derived carbon electrodes. *Advanced Materials Interfaces*, 3 (11), 1600057.

Yap, M., Mubarak, N., Sahu, J., Abdullah, E. (2017). Microwave induced synthesis of magnetic biochar from agricultural biomass for removal of lead and cadmium from wastewater. *Journal of Industrial and Engineering Chemistry*, 45, 287-295.

Yavari, S., Malakahmad, A., Sapari, N., Yavari, S. (2017). Sorption properties optimization of agricultural wastes-derived biochars using response surface methodology. *Process Safety and Environmental Protection*, 109, 509-519.

Ye, Q., Li, Q., Li, X. (2022). Removal of heavy metals from wastewater using biochars: adsorption and mechanisms. *Environmental Pollutants & Bioavailability*, 34, 385-394.

Yek, P., Osman, M., Wong, C., Wong, C., Liew, R. (2020a). Microwave wet torrefaction: a catalytic process to convert waste palm shell into porous biochar. *Materials Science for Energy Technologies*, 3, 742-747.

Yek, P., Peng, W., Wong, C., Liew, R., Ho, Y., Mahari, W., Azwar, E., Yuan, T., Tabatabaei, M., Aghbashlo, M. (2020b). Engineered biochar via microwave CO_2 and steam pyrolysis to treat carcinogenic congo red dye. *Journal of Hazardous Materials*, 395, 122636.

Yin, Y., Gao, Y., Li, A. (2018). Self-activation of biochar from furfural residues by recycled pyrolysis gas. *Waste Management*, 77, 312-321.

Yu, J., Paterson, N., Blamey, J., Millan, M. (2017). Cellulose, xylan and lignin interactions during pyrolysis of lignocellulosic biomass. *Fuel*, 191, 140-149.

Yu, K., Chen, W., Sheen, H., Chang, J., Lin, C., Ong, H., Show, P., Ng, E., Ling, T. (2020). Production of microalgal biochar and reducing sugar using wet torrefaction with microwave-assisted heating and acid hydrolysis pretreatment. *Renewable Energy*, 156, 349-360.

Yuan, J., Xu, R. (2011a). The amelioration effects of low temperature biochar generated from nine crop residues on an acidic ultisol. *Soil Use and Management*, 27, 110-115.

Yuan, J., Xu, R., Zhang, H. (2011b). The forms of alkalis in the biochar produced from crop residues at different temperatures. *Bioresource Technology*, 102, 3488-3497.

Zhang, C., Wang, C., Cao, G., Chen, W., Ho, S. (2019a). Comparison and characterization of property variation of microalgal biomass with non-oxidative and oxidative torrefaction. *Fuel*, 246, 375-385.

Zhang, F., Nriagu, J., Itoh, H. (2005). Mercury removal from water using activated carbons derived from organic sewage sludge. *Water Research*, 39, 389-395.

Zhang, H., Voroney, R., Price, G. (2015a). Effects of temperature and processing conditions on biochar chemical properties and their influence on soil C and N transformations. *Soil Biology & Biochemistry*, 83, 19-28.

Zhang, J., Liu, J., Liu, R. (2015b). Effects of pyrolysis temperature and heating time on biochar obtained from the pyrolysis of straw and lignosulfonate. *Bioresource Technology*, 176, 288-291.

Zhang, J., Nolte, M., Shanks, B. (2014). Investigation of primary reactions and secondary effects from the pyrolysis of different celluloses. *ACS Sustainable Chemistry & Engineering*, 2, 2820-2830.

Zhang, J., Wang, Q. (2016). Sustainable mechanisms of biochar derived from brewers' spent grain and sewage sludge for ammonia-nitrogen capture. *Journal of Cleaner Production*, 112, 3927-3934.

Zhang, M., Hu, Y., Wang, H., Li, H., Han, X., Zeng, Y., Xu, C. (2021). A review of bio-oil upgrading by catalytic hydrotreatment: advances, challenges, and prospects. *Molecular Catalysis*, 504, 111438.

Zhang, S., Dong, Q., Zhang, L., Xiong, Y., Liu, X., Zhu, S. (2015c). Effects of water washing and torrefaction pretreatments on rice husk pyrolysis by microwave heating. *Bioresource Technology*, 193, 442448.

Zhang, T., Walawender, W., Fan, L., Fan, M., Daugaard, D., Brown, R. (2004). Preparation of activated carbon from forest and agricultural residues through CO_2 activation. *Chemical Engineering Journal*, 105, 53-59.

Zhang, W., Meng, J., Wang, J., Fan, S., Chen, W. (2013). Effect of biochar on root morphological and physiological characteristics and yield in rice. *Acta Agronomica Sinica*, 39, 1445-1451.

Zhang, X., Yang, X., Yuan, X., Tian, S., Wang, X., Zhang, H., Han, L. (2022a). Effect of pyrolysis temperature on composition, carbon fraction and abiotic stability of straw biochars: correlation and quantitative analysis. *Carbon Research*, 1, 17.

Zhang, X., Zhang, P., Yuan, X., Li, Y., Han, L. (2020a). Effect of pyrolysis temperature and correlation analysis on the yield and physicochemical properties of crop residue biochar. *Bioresource Technology*, 296.

Zhang, Y., Chen, P., Liu, S., Peng, P., Min, M., Cheng, Y., Anderson, E., Zhou, N., Fan, L., Liu, C., Chen, G., Liu, Y., Lei, H., Li, B., Ruan, R. (2017). Effects of feedstock characteristics on microwave-assisted pyrolysis-a review. *Bioresource Technology*, 230, 143-145.

Zhang, Y., Cui, Y., Liu, S., Fan, L., Zhou, N., Peng, P., Wang, Y., Guo, F., Min, M., Cheng, Y., Liu, Y., Lei, H., Chen, P., Li, B., Ruan, R. (2020b). Fast microwave-assisted pyrolysis of wastes for biofuels production-a review. *Bioresource Technology*, 297, 122480.

Zhang, Y., He, M., Wang, L., Yan, J., Ma, B., Zhu, X., Ok, Y., Mechtcherine, V., Tsang, D. (2022b). Biochar as construction materials for achieving carbon neutrality. *Biochar*, 4, 59.

Zhang, Y., Ke, C., Gao, Y., Liu, S., Pan, Y., Zhou, N., Wang, Y., Fan, L., Peng, P., Li, B., Ruan, R. (2019b). Syngas production from microwave-assisted air gasification of biomass: part 2 model validation. *Renewable Energy*, 140, 625-632.

Zhang, Z., Zhu, Z., Shen, B., Liu, L. (2019c). Insights into biochar and hydrochar production and applications: a review. *Energy*, 171, 581-598.

Zhao, B., O'Connor, D., Zhang, J., Peng, T., Shen, Z., Tsang, D., Hou, D. (2018a). Effect of pyrolysis temperature, heating rate, and residence time on rapeseed stem derived biochar. *Journal of Cleaner Production*, 174, 977-987.

Zhao, C., Huang, Y., Zhao, C., Shao, X., Zhu, Z. (2018b). Rose-derived 3D carbon nanosheets for high cyclability and extended voltage supercapacitors. *Electrochimica Acta*, 291, 287-296.

Zhao, X., Song, Z., Liu, H., Li, Z., Li, L., Ma, C. (2010). Microwave pyrolysis of corn stalk bale: a promising method for direct utilization of large-sized biomass and syngas production. *Journal of Analytical and Applied Pyrolysis*, 89, 87-94.

Zheng, N., Lee, M., Lin, Y. (2020). Co-processing textile sludge and lignocellulose biowaste for biofuel production through microwave-assisted wet torrefaction. *Journal of Cleaner Production*, 268, 122200.

Zhou, C., Zhang, Y., Liu, Y., Deng, Z., Li, X., Wang, L., Dai, J., Song, Y., Jiang, Z., Qu, J., Siyal, A. (2021). Co-pyrolysis of textile dyeing sludge and red wood waste in a continuously operated auger reactor under microwave irradiation. *Energy*, 218, 119398.

Zhou, J., Liu, S., Zhou, N., Fan, L., Zhang, Y., Peng, P., Anderson, E., Ding, K., Wang, Y., Liu, Y., Chen, P., Ruan, R. (2018). Development and application of a continuous fast microwave pyrolysis system for sewage sludge utilization. *Bioresource Technology*, 256, 295-301.

Zhou, J., Yan, P., Zhang, W., Zheng, F., Cheng, X., Chen, W. (2017). Effect of biochar on root morphogenesis and anatomical structure of rice cultivated in cold region of northeast China. *Acta Agronomica Sinica*, 43, 72-81.

Zhou, N., Zhou, J., Dai, L., Guo, F., Wang, Y., Li, H., Deng, W., Lei, H., Chen, P., Liu, Y., Ruan, R. (2020). Syngas production from biomass pyrolysis in a continuous microwave assisted pyrolysis system. *Bioresource Technology*, 314, 123756.

Zhou, R., Lei, H., Julson, J. (2013). The effects of pyrolytic conditions on microwave pyrolysis of prairie cordgrass and kinetics. *Journal of Analytical and Applied Pyrolysis*, 101, 172-176.

Zhu, L., Lei, H., Wang, L., Yadavalli, G., Zhang, X., Wei, Y., Liu, Y., Yan, D., Chen, S., Ahring, B. (2015). Biochar of corn stover: Microwave-assisted pyrolysis condition induced changes in surface functional groups and characteristics. *Journal of Analytical and Applied Pyrolysis*, 115, 149-156.

Zhu, X., Li, C., Li, J., Xie, B., Lv, J., Li, Y. (2018). Thermal treatment of biochar in the air/nitrogen atmosphere for developed mesoporosity and enhanced adsorption to tetracycline. *Bioresource Technology*, 263, 475-482.

Zornoza, R., Moreno-Barriga, E., Acosta, J., Muñoz, M., Faz, A. (2016). Stability, nutrient availability and hydrophobicity of biochars derived from manure, crop residues, and municipal solid waste for their use as soil amendments. *Chemosphere*, 144, 122-130.

Zou, J., Qian, W., Li, Y., Yu, Q., Yu, Y., Chen, S., Qu, F., Gao, Y., Lu, L. (2021). Multilayer activated biochar/UiO-66-NH_2 film as intelligent sensing platform for ultra-sensitive electrochemical detection of Pb^{2+} and Hg^{2+}. *Applied Surface Science*, 569, 151006.

Chapter 7

Microwave Irradiation for the Gasification of Organic Wastes

Wenming Fu
Wenke Zhao
Yaning Zhang*
and Bingxi Li
School of Energy Science and Engineering, Harbin Institute of Technology, Harbin, China

Abstract

Syngas produced from biomass is a carbon-neutral energy source, making it crucial in efforts to reduce carbon emissions. Beyond its energy-generating capabilities, syngas components are also useful in methane synthesis and the production of various chemicals, including ammonia, methanol, and ethylene. Microwave irradiation heating, which is known for its rapid heating rates, precise control, and enhanced thermal efficiency, is a highly effective heating technique for reducing tar formation during the gasification process. This chapter introduces the microwave irradiation gasification of biomass and plastics for syngas production. It discusses in detail the effects of gasification temperature, retention time, equivalence ratio, and silicon carbide load on syngas production and its components. Moreover, the simulation of microwave irradiation for gasification is introduced as well.

* Corresponding Author's Email: ynzhang@hit.edu.cn.

In: Characteristics and Applications of Microwave Irradiation
Editors: Yaning Zhang and Wenke Zhao
ISBN: 979-8-89530-170-8

Keywords: microwave irradiation, organic wastes, biomass, plastics, gasification

1. Introduction of Syngas

Syngas, a mixture primarily composed of carbon monoxide and hydrogen, has versatile applications ranging from power generation to chemical synthesis (Toledo et al. 2023; Solarte-Toro et al. 2018). Its properties vary depending on composition, typically being lighter than air, colorless, and odorless due to its high hydrogen and carbon monoxide content. The combustion of syngas releases a substantial amount of heat, and in an oxygen-rich environment, it results in the generation of carbon dioxide and water vapor, demonstrated by the following chemical reactions: $2CO + O_2 \rightarrow 2CO_2$, and $2H_2 + O_2 \rightarrow 2H_2O$. Carbon monoxide in syngas can also undergo the Water Gas Shift reaction with water, producing more hydrogen and carbon dioxide ($CO + H_2O \rightarrow CO_2 + H_2$) (Kryca et al. 2018). The higher heating value (HHV) of syngas varies based on its specific composition, with hydrogen having an HHV of 18.8 MJ/Nm^3 and carbon monoxide 12.64 MJ/Nm^3. In addition to providing energy, syngas components are useful for methane synthesis and the production of various chemicals such as ammonia, methanol, and ethylene ($CO + 3H_2 \rightarrow CH_4 + H_2O$, $N_2 + 3H_2 \rightarrow 2NH_3$, $CO + 2H_2 \rightarrow CH_3OH$, $2CO + 2H_2 \rightarrow C_2H_4 + H_2O$) (Makwana et al. 2023; Sepahi et al. 2023). The industrial significance of syngas lies in its diverse chemical applications, determined by its specific composition and component ratios.

Syngas is primarily derived from the gasification of biomass, plastics, and other carbonaceous materials (Gomaa et al. 2020). The gasification process occurs in a gasifier, a high-temperature reactor where the carbonaceous feedstock undergoes several chemical reactions with controlled amounts of oxygen, steam, or carbon dioxide (Fu et al. 2023; Susastriawan et al. 2019). The gasification process typically involves drying, pyrolysis, and gasification stages (Nunes, 2022). However, gasification is often accompanied by tar formation, leading to serious issues in downstream equipment such as coolers, compressors and gas turbines (Li et al. 2021; Zhang et al. 2018). Tar yield during gasification is influenced by several factors (Kong et al. 2023), including gasification temperature (Wang et al. 2024), equivalence ratio (Harb et al. 2020), and catalyst type (Galadima et al. 2022). Zhou et al. (2018) studied the gasification of lignin and cellulose extracted from Camellia sinensis branches, finding that the tar compounds concentration increased with temperature, with the major compounds in tar being lignin-derived phenolics. Microwave-assisted heating is proposed as an effective method for mitigating tar formation due to its unique advantages, including faster heating rates, better control, and higher thermal efficiency (Arpia et al. 2021; Huang et al. 2020).

Recent studies have investigated the microwave-assisted gasification of carbon-rich materials. Zhang et al. (2023) studied the microwave-assisted air gasification of corn stover for syngas production in a fixed bed, finding a syngas yield of 38.90-76.14 wt.% and LHV of 4.95-6.34 MJ/m^3 at gasification temperatures of 500-700 °C, retention time of 10-30 minutes, and an air flow rate of 200-600 ml/min. Fu et al. (2023) explored microwave-assisted gasification for rice husk in a fluidized bed reactor. They found that operating at a gasification temperature of 900 °C resulted in optimal outcomes, including the highest syngas yield (78.2 wt.%), elevated heating value (6.3 MJ/Nm^3), and enhanced cold gas efficiency (81.8 %). Furthermore, microwave-assisted heating had a beneficial influence on tar reduction, as evidenced by a lower tar content (2.1 wt.%) compared to 3.2 wt.% observed during catalytic electrical air gasification of rice husk at 950 °C. In a separate study, Ellison et al. (2022) investigated the CO_2 gasification process of four coal types using both microwave-assisted and electrical heating methods. They found superior outcomes with microwave-assisted heating across all coal types, including higher syngas yields, improved carbon conversion efficiencies, and enhanced cold gas efficiencies. Additionally, certain coal varieties showed lower tar yields under microwave-assisted heating conditions. Overall, microwave-assisted gasification demonstrates excellent performance in carbon conversion efficiency and tar reduction.

The microwave-assisted gasification process is affected by various parameters (Faizan et al. 2023; State et al. 2019). Many organic materials, such as biomass and plastics, have weak microwave absorption capabilities, requiring the addition of microwave absorbents to enhance absorption efficiency (Li et al. 2024). The type and amount of microwave absorbents significantly affects the heating rate, thereby influencing the gasification process. Gasification temperature affects the rupture of chemical bonds and ensuing chemical reactions, owing to the diverse heat-absorbing and heat-releasing reactions occurring during the gasification procedure (Wang et al. 2023). Equivalence ratio, representing the relationship between the actual combustible-to-oxidant proportion and the stoichiometric combustible-to-oxidant ratio, plays a significant role in determining syngas components and carbon conversion efficiency (Wang et al. 2023). Optimal retention time balances energy consumption and syngas yield, thereby maximizing economic benefits (Zhang et al. 2023).

Gasification, recognized as one of the key technologies for carbonaceous materials, holds promising prospects for resource recovery from solid waste. In this chapter, the microwave-assisted conversion of biomass and plastics for syngas production is examined. The discussion encompasses the influence of various parameters (gasification temperatures, silicon carbide loading, equivalence ratio, and retention time) on syngas yields and components, as well as the incorporation of microwave irradiation simulation.

2. Microwave Irradiation for Gasification of Biomass

Microwave irradiation for biomass gasification has been extensively studied and shown to be effective in reducing tar yield (Arpia et al. 2021; Huang et al. 2020). Zhang et al. (2023) investigated microwave-assisted technology for corn stover gasification using a fixed bed reactor under oxygen-deficient conditions. They detailed the effects of gasification temperature (500, 700, and 900 °C) and retention time (10, 20, and 30 min) on product distribution and syngas composition. Fu et al. (2023) investigated microwave irradiation for rice husk gasification using a fluidized bed reactor with air as the gasification agent. They examined the effects of gasification temperature (700, 750, 800, 850, and 900 °C), silicon carbide load (0, 10, 20, 30, and 40 g), and equivalence ratio (0.20, 0.25, 0.3, 0.35 and 0.40).

Table 1. Proximate and ultimate analyses of rice husk

Characteristics	Corn stover	Rice husk	Unit
Proximate analysis[a, d]			
Moisture content	8.67	6.9	wt. %
Volatile matter	74.29	60.9	wt. %
Ash content	7.78	17.0	wt. %
Fixed carbon	9.26	15.2	wt. %
Ultimate analysis[b]			
Carbon	47.01	37.4	wt. %
Hydrogen	6.06	4.4	wt. %
Nitrogen	0.40	0.2	wt. %
Sulfur	0.12	0.1	wt. %
Oxygen	37.89	34.1	wt. %

[a]: on as received basis, [b]: on dry basis, [c]: calculated by difference.

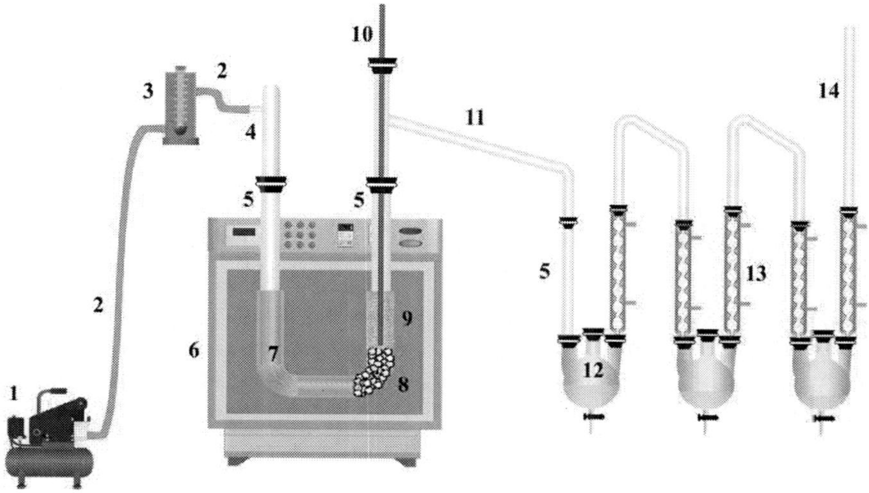

1. air compressor, 2. gas hose, 3. flow meter, 4. adapter, 5. connector, 6. microwave oven, 7. reactor, 8. ceramic wool, 9. corn stover, 10. thermocouple, 11. connector, 12. tar collection, 13. cooling, 14. Outlet.

Figure 1. Experimental system developed for the microwave-assisted gasification of corn stover.

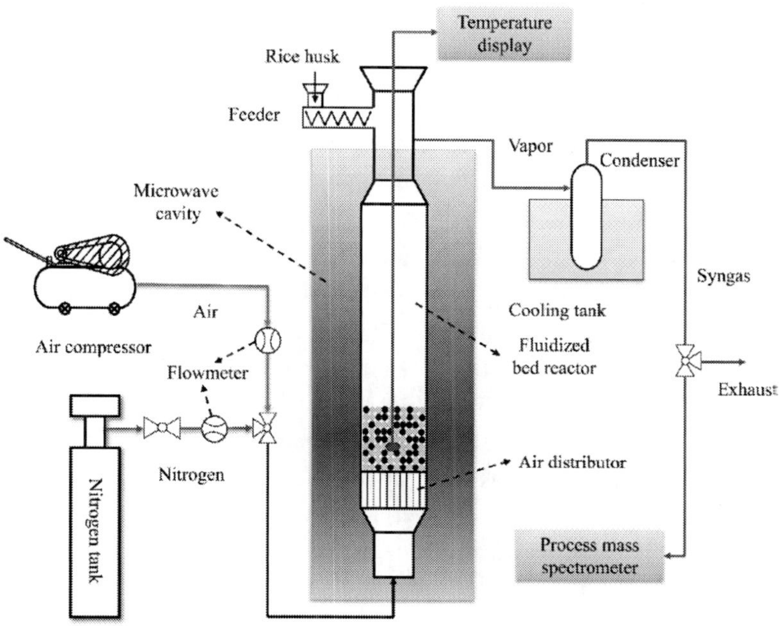

Figure 2. Schematic diagram of microwave-assisted fluidized bed gasification system.

Table 1 presents the proximate and ultimate analyses of corn stover and rice husk. Figure 1 illustrates the experimental setup for microwave-assisted gasification of corn stover. The system includes an air compressor (1), an airflow meter (3), an air adapter (4), a microwave oven (6), a U-type quartz reactor (7), a K-type thermocouple (10), a cooling line (13), a syngas outlet (14), three collection flasks (12), gas hoses (2), and quartz connectors (5, 11). Unlike setups for microwave-assisted pyrolysis (Liu et al. 2017; Fan et al. 2017), this system features a U-type quartz reactor (7) and an air supply line (1, 2, 3, and 4) (Liu et al. 2017; Fan et al. 2017). Figure 2 shows the schematic diagram of the microwave-assisted fluidized bed gasification system, which includes a nitrogen tank for adjusting equivalence ratio and fluidized velocity simultaneously.

2.1. Effect of Gasification Temperature

Figure 3 shows the product yields from the gasification of (a) corn stover and (b) rice husk at various gasification temperatures (Zhang et al. 2023; Fu et al. 2023). For corn stover, the syngas yield ranged from 51.91 to 69.56 wt.%, while the tar yield ranged between 20.07 and 22.27 wt.% at gasification temperatures of 500-700°C in a fixed bed reactor. For rice husk, the syngas yield ranged from 61.13 to 78.15 wt.%, while the tar yield varied from 2.5 to 3.4 wt.% at temperatures of 700 to 900°C in a fluidized-bed reactor. The findings indicated that elevated temperatures favored increased syngas yield but reduced tar and char yields. This can be explained by the promotion of biomass decomposition (Biomass → Char + Tar + H_2O + Syngas) and macromolecular breakdown, as well as the facilitation of endothermic reactions ($C + O_2 \rightarrow CO$, $C + H_2O \rightarrow CO + H_2$) (Gao et al. 2016; Kombe et al. 2022) at elevated temperatures. The tar yield from microwave-assisted gasification of rice husk at 900°C, reaching a minimum of 2.5 wt.%, was lower than that from electric gasification at 950°C, even with calcined dolomite as the catalyst (Zhang et al. 2018). Although calcined dolomite had a beneficial effect on tar cracking (Kook et al. 2016), the microwave gasification process still resulted in significantly lower tar content. This was attributed to the formation of hot spots caused by the uneven distribution of temperature within the fluidized bed reactor during the microwave-assisted heating process. These hot spots provided additional heat for further tar decomposition, significantly reducing tar production (Li et al. 2021).

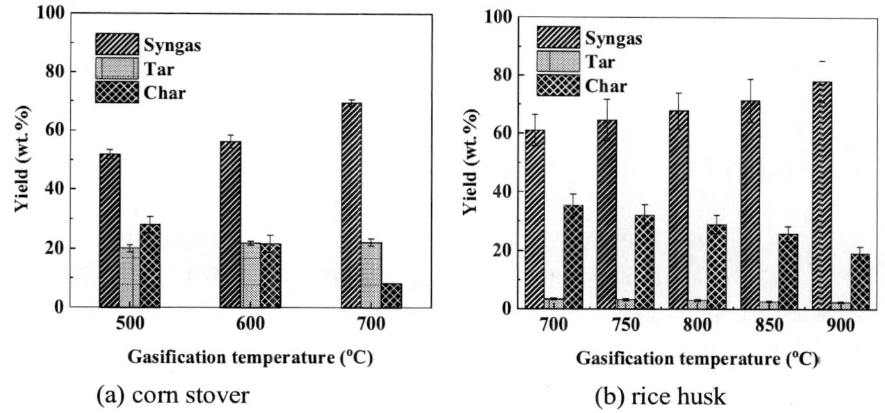

Figure 3. Product yields obtained from gasification of biomass at different gasification temperatures.

Table 2. Syngas components obtained from gasification of (a) corn stover and (b) rice husk at different gasification temperatures

Gasification temperature	H_2 vol.%	CO vol.%	CO_2 vol.%	N_2 vol.%	CH_4 vol.%	C_2H_4 vol.%	C_2H_6 vol.%	C_3H_6 vol.%	C_3H_8 vol.%
CS									
500 °C	6.40	6.55	17.84	55.2	3.26	1.15	1.13	1.06	0.02
600 °C	7.83	6.55	19.51	54.92	3.78	1.21	1.19	1.41	0.02
700 °C	8.45	9.55	19.18	50.92	3.92	1.28	1.22	1.16	0.03
RH									
700 °C	7.23	9.12	23.54	55.43	3.11	0.81	0.76	/	/
750 °C	8.95	8.83	21.87	54.95	3.69	0.91	0.80	/	/
800 °C	10.95	10.58	19.89	52.67	4.21	0.85	0.86	/	/
850 °C	11.23	11.94	18.25	52.36	4.14	0.95	1.13	/	/
900 °C	11.89	12.38	17.11	51.87	4.58	1.17	0.98	/	/

Table 2 presents the syngas components obtained from the gasification of (a) corn stover and (b) rice husk at various temperatures. Besides N_2 (primarily from air), the main components of syngas were H_2, CO, and CO_2. For corn stover, H_2 concentration ranged from 6.40 to 8.45 vol.%, CO concentration from 6.55 to 9.55 vol.%, and CO_2 concentration from 17.84 to 19.51 vol.% at 500-700 °C. For rice husk, H_2 concentration ranged from 7.23 to 11.89 vol.%, CO concentration from 8.83 to 12.38 vol.%, and CO_2 concentration from 17.11 to 23.54 vol.% at 700-900 °C. The findings indicated that, for both rice husk and corn stover, higher temperatures resulted in higher H_2 concentrations. This was attributed to higher temperatures promoting heterogeneous water-gas shift reactions and methane reforming, thereby increasing H_2 production (Liu et al. 2018).

2.2. Effect of Silicon Carbide Load

Figure 4 shows the product outputs from the rice husk gasification process at different silicon carbide loads (0, 10, 20, 30, and 40 g), while maintaining a gasification temperature at 900 °C and an equivalence ratio of 0.35. The syngas yield ranged from 48.45 to 78.15 wt.%, the tar yield varied from 2.1 to 13.3 wt.%, and the char yield ranged from 19.3 to 38.3 wt.%. The syngas yield initially rose sharply before reaching a steady state, while the char yield initially dropped significantly before stabilizing. Conversely, the tar yield showed an abrupt initial decline followed by a more gradual change. The highest syngas yield (78.15 wt.%) was observed with a silicon carbide load of 30 g, while the lowest yield (48.45 wt.%) occurred without silicon carbide. Without silicon carbide, rice husk exhibited weak microwave absorption, preventing the attainment of the desired gasification temperature. Consequently, this led to the lowest syngas output and the highest char and tar production. With the addition of silicon carbide, microwave energy absorption increased, leading to higher heating rates. As the silicon carbide content increased from 10 to 30 g, heat distribution within the fluidized-bed reactor improved. This improvement led to higher mean temperatures and greater conversion of tar and char, resulting in increased syngas yields (AlNouss et al. 2020).

Table 3 shows the syngas components obtained from rice husk gasification at different silicon carbide loads (0, 10, 20, 30, and 40 g) while maintaining a gasification temperature of 900 °C and an equivalence ratio of 0.35. Besides N_2 (mainly from air), the main components of syngas were H_2, CO, and CO_2. The volume concentration of H_2 was in the range of 2.32-11.92 vol.%, CO was in the range of 10.35-12.97 vol.%, and CO_2 was in the range of 17.01-20.38 vol.% at silicon carbide loads of 0-40 g. The results indicated that increasing the silicon carbide load led to higher H_2 concentrations, attributed to the higher heating rates and increased mean temperatures facilitated by the introduction of more silicon carbides. The higher heating rates and increased mean temperatures promoted endothermic reactions such as biomass decomposition (Biomass → Char + Tar + H_2O + Syngas) and the water-gas shift reaction ($C + H_2O \rightarrow CO + H_2$) (Zha et al. 2022).

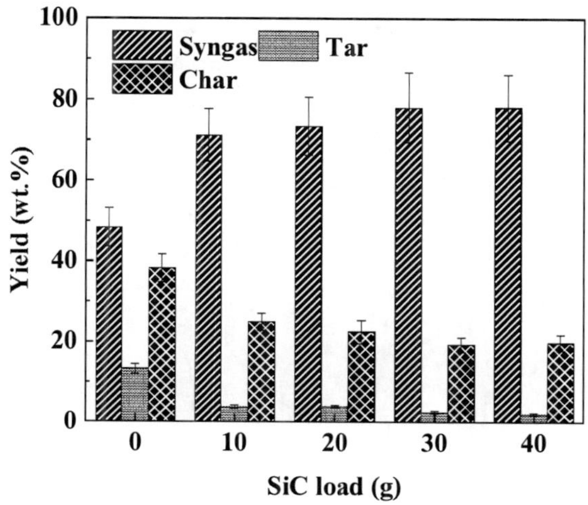

Figure 4. Product yields obtained from fluidized bed at different silicon carbide loads.

Table 3. Syngas components obtained from fluidized bed at different silicon carbide loads

SiC load	H_2 vol.%	CO vol.%	CO_2 vol.%	N_2 vol.%	CH_4 vol.%	C_2H_4 vol.%	C_2H_6 vol.%
0 g	2.32	19.38	19.23	57.79	1.28	0	0
10 g	8.23	10.35	20.38	55.57	3.84	0.79	0.83
20 g	9.63	11.38	19.24	53.62	4.13	1.08	0.92
30 g	11.89	12.38	17.11	51.87	4.58	1.17	0.98
40 g	11.92	12.97	17.01	51.73	4.38	1.10	0.89

2.3. Effect of Equivalence Ratio

Figure 5 shows the product yields from rice husk gasification at different equivalence ratios (0.20, 0.25, 0.30, 0.35 and 0.40) while maintaining a gasification temperature of 900°C and adding 30 g of silicon carbide in a fluidized bed. The syngas yield ranged from 69.25 to 78.15 wt.%, and the tar yield varied from 2.5 to 3.9 wt.%. The highest syngas yield (78.15 wt.%) and the lowest tar yield (2.5 wt.%) were achieved at an equivalence ratio of 0.35. In contrast, the lowest syngas yield (69.25 wt.%) was observed at an equivalence ratio of 0.20. As the equivalence ratio increased from 0.20 to 0.35, the oxygen content in the input gas rose while the nitrogen concentration declined, fostering char oxidation ($C + O_2 \rightarrow CO_2$, $C + 1/2\ O_2 \rightarrow CO$) and tar decomposition (Zhu et al. 2024; Han et al. 2022). Consequently, char and tar yields decreased while syngas yield increased. However, with a further increase in the equivalence ratio from 0.35 to 0.40, there were slight increases in the yields of tar and char. This can be attributed to higher CO_2 production and lower CO production, necessitating increased oxygen consumption for char oxidation, thereby elevating the char content.

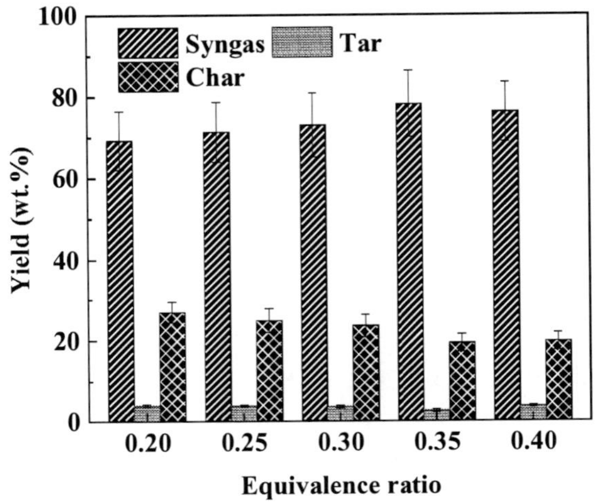

Figure 5. Product outputs obtained from the fluidized bed reactor under varying equivalence ratios.

Table 4. Syngas components obtained from the fluidized bed reactor under varying equivalence ratios

Equivalence ratio	H_2 vol.%	CO vol.%	CO_2 vol.%	N_2 vol.%	CH_4 vol.%	C_2H_4 vol.%	C_2H_6 vol.%
0.20	9.04	8.24	21.33	57.38	2.38	0.85	0.78
0.25	10.21	9.16	18.33	56.66	3.87	0.93	0.84
0.30	11.34	10.33	17.22	54.87	4.26	1.02	0.96
0.35	11.89	12.38	17.11	51.87	4.58	1.19	0.98
0.40	9.23	9.26	21.33	54.76	3.58	0.95	0.87

Table 4 shows the syngas components obtained from rice husk gasification at different equivalence ratios while maintaining a gasification temperature of 900 °C and a silicon carbide load of 30 g in a fluidized bed. Besides N_2 (mainly from air), the main components of syngas were H_2, CO, and CO_2. The volume concentration of H_2 was in the range of 9.04-11.89

vol.%, CO was in the range of 8.24-12.38 vol.%, and CO_2 was in the range of 17.11-21.33 vol.% at equivalence ratios of 0.20-0.40. The peak H_2 and CO concentrations were achieved at an equivalence ratio of 0.35, while the lowest levels of both were observed at an equivalence ratio of 0.20. As the equivalence ratio increased from 0.20 to 0.35, there was a decrease in CO_2 concentration and a simultaneous increase in the concentrations of combustible gases such as CO, H_2, and C_nH_m. This change is attributed to the higher O_2 content, which promoted char oxidation processes ($C + O_2 \rightarrow CO_2$, $C + 1/2\ O_2 \rightarrow CO$), leading to the release of significant heat (de Sales et al. 2017). This increased heat favored endothermic reactions ($C + CO_2 \rightarrow 2\ CO$, $C + H_2O \rightarrow CO + H_2$), leading to elevated concentrations of H_2 and CO and a decline in CO_2 content (Shahbaz et al. 2020). With the increase in the equivalence ratio from 0.35 to 0.40, there was an increase in CO_2 concentrations and a corresponding decline in the levels of combustible gases (CO, H_2, and C_nH_m). This change occurred because the higher O_2 concentration favored reactions such as $C + O_2 \rightarrow CO_2$, $CO + 1/2\ O_2 \rightarrow CO_2$, and $CH_4 + 2\ O_2 \rightarrow CO_2 + 2\ H_2O$, which are pathways for CO_2 production.

2.4. Effect of Retention Time

Figure 6 illustrates the product yields from corn stover gasification at different retention times (10, 20, 30 min) in a fixed bed reactor, with an airflow rate of 400 ml/min and a gasification temperature of 700 °C. The syngas yield ranged from 69.56 to 72.79 wt.%, while the tar yield varied from 20.61 to 22.27 wt.%. The highest syngas yield was obtained at a retention time of 30 min, while the lowest syngas yield was obtained at retention time of 10 min. Generally, extended retention time resulted in higher conversion of char and tar into syngas constituents, as exemplified by reactions such as $C + O_2 \rightarrow CO_2$ (de Sales et al. 2017), $Tar + O_2 \rightarrow H_2O + CO$ (Yao et al. 2018), $C_aH_bO_c + (a + b/2 - c)\ O_2 \rightarrow a\ CO + b/2\ H_2O$ (Gao et al. 2016), among others. Consequently, increasing retention time led to greater syngas production and reduced tar and char yields.

Table 5 shows the syngas components from corn stover gasification at different retention times. Besides N_2 (mainly from air), the main components of syngas were H_2, CO, and CO_2. The volume concentration of H_2 was in the range of 6.48-8.55 vol.%, CO was in the range of 6.30-9.55 vol.%, and CO_2

was in the range of 19.18-20.46 vol.% at retention time of 10-30 minutes. Extending retention time typically led to higher N_2 influx, a more pronounced decrease in combustible gases (H_2, CO, and C_nH_m), and increased CO_2 production. As a result, the concentrations of N_2 and CO_2 rose, while those of H_2, CO, and C_nH_m decreased due to the increased air introduction into the reactor. The highest hydrogen concentration was obtained at a retention time of 20 min, while the lowest was observed at 30 min. The highest CO concentration was obtained at a retention time of 10 min, while the lowest was observed at 30 min.

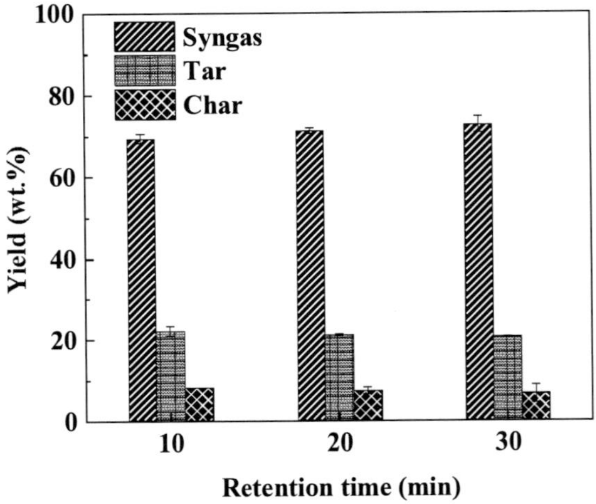

Figure 6. Product yields obtained from fluidized bed at different retention times.

Table 5. Syngas components obtained from fluidized bed at different retention times

Retention time	H_2 vol.%	CO vol.%	CO_2 vol.%	N_2 vol.%	CH_4 vol.%	C_2H_4 vol.%	C_2H_6 vol.%	C_3H_6 vol.%	C_3H_8 vol.%
10 min	8.45	9.55	19.18	50.92	3.92	1.28	1.22	1.16	0.03
20 min	8.55	7.59	20.01	51.11	3.69	1.16	1.12	1.07	0.03
30 min	6.48	6.30	20.46	52.24	3.08	1.16	1.12	1.07	0.02

3. Microwave Irradiation for Gasification of Plastics

Plastics play a crucial role in modern society, providing unparalleled convenience in our daily lives (Kaing et al. 2023). However, global plastic production has been steadily increasing, reaching 400.3 million tons in 2022 (Plastics Europe, 2023). Despite their widespread use, many plastics have short lifespans and are disposable (Fu et al. 2021). Combined with their resistance to decomposition, this leads to significant accumulation of waste plastics, resulting in extensive environment pollution (Vuppaladadiyam et al. 2023). Fortunately, plastics are derived from petroleum, presenting a potential energy source (Plastics Europe, 2023). However, burning plastics can release harmful gases, including dioxins, which pose health risks (Guo et al. 2023). To address this issue, researchers have extensively studied plastic gasification as a means to recycle waste and reduce environmental pollution (Dogu et al. 2021). Unlike biomass gasification, plastic gasification yields syngas, liquid oil, and wax without producing char (Saebea et al. 2020). Common plastics, including polypropylene, polyethylene, polyethylene terephthalate, and polystyrene, are mainly composed of carbon, hydrogen, and oxygen atoms. They can be gasified to produce syngas primarily consisting of carbon monoxide, hydrogen, and various hydrocarbons (Lopez et al. 2018). In this section, microwave irradiation is introduced as a novel approach for plastic gasification, an approach that has received relatively little attention in the research community. The effects of key parameters such as gasification temperature, equivalence ratio, and microwave power on the gasification process are examined.

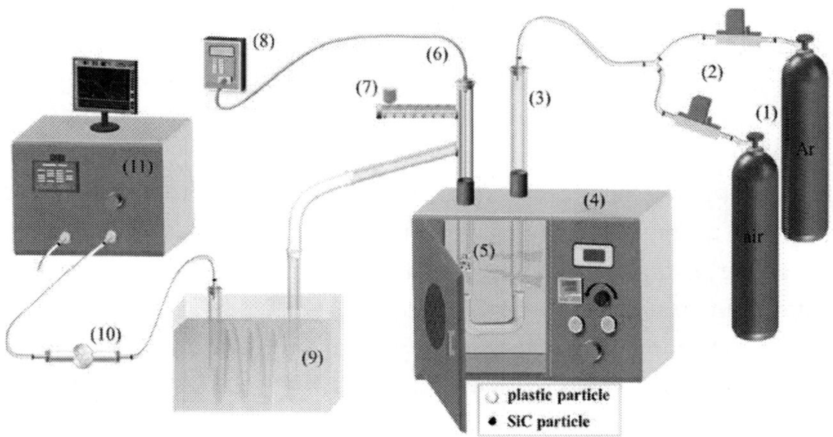

Figure 7. Experimental systematic diagram for the microwave assisted air gasification of PP. (1) gas cylinder, (2) gas flow meter, (3) connecting tube, (4) microwave assisted heating oven, (5) fluidized bed, (6) thermocouple, (7) screw feeder, (8) temperature monitor, (9) cooling tower, (10) syngas filtration system, (11) mass spectrometer.

Figure 7 illustrates the experimental setup for the microwave-assisted air gasification of PP particles in a fluidized bed. The setup consists of four primary components: (a) the gas intake system, which includes argon and air tanks connected to a gas mass flow meter for controlling flow rate and adjusting fluidization velocity and air-to-fuel ratio; (b) the microwave heating system, which features a microwave oven with a power range of 200-1000W; (c) the reaction system, comprising a fluidized bed reactor made of quartz glass and a K-type thermocouple for monitoring gasification temperature; and (d) the testing system, which includes a cooling tower and a gas mass spectrometer (LC-D200M, Tilon).

3.1. Effect of Gasification Temperature

Similar to biomass gasification, the gasification temperature in plastic gasification influences endothermic and exothermic reactions, affecting product distribution (Bai et al. 2020). Effects of syngas temperature on product yields and syngas components were systematically investigated by researchers (Cui et al. 2023).

Figure 8 shows the yields of syngas, liquid oil, and wax from the gasification of PP at various gasification temperatures (700, 750, 800, 850, and 900 °C), while keeping the equivalence ratio and microwave power constant at 0.15 and 800 W. The yields of syngas, liquid oil and wax varied in the ranges of 56.3-85.6 wt.%, 1-20.8 wt.%, and 13.4-22.9 wt.%, respectively. As the gasification temperature increased from 700 to 900 °C, the gas yield rose, while the yields of liquid oil and wax declined. This trend can be attributed to the intensified cracking reaction of PP (PP → gas + liquid oil + wax) with rising temperatures (Xiao et al. 2007). Higher temperatures also facilitated the breakdown of macromolecules (liquid oil/wax + O_2 → oxygen-containing organic compounds + CO + CO_2 + H_2) and the progression of endothermic reactions ($C + H_2O \rightarrow CO + H_2$) (Xiao et al. 2007; Kombe et al. 2022), leading to higher conversion of wax and oil. The effect of gasification temperature on gas yield was similar to that observed in biomass gasification. At gasification temperatures exceeding 850 °C, the liquid oil yield significantly decreased from 12.3 wt.% to 2 wt.%. Despite this decline, the wax yield remained relatively high at about 14.1 wt.%. The persistence in wax yield could be attributed to the presence of inert gases like N_2 and CO_2, which favor the production of wax over non-condensable gases (Li et al. 2017).

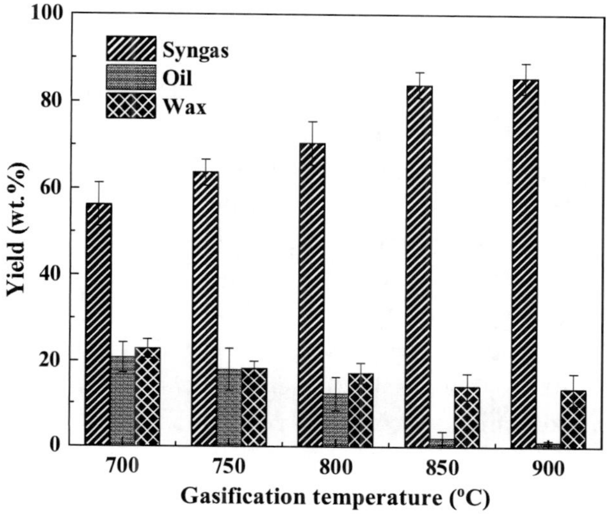

Figure 8. Product yields obtained from gasification of PP at different gasification temperatures.

Table 6. Syngas components obtained from gasification of PP at different gasification temperatures

Gasification temperature	H_2 Nm^3/kg_{PP}	CO Nm^3/kg_{PP}	CO_2 Nm^3/kg_{PP}	CH_4 Nm^3/kg_{PP}	C_2H_4 Nm^3/kg_{PP}	C_3H_5 Nm^3/kg_{PP}
700 °C	0.78	1.89	0.58	0.51	0.31	1.14
750 °C	0.80	2.23	0.70	0.52	0.25	1.06
800 °C	0.84	3.01	1.14	0.63	0.14	0.79
850 °C	1.01	3.70	0.93	0.74	0.17	1.13
900 °C	1.28	3.90	0.59	0.73	0.10	0.93

Table 6 shows the syngas components derived from the gasification of PP at different temperatures. The primary constituents of the syngas included H_2, CO, CO_2, CH_4, C_2H_4, and C_3H_6, with concentrations ranging from 0.78-1.28, 1.89-3.90, 0.58-1.14, 0.51-0.73, 0.10-0.31, and 0.79-1.14 Nm^3/kg_{PP}, respectively. The gasification mechanism of PP involves several key reactions (Xiao et al. 2007): (1) thermal degradation of PP particles at high temperatures in the absence of oxygen, yielding smaller hydrocarbon molecules such as methane, ethane, and propane (PP → gas + liquid oil + wax); (2) cracking reactions of these smaller hydrocarbon molecules, resulting in the formation of smaller species like hydrogen, carbon monoxide, and light hydrocarbons (liquid oil/wax + O_2 → oxygen-containing organic compounds + CO + CO_2 +

H_2); (3) subsequent reactions of these hydrocarbons with oxygen or steam, producing carbon dioxide, water, and additional hydrogen ($CH_4 + H_2O \leftrightarrow CO + 3H_2$, $C + H_2O \rightarrow CO + H_2$, $C + 1/2\ O_2 \rightarrow CO$, $C + O_2 \rightarrow CO_2$) (Shahbaz et al. 2020; Liu et al. 2018). Overall, higher gasification temperatures favored the production of CO, CH_4, and H_2. As the gasification temperature increased from 700 to 900°C, the yield of CO rose from 1.89 to 3.90 Nm^3/kg_{PP}, while H_2 and CH_4 yields saw slight increases. However, the CO_2 yield initially increased from 0.58 to 1.14 Nm^3/kg_{PP} and then declined to 0.59 Nm^3/kg_{PP} as the gasification temperature increased. This trend can be attributed to the promotion of endothermic reactions ($C + CO_2 \rightarrow 2CO$) at higher temperatures (Pinto et al. 2002). Conversely, higher gasification temperatures led to a reduction in hydrocarbons such as C_2H_4 and C_3H_6.

3.2. Effect of Equivalence Ratio

Figure 9 illustrates the yields of syngas, liquid oil, and wax resulting from the gasification of PP at various equivalence ratios (0.05, 0.10, 0.15, 0.20, and 0.25), with the gasification temperature and microwave power held constant at 850 °C and 800 W, respectively. The syngas, liquid oil, and wax yields ranged from 78.2 to 87.2 wt.%, 2 to 10 wt.%, and 9.6 to 11.8 wt.%, respectively. As the equivalence ratio increased from 0.05 to 0.25, the syngas yield increased, the oil yield decreased, and the wax yield initially rose before declining. At an equivalence ratio of 0.25, the syngas yield peaked, while the yields of oil and wax reached their lowest points. Increasing equivalence ratio from 0.05 to 0.25 supplied more oxidant during gasification, promoting the oxidation of oil and wax (Pinto et al. 2002), thereby reducing their production and boosting the syngas yield. As the equivalence ratio increased from 0.05 to 0.15, more oil underwent oxidation, contributing to increased gas yield, with less oxidation observed for wax. With further increases in the equivalence ratio to 0.25, oil oxidation remained relatively stable, while more wax underwent oxidation, leading to increased gas production. This phenomenon may be attributed to the increased primary pyrolysis rate of PP particles at higher equivalence ratios (0.05 to 0.15), resulting in shorter residence times for volatiles, thereby promoting wax production with reduced pyrolysis and oxidation. Additionally, oxygen in the air tended to react more readily with oils and small hydrocarbon molecules than with wax, resulting in a slight increase in wax yield. However, with further increases in the equivalence ratio

(0.15 to 0.25), more wax underwent oxidation by the air, consequently decreasing the wax yield as it was converted into syngas.

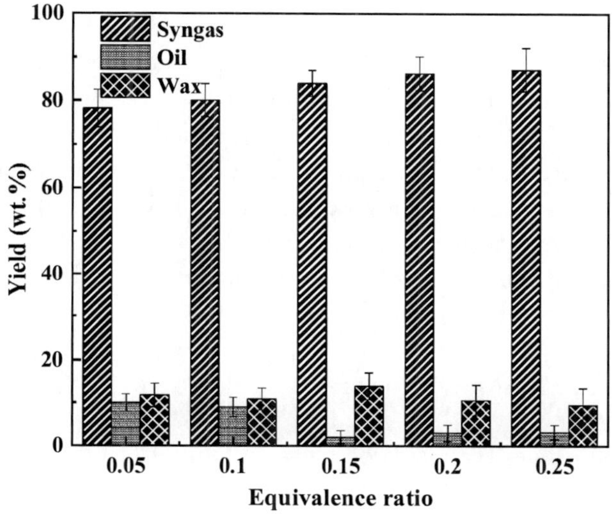

Figure 9. Product yields obtained from gasification of PP at different equivalence ratios.

Table 7. Syngas components obtained from gasification of PP at different equivalence ratios

Equivalence ratio	H_2 Nm^3/kg_{PP}	CO Nm^3/kg_{PP}	CO_2 Nm^3/kg_{PP}	CH_4 Nm^3/kg_{PP}	C_2H_4 Nm^3/kg_{PP}	C_3H_6 Nm^3/kg_{PP}
0.05	1.75	2.48	0.67	0.96	0.18	1.14
0.1	1.56	3.02	0.74	0.95	0.23	1.16
0.15	1.01	3.70	0.93	0.74	0.17	1.13
0.2	0.98	3.64	1.74	0.73	0.16	0.78
0.25	0.74	3.28	2.63	0.70	0.13	0.80

Table 7 delineates the syngas components observed across equivalence ratios. As the equivalence ratio increased from 0.05 to 0.15, the CO yield initially rose from 2.47 to 3.70 Nm^3/kg_{PP} before decreasing to 3.28 Nm^3/kg_{PP}. This trend is attributed to the increased presence of oxidant, promoting the partial oxidation of wax (Zhang et al. 2021) and facilitating endothermic reactions ($C + H_2O \rightarrow CO + H_2$, $C + CO_2 \rightarrow 2CO$) (Ding et al. 2019). As the equivalence ratio increased, more CO_2 was produced, while the proportions of combustible gas components (H_2, CO, and C_nH_m) decreased. This shift was a result of reactions such as $CO + 1/2\ O_2 \rightarrow CO_2$, $C_nH_m + (n+m/4)\ O_2 \rightarrow nCO_2 + m/2\ H_2O$, and $H_2 + 1/2\ O_2 \rightarrow H_2O$ being augmented to produce more CO_2 and H_2O. Yang et al. (2018) also noted similar trends, with reductions in H_2, CO, CH_4, and C2-C5 yields alongside an increase in CO_2 yield with rising equivalence ratio. Xiao et al. (2007) studied the air gasification of PP in a fluidized bed gasifier, analyzing the influence of equivalence ratio on gas composition. Their findings revealed an initial increase followed by a subsequent decrease in CO content, along with a continuous decrease in CO_2 content as the equivalence ratio rose, mirroring the observations in this study. These trends were attributed to the thermal cracking of hydrocarbons and enhanced partial combustion of other gas components.

3.3. Effect of Microwave Power

Figure 10 illustrates the yields of syngas, liquid oil, and wax resulting from the gasification of PP at different microwave powers (600, 700, 800, 900, and 1000 W), with a constant gasification temperature of 850 °C and an equivalence ratio of 0.15. The yields of syngas, liquid oil, and wax ranged from 68.9 to 84 wt.%, 2 to 10.3 wt.%, and 11.7 to 28.1 wt.%, respectively. Increasing the microwave power from 600 W to 1000 W resulted in an initial rise in syngas yield from 78.2 wt.% to 87.2 wt.% followed by a decrease to 68.9 wt.%. Concurrently, the oil yield decreased from 10.3 wt.% to 3.1 wt.%, while the wax yield increased from 11.7 wt.% to 28.1 wt.%. This phenomenon was attributed to secondary oil decomposition due to non-uniform temperature distribution caused by generated hot spots during microwave-assisted heating, leading to a significant decrease in oil yield (Ding et al. 2019). Additionally, microwave-induced plasma also contributed to chemical bond breakage and promoted tar degradation.

Table 8 presents the syngas components observed at different gasification temperatures. Increasing the microwave power resulted in enhanced production of H_2 and CH_4. Specifically, as the microwave power increased from 600 W to 1000 W, the yield of CO initially increased from 3.10 Nm^3/kg_{PP} to 3.70 Nm^3/kg_{PP} before decreasing to 3.20 Nm^3/kg_{PP}. Concurrently, the yield of H_2 rose from 0.81 Nm^3/kg_{PP} to 1.03 Nm^3/kg_{PP} initially and then decreased to 0.93 Nm^3/kg_{PP}, while the yield of CH_4 initially increased from 0.49 Nm^3/kg_{PP} to 0.80 Nm^3/kg_{PP} before decreasing to 0.79 Nm^3/kg_{PP}. Conversely, the yield of C_3H_6 decreased from 1.12 Nm^3/kg_{PP} to 0.78 Nm^3/kg_{PP}. At a microwave power of 800 W, the gas composition was as follows: CO (3.70 Nm^3/kg_{PP}), C_3H_6 (1.13 Nm^3/kg_{PP}), H_2 (1.01 Nm^3/kg_{PP}), CO_2 (0.93 Nm^3/kg_{PP}), CH_4 (0.74 Nm^3/kg_{PP}), and C_2H_4 (0.17 Nm^3/kg_{PP}).

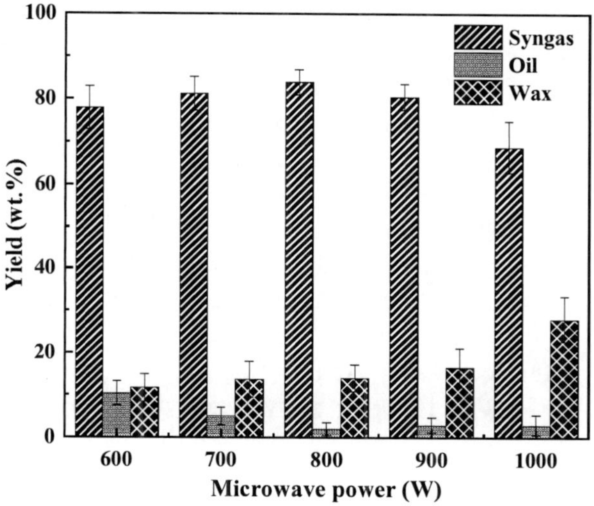

Figure 10. Product yields obtained from gasification of PP at different microwave powers.

Table 8. Syngas components obtained from gasification of PP at different microwave powers

Microwave power	H_2 Nm^3/kg_{PP}	CO Nm^3/kg_{PP}	CO_2 Nm^3/kg_{PP}	CH_4 Nm^3/kg_{PP}	C_2H_4 Nm^3/kg_{PP}	C_3H_6 Nm^3/kg_{PP}
600 W	0.81	3.10	0.90	0.49	0.15	1.13
700 W	0.90	3.29	0.89	0.60	0.20	1.17
800 W	1.01	3.70	0.93	0.74	0.17	1.13
900 W	1.03	3.62	0.85	0.80	0.15	0.97
1000 W	0.93	3.20	0.75	0.79	0.10	0.78

3.4. Effect of Catalyst

Abedin et al. (2023) investigated the microwave-assisted gasification of plastics (PET/PP/PS = 1:1:1) using air as the gasifying agent at a gasification temperature of 700°C, an air flow rate of 200 sccm, and a feedstock of 6 g (Plastics: SiC = 1:1). The results revealed gas product yields of 52 wt.% and tar product yields of 9 wt.%. The concentrations of hydrogen and carbon monoxide were 20 vol.% and 46 vol.%, respectively, and the total hydrogen yield was 4 mmol/$g_{plastic}$. These yields were relatively low, likely due to the use of air as the gasification agent and the inclusion of PET among the feedstocks.

Zhang et al. (2024) investigated the microwave-assisted CO_2 gasification of PE for hydrogen production using various metal oxides ($AlCoO_x$, $AlNiO_x$, $AlFeO_x$, $CaCoO_x$, $CaNiO_x$, $CaFeO_x$, $MgCoO_x$, $MgNiO_x$, $MgFeO_x$, $LaCoO_x$, $LaNiO_x$, and $LaFeO_x$) as catalysts. Figure 11 illustrates the carbon conversion and hydrogen yield from the gasification of PE using different catalysts. The hydrogen yield ranged from 78.5 to 431.8 mmol·$g^{-1}H_{plastic}$. The highest hydrogen yield was achieved with the introduction of $AlNiO_x$, reaching more than 90% of the theoretical yield of hydrogen. The carbon conversion efficiency ranged from 4.3% to 65.72%, with the highest value achieved using $AlCoO_x$. Microwave power, plastic-to-mass ratio, and CO_2 concentration significantly influenced the hydrogen yield. When microwave power was increased from 800 W to 1000 W, the hydrogen yield rose from 298 mmol·$g^{-1}H_{plastic}$ to 415 mmol·$g^{-1}H_{plastic}$. An increase in the plastic-to-mass ratio decreased the hydrogen yield, while an increase in CO_2 concentration reduced the carbon conversion efficiency.

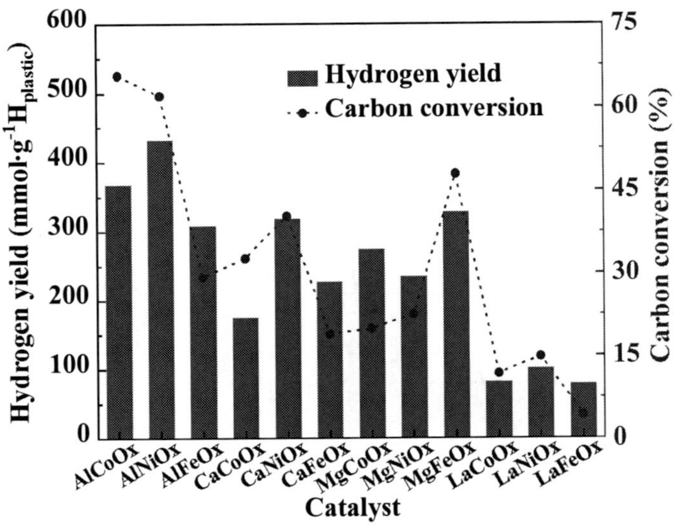

Figure 11. Catalyst effects on hydrogen yield and carbon conversion efficiency in microwave irradiation for gasification of PE.

4. Simulation of Microwave Irradiation for Gasification

Ke et al. (2019) developed a Computational Fluid Dynamics (CFD) framework to simulate microwave-assisted air gasification of biomass. This model incorporated an energy source term to capture the influence of microwave irradiation, using the Euler-Euler method within the Ansys Fluent software platform. The governing equations were solved using the standard k-ε model, and reaction rates between gas and solid phases were determined using the finite rate/eddy dissipation model. The transient dynamics of microwave heating were integrated into the gasification model via User-Defined Function (UDF) compilation. The biomass particles were assumed to be uniformly sized and spherical, and the incoming air was represented as an ideal gas mixture containing 21% O_2 and 79% N_2. The gas composition during the reaction included CO, CO_2, H_2, CH_4, C_2H_4, H_2O, and N_2, excluding gases such as C_3H_6. Sulfur and nitrogen content in the raw materials, along with related reactions, were excluded from consideration. Microwave power was exclusively absorbed by solid biomass particles, and its effect on gas components ignored. Furthermore, the reactor wall was assumed to be adiabatic.

Figure 12 illustrates the absorption and conversion of energy within each control unit. Both gas and solid phases coexist, each governed by separate equations. Mass and energy transfer occur between these phases, with transfer coefficients calculated using a two-fluid model based on the Euler-Euler method. The temperature of the solid phase increases with continuous microwave input, leading to successive water evaporation and volatile matter release. Concurrently, heat begins to transfer to the gas phase. As the temperature rises further, some volatiles and fixed carbon undergo combustion. During this oxidation process, chemical energy in the reactants is converted to heat energy. As combustion progresses, reduction reactions occur between components after the oxygen content is depleted, with some heat absorbed in this process. Chemical reactions include both heterogeneous reactions between gaseous and solid phases and homogeneous reactions among gaseous components. The respective chemical reaction rates are determined using a finite-rate chemistry model.

Biomass gasification encompasses intricate physical and chemical transformations, such as water evaporation, volatile matter release, combustion of volatile matter and fixed carbon, and reduction reactions involving gases and fixed carbon. Table 9 primarily illustrates the homogeneous reactions among gases and the heterogeneous reactions between char and gas.

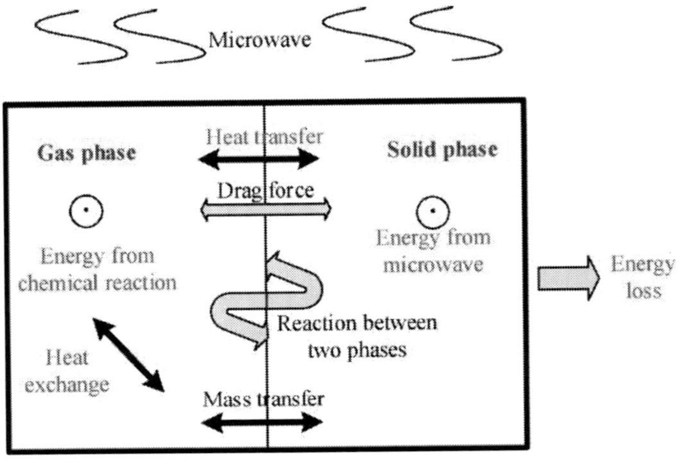

Figure 12. A control zone within the reactor.

Table 9. Kinetic parameters of chemical reactions

Reaction equation	Activation energy E (J/kg. mol)	Pre-exponential factor A	Reference
$CO + 0.5O_2 \rightarrow CO_2$	1.33×10^8	1.78×10^{14}	Xie et al. 2012
$H_2 + 0.5O_2 \rightarrow H_2O$	1.25×10^8	1.08×10^{13}	Xie et al. 2012
$CH_4 + 1.5O_2 \rightarrow CO + 2H_2O$	2.03×10^8	2.80×10^9	Xie et al. 2012
$C_2H_4 + 2O_2 \rightarrow 2CO + 2H_2O$	1.73×10^8	1.00×10^{12}	Zhou et al. 2009
$CO + 0.5H_2O \rightleftharpoons CO_2 + H_2$	3.046×10^8	3.43×10^9	Ke et al. 20019
	3.264×10^8	6.4×10^{12}	
$C_xH_yO_z + (x+0.5y-z)O_2 \rightarrow xCO + 0.5yH_2O$	0.802×10^8	9.17×10^6	Bustamante et al. 2005
$C(s) + 0.5O_2 \rightarrow CO$	1.79×10^8	1.00×10^9	Bustamante et al. 2004
$C(s) + CO_2 \rightarrow 2CO$	2.45×10^8	1.20×10^8	Bustamante et al. 2004
$C(s) + H_2O \rightarrow CO + H_2$	1.70×10^8	3.55×10^5	Bustamante et al. 2004

Zhang et al. (2019) verified the model using experimental results obtained from the gasification of rice husk under microwave irradiation in a fixed-bed reactor at temperatures ranging from 500 to 700 °C. Figure 13 presents the simulated values of volume fractions of H_2, CO, CO_2, N_2, CH_4, and C_2H_4 alongside the corresponding experimental values at 500-700 °C. Figure 14 illustrates the simulated and experimental HHVs at gasification temperatures ranging from 500 to 700 °C.

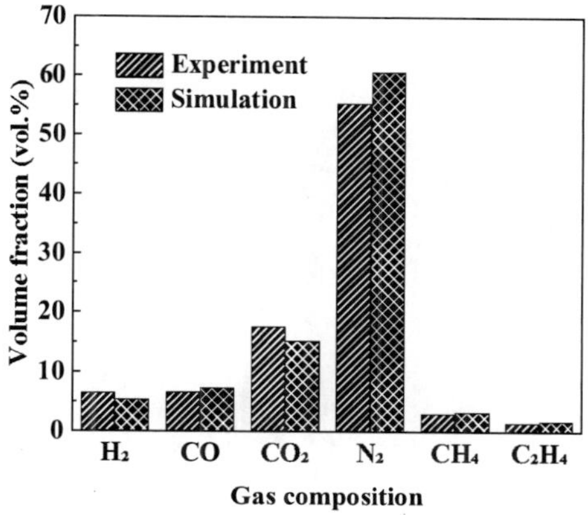

(a) 500 °C

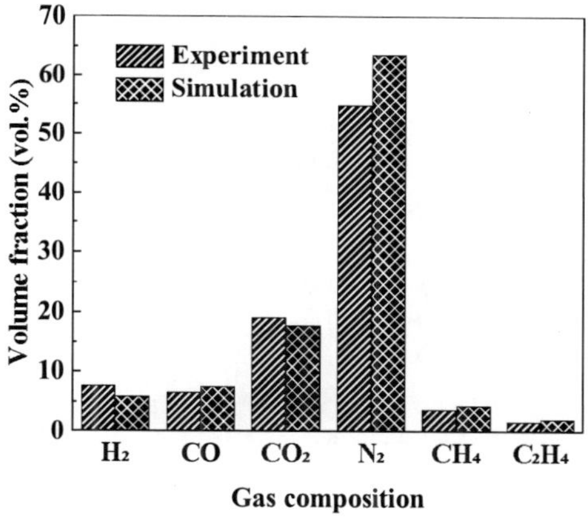

(b) 600 °C

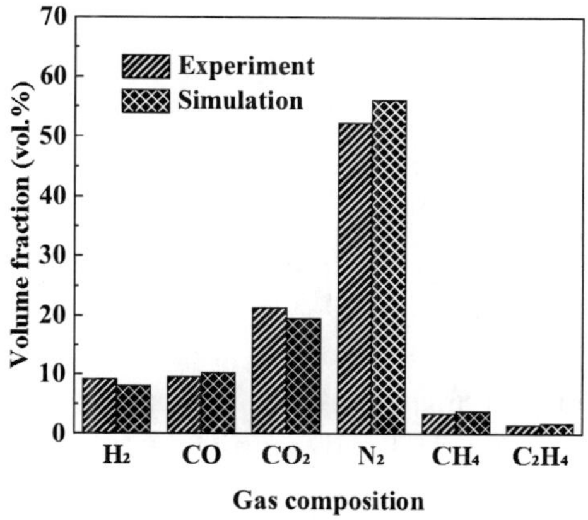

(c) 700 °C

Figure 13. Volume fractions of gas compositions obtained from experiment and simulation at different gasification temperatures.

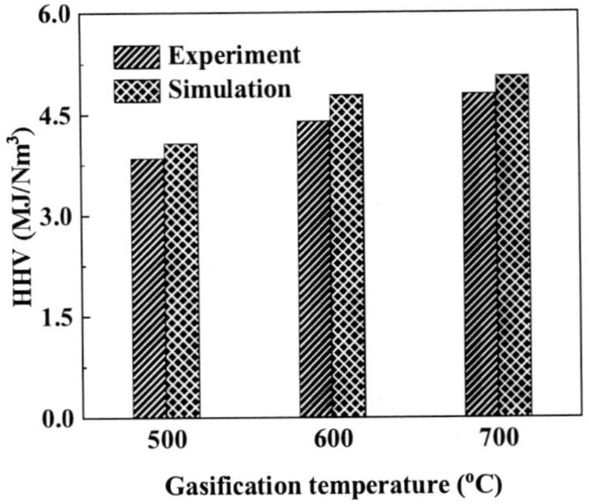

Figure 14. The HHVs of syngas obtained from simulation and experiment at gasification temperatures of 500-700 °C.

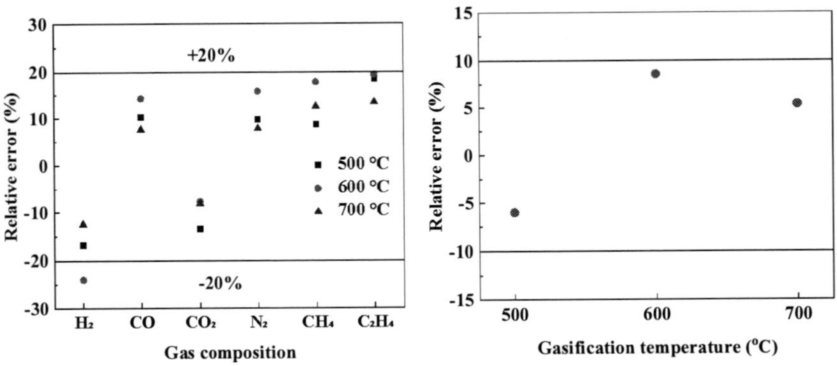

Figure 15. The discrepancies between simulated and experimental values at different gasification temperatures.

Figure 15 depicts the relative discrepancies between the simulated and experimental gas composition values at various gasification temperatures. The findings indicated that the majority (94.44%) of the relative errors, ranging from 7.84% to 19.34% (absolute values), were within the 20% threshold for both simulated and experimental syngas compositions (CO, CO_2, H_2, CH_4, C_2H_4, and N_2). Moreover, the relative errors for simulated and experimental syngas HHVs ranged from 3.44% to 8.62%.

Conclusion

This chapter introduces microwave irradiation for the gasification of plastics and biomass, exploring the effects of gasification temperature, silicon carbide load, equivalence ratio, and retention time on syngas components.

The silicon carbide load affected the heating rate and mean gasification temperature, while the equivalence ratio and retention time influenced the amount of oxygen introduced. Gasification temperature, in turn, affected the decomposition of feedstocks and the endothermic reactions during the gasification process. Microwave irradiation can enhance tar decomposition due to the generation of hot spots.

Microwave irradiation for plastic gasification showed promise for hydrogen production. Microwave power, plastic-to-mass ratio, catalyst, and gasification agent had a significant influence on the hydrogen yield. A yield of 415 mmol·$g^{-1}H_{plastic}$ was achieved from the CO_2 gasification of PE at a microwave power of 1000 W using $AlCoO_x$ as the catalyst.

The analysis of microwave irradiation simulations for gasification revealed that 94.44% of the relative errors were within 20% for both simulated and experimental syngas compositions and syngas HHV.

Acknowledgments

The following financial supports are acknowledged: Financial support was provided by the National Natural Science Foundation of China (52076049, 52476005), Heilongjiang Province "Double First-class" Discipline Collaborative Innovation Achievement Project (LJGXCG2023-080), Heilongjiang Provincial Key R&D Program (2023ZX02C05), and Heilongjiang Provincial Key R&D Program "Unveiling the Leader" Project (2023ZXJ02C04).

References

Abedin, A., Bai, X., Smith, M., Muley, P. (2023). Microwave-assisted co-gasification of mixed plastics and corn stover: A synergistic approach to produce clean hydrogen. *Energy Conversion and Management*, 280, 116774.

Arpia, A. A., Chen, W. H., Lam, S. S., Rousset, P. De Luna, M. D. G. (2021). Sustainable biofuel and bioenergy production from biomass waste residues using microwave-assisted heating: A comprehensive review. *Chemical Engineering Journal*, 403, 126233.

Bai, B., Wang, W., Jin, H. (2020). Experimental study on gasification performance of polypropylene (PP) plastics in supercritical water. *Energy*, 191, 116527.

Bustamante, F., Enick, R. M., Cugini, A. V., Killmeyer, R. P., Howard, B. H., Rothenberger, K. S., Morreale, B. D., Chattopadhyay, S., Shi, S. (2004). High-temperature kinetics of the homogeneous reverse water–gas shift reaction. *AIChE Journal*, 50 (5), 1028-1041.

Bustamante, F., Enick, R. M., Killmeyer, R. P., Howard, B. H., Rothenberger, K. S., Cugini, A. V., Morreale, B. D., Ciocco, M. V. (2005). Uncatalyzed and wall-catalyzed forward water–gas shift reaction kinetics. *AIChE journal*, 51 (5), 1440-1454.

Cui, Y., Shi, H., Zhang, Y., Cui, L., Zhao, W., Liu, W. (2023). Experimental investigation on air gasification of polypropylene particles in a microwave fluidized bed reactor. *Industrial & Engineering Chemistry Research*, 62 (42), 17306-17313.

De Sales, C. A. V. B., Maya, D. M. Y., Lora, E. E. S., Jaén, R. L., Reyes, A. M. M., González, A. M., Andrade, R. V., Martínez, J. D. (2017). Experimental study on biomass (eucalyptus spp.) gasification in a two-stage downdraft reactor by using mixtures of air, saturated steam and oxygen as gasifying agents. *Energy Conversion and Management*, 145, 314-323.

Ding, K., Liu, S., Huang, Y., Liu, S., Zhou, N., Peng, P., Wang, Y., Chen, P., Ruan, R. (2019). Catalytic microwave-assisted pyrolysis of plastic waste over NiO and HY for gasoline-range hydrocarbons production. *Energy Conversion and Management*, 196, 1316-1325.

Dogu, O., Pelucchi, M., Van de Vijver, R., Van Steenberge, P. H., D'hooge, D. R., Cuoci, A., Mehl, M., Frassoldati, A., Faravelli, T. Van Geem, K. M. (2021). The chemistry of chemical recycling of solid plastic waste via pyrolysis and gasification: State-of-the-art, challenges, and future directions. *Progress in Energy and Combustion Science*, 84, 100901.

Ellison, C., Abdelsayed, V., Smith, M., Shekhawat, D. (2022). Comparative evaluation of microwave and conventional gasification of different coal types: Experimental reaction studies. *Fuel*, 321, 124055.

Faizan, M., Song, H. (2023). Critical review on catalytic biomass gasification: State-of-Art progress, technical challenges, and perspectives in future development. *Journal of Cleaner Production*, 137224.

Fan, L., Zhang, Y., Liu, S., Zhou, N., Chen, P., Liu, Y., Wang, Y., Peng, P., Cheng, Y., Addy, M., Lei, H., Ruan, R. (2017). Ex-situ catalytic upgrading of vapors from microwave-assisted pyrolysis of low-density polyethylene with MgO. *Energy Conversion and Management*, 149, 432-441.

Fu, W., Dai, J., Zhang, Y., Guang, M., Liu, Y. Li, B. (2021). Heating performances of high density polyethylene (HDPE) plastic particles in a microwave chamber. *Sustainable Energy Technologies and Assessments*, 48, 101581.

Fu, W., Zhang, Y., Cui, L., Liu, H., Maqsood, T. (2023). Experimental microwave-assisted air gasification of biomass in fluidized bed reactor. *Bioresource Technology*, 369, 128378.

Galadima, A., Masudi, A. Muraza, O. (2022). Catalyst development for tar reduction in biomass gasification: Recent progress and the way forward. *Journal of Environmental Management*, 305, 114274

Gao, X., Zhang, Y., Li, B., Yu, X. (2016). Model development for biomass gasification in an entrained flow gasifier using intrinsic reaction rate submodel. *Energy Conversion and Management*, 108, 120-131.

Gomaa, M. R., Mustafa, R. J., Al-Dmour, N. (2020). Solar thermochemical conversion of carbonaceous materials into syngas by Co-Gasification. *Journal of Cleaner Production*, 248, 119185.

Guo, J., Bo, X., Xie, Y., Tang, L., Xu, J., Zhang, Z., Wan, R., Xu, H. Mi, Z. (2023). Health effects of future dioxins emission mitigation from Chinese municipal solid waste incinerators. *Journal of Environmental Management*, 345, 118805.

Han, S. W., Lee, J. J., Tokmurzin, D., Lee, S. H., Nam, J. Y., Park, S. J., Ra, H. W., Mun, T. Y., Yoon, S. J., Yoon, S. M., Moon, J. H., Lee, J. G., Kim, Y. M., Rhee, Y. W., Seo, M. W. (2022). Gasification characteristics of waste plastics (SRF) in a bubbling fluidized bed: Effects of temperature and equivalence ratio. *Energy*, 238, 121944.

Harb, R., Rivera-Tinoco, R., Nemer, M., Zeghondy, B., Bouallou, C. (2020). Towards synthetic fuels production from biomass gasification: Tar content at low temperatures. *Biomass & Bioenergy*, 137, 105540.

Huang, J., Xu, G., Liang, Y., Hu, G. Chang, P. (2020). Improving coal permeability using microwave heating technology—A review. *Fuel*, 266, 117022.

Kaing, V., Guo, Z., Sok, T., Kodikara, D., Breider, F. Yoshimura, C. (2023). Photodegradation of biodegradable plastics in aquatic environments: Current understanding and challenges. *Science of the Total Environment*, 168539.

Kombe, E. Y., Lang'at, N., Njogu, P., Malessa, R., Weber, C. T., Njoka, F., Krause, U. (2022). Process modeling and evaluation of optimal operating conditions for production of hydrogen-rich syngas from air gasification of rice husks using aspen plus and response surface methodology. *Bioresource Technology*, 361, 127734.

Kong, G., Zhang, X., Wang, K., Li, J., Zhou, L., Wang, J., Zhang, X., Han, L. (2023). Coupling biomass gasification and inline co-steam reforming: Synergistic effect on promotion of hydrogen production and tar removal. *Fuel Processing Technology*., 243, 107689.

Kook, J. W., Choi, H. M., Kim, B. H., Ra, H. W., Yoon, S. J., Mun, T. Y., Kim, J. H., Kim, Y. K., Lee, J. G., Seo, M. W. (2016). Gasification and tar removal characteristics of rice husk in a bubbling fluidized bed reactor. *Fuel*, 181, 942-950.

Kryca, J., Priščák, J., Łojewska, J., Kuba, M., Hofbauer, H. (2018). Apparent kinetics of the water-gas-shift reaction in biomass gasification using ash-layered olivine as catalyst. *Chemical Engineering Journal*, 346, 113-119.

Li, J., Lin, L., Ju, T., Meng, F., Han, S., Chen, K., Jiang, J. (2024). Microwave-assisted pyrolysis of solid waste for production of high-value liquid oil, syngas, and carbon solids: A review. *Renewable & Sustainable Energy Reviews*, 189, 113979.

Li, J., Tao, J., Yan, B., Jiao, L., Chen, G., Hu, J. (2021). Review of microwave-based treatments of biomass gasification tar. *Renewable & Sustainable Energy Reviews*, 150, 111510.

Li, Z., Li, X., Zong, Y., Tan, G., Sun, Y., Lan, Y., He, M., Ren, Z., Zheng, X. (2017). Solvothermal synthesis of nitrogen-doped graphene decorated by superparamagnetic Fe_3O_4 nanoparticles and their applications as enhanced synergistic microwave absorbers. *Carbon*, 115, 493-502.

Liu, L., Huang, Y., Cao, J., Liu, C., Dong, L., Xu, L., Zha, J. (2018). Experimental study of biomass gasification with oxygen-enriched air in fluidized bed gasifier. *Science of the Total Environment*, 626, 423-433.

Liu, S., Zhang, Y., Fan, L., Zhou, N., Tian, G., Zhu, X., Cheng, Y., Wang, Y., Liu, Y., Chen, P., Ruan, R. (2017). Bio-oil production from sequential two-step catalytic fast microwave-assisted biomass pyrolysis. *Fuel*, 196, 261-268.

Lopez, G., Artetxe, M., Amutio, M., Alvarez, J., Bilbao, J. Olazar, M. (2018). Recent advances in the gasification of waste plastics. A critical overview. *Renewable & Sustainable Energy Reviews*, 82, 576-596.

Makwana, J., Dhass, A. D., Ramana, P. V., Sapariya, D., Patel, D. (2023). An analysis of waste/biomass gasification producing hydrogen-rich syngas: A review. *International Journal of Thermofluids*, 100492.

Nunes, L. J. (2022). Biomass gasification as an industrial process with effective proof-of-concept: A comprehensive review on technologies, processes and future developments. *Results in Engineering*, 14, 100408.

Pinto, F., Franco, C., André, R., Miranda, M., Gulyurtlu, I., Cabrita, I. (2002). Co-gasification study of biomass mixed with plastic wastes. *Fuel*, 81 (3), 291-297.

Plastics Europe, Plastics - the facts 2023. https://plasticseurope.org/knowledge-hub/plastics-the-fast-facts-2023/, 2023.

Saebea, D., Ruengrit, P., Arpornwichanop, A., Patcharavorachot, Y. (2020). Gasification of plastic waste for synthesis gas production. *Energy Reports*, 6, 202-207.

Sepahi, S., Rahimpour, M. R. (2023). Methanol production from syngas. In Advances in Synthesis Gas: Methods, Technologies and Applications, 111-146.

Shahbaz, M., Taqvi, S. A. A., Inayat, M., Inayat, A., Sulaiman, S. A., McKay, G., Al-Ansari, T. (2020). Air catalytic biomass (PKS) gasification in a fixed-bed downdraft gasifier using waste bottom ash as catalyst with NARX neural network modelling. *Computers & Chemical Engineering*, 142, 107048.

Solarte-Toro, J. C., Chacón-Pérez, Y., Cardona-Alzate, C. A. (2018). Evaluation of biogas and syngas as energy vectors for heat and power generation using lignocellulosic biomass as raw material. Electron. *Journal of Biotechnology*, 33, 52-62.

State, R. N., Volceanov, A., Muley, P., Boldor, D. (2019). A review of catalysts used in microwave assisted pyrolysis and gasification. *Bioresource Technology*, 277, 179-194.

Susastriawan, A. A. P., Saptoadi, H. (2019). Comparison of the gasification performance in the downdraft fixed-bed gasifier fed by different feedstocks: Rice husk, sawdust, and their mixture. *Sustainable Energy Technologies and Assessments*, 34, 27-34.

Toledo, M., Arriagada, A., Ripoll, N., Salgansky, E. A., Mujeebu, M. A. (2023). Hydrogen and syngas production by hybrid filtration combustion: Progress and challenges. *Renewable & Sustainable Energy Reviews*, 177, 113213.

Vuppaladadiyam, S. S. V., Vuppaladadiyam, A. K., Sahoo, A., Urgunde, A., Murugavelh, S., Šrámek, V., Pohořelý, M., Trakal, L., Bhattacharya, S., Sarmah, A. K. Shah, K.

(2023). Waste to energy: Trending key challenges and current technologies in waste plastic management. *Science of the Total Environment*, 169436.

Wang, B., Gupta, R., Bei, L., Wan, Q., Sun, L. (2023). A review on gasification of municipal solid waste (MSW): Syngas production, tar formation, mineral transformation and industrial challenges. *International Journal of Hydrogen Energy*, 48, 26676-26706.

Wang, C., Liu, F., Zhu, L., Wang, X., Zhang, C., Zhang, M., Han, Z., Jia, X., Guan, G., Zeng, X. Wang, F. (2024). Two-stage fluidized bed gasification enhanced with oxidative pyrolysis for low-tar producer gas from biomass. *Fuel*, 363, 130839.

Xiao, R., Jin, B., Zhou, H., Zhong, Z., Zhang, M. (2007). Air gasification of polypropylene plastic waste in fluidized bed gasifier. *Energy Conversion and Management*, 48 (3), 778-786.

Xie, J., Zhong, W., Jin, B., Shao, Y., Liu, H. (2012). Simulation on gasification of forestry residues in fluidized beds by Eulerian–Lagrangian approach. *Bioresource Technology*, 121, 36-46.

Yang, R.-X., Chuang, K.-H., Wey, M.-Y. (2018). Effects of temperature and equivalence ratio on carbon nanotubes and hydrogen production from waste plastic gasification in fluidized bed. *Energy Fuels*, 32 (4), 5462-5470.

Yao, Z., You, S., Ge, T., Wang, C. H. (2018). Biomass gasification for syngas and biochar co-production: Energy application and economic evaluation. *Applied Energy*, 209, 43-55.

Zha, Z., Wang, K., Ge, Z., Zhou, J., Zhang, H. (2022). Morphological and heat transfer characteristics of biomass briquette during steam gasification process. *Bioresource Technology*, 356, 127334.

Zhang, G., Liu, H., Wang, J., Wu, B. (2018). Catalytic gasification characteristics of rice husk with calcined dolomite. *Energy*, 165, 1173-1177.

Zhang, P., Liang, C., Wu, M., Li, Y., Chen, X., Liu, D., Ma, J. (2024). Sustainable microwave-driven CO_2 gasification of plastic waste for high-yield H_2 and CO production. *Applied Catalysis B: Environmental Energy*, 345, 123718.

Zhang, X., Yu, Z., Lu, X., Ma, X. (2021). Catalytic co-pyrolysis of microwave pretreated chili straw and polypropylene to produce hydrocarbons-rich bio-oil. *Bioresource Technology*, 319, 124191.

Zhang, Y., Fu, W., Cui, L., Maqsood, T., Li, B. (2023). Experimental microwave-assisted air gasification of biomass for syngas production. *Fuel*, 339, 126954.

Zhou, B., Dichiara, A., Zhang, Y., Zhang, Q., Zhou, J. (2018). Tar formation and evolution during biomass gasification: An experimental and theoretical study. *Fuel*, 234, 944-953.

Zhou, J., Chen, Q., Zhao, H., Cao, X., Mei, Q., Luo, Z., Cen, K. (2009). Biomass–oxygen gasification in a high-temperature entrained-flow gasifier. *Biotechnology Advances*, 27 (5), 606-611.

Zhu, D., Wang, Q., Xie, G., Ye, Z., Zhu, Z., Ye, C. (2024). Effect of air equivalence ratio on the characteristics of biomass partial gasification for syngas and biochar co-production in the fluidized bed. *Renewable Energy*, 222, 119881.

Chapter 8

Microwave Irradiation for Organic Synthesis

Deepika Dwarakanath
and Santosh L. Gaonkar*
Department of Chemistry, Manipal Institute of Technology,
Manipal Academy of Higher Education, Manipal, Karnataka, India

Abstract

The field of organic chemistry is a colossal branch of chemistry dealing with synthesis, which is one of the subdisciplines of the former, resulting in simple to complex molecules. The evolution of organic chemistry, starting with extraction from plant resources, distillation, and conventional methods to microwave synthesis, as an approach to green chemistry, has been a tremendous breakthrough in the scientific world. Most scientists across the world aim to adopt environmentally friendly processes to accomplish their synthesis; one such favoured approach is microwave irradiation. Since its introduction, the microwave irradiation technique has been a transformational change and is exploited by researchers for its reduced energy consumption, quick reaction span, better yields and no or less solvent in some cases, which makes it an ideal route for green synthesis. This technique makes the synthesis of organic molecules with stereochemistry simpler and easier. A number of one-pot multicomponent reactions can be achieved via microwave synthesis, and the resulting outcome can be applied in medicine and pharmaceuticals. Nevertheless, its shortcomings are infeasible to monitor the reaction, expensive instrumentation and poor scalability in large quantities. This chapter discusses the organic synthesis of various reactions under microwave irradiation and its pros and cons.

---

* Corresponding Author's Email: sl.gaonkar@manipal.edu.

In: Characteristics and Applications of Microwave Irradiation
Editors: Yaning Zhang and Wenke Zhao
ISBN: 979-8-89530-170-8

Keywords: microwave irradiation, microwave heating, organic synthesis

1. Introduction

The history of microwave irradiation applications dates back to the 1940s when it was initially used to heat food, which was fortuitously discovered by Dr. Percy Spencer, an autodidact. The observations began when the chocolate bar in Dr. Spencer's pocket melted while researching a radar-related project. He further exposed popcorn and an egg to low-density microwave energy where the kernels were sputtered, and the hot yolk was spluttered over the laboratory during individual experiments. Since then, microwave cooking has paced since its launch in the market (Adhikari et al. 2022). However, this innovation in microwave energy has led to the exploration of various applications in different fields, such as waste treatment, hydrolysis of proteins, polymer technology, drug targeting, preparation of samples, and so on. The reason why microwave energy is superior to X-ray and gamma-ray energy is that it is non-ionizing and imparts thermal activation without altering the molecular construct. Consequently, chemical reactions using microwave irradiation have become increasingly popular. In a microwave system, the heating system used to transform organic reactants to products is associated with dielectric polarization. Microwave energy is an electromagnetic energy, so the ability of a material to convert this energy to thermal energy will depend on its dielectric constant. A larger dielectric constant corresponds to greater coupling with microwaves. For example, solvents such as methanol, N-N-dimethylformamide (DMF), acetone, water, acetic acid and ethyl acetate have higher dielectric constants, whereas solvents such as hexane, carbon tetrachloride (CCl_4), and toluene have low dielectric constants and hence are not rapidly heated when irradiated with microwaves. The use of organic synthesis methods has greatly advanced from conventional methods to synthesis methods under microwave irradiation, as there are added advantages, such as reactions take place quicker, environmentally friendly (green chemistry), safer and higher yields (Sanghi, 2000). This was one of the major breakthroughs in chemistry since 1986 by Gedye and Giguere, who were the first to contribute to this field, and the number of advantages outweighs the number of disadvantages.

The heating ranges obtained in microwave techniques are multiple times greater than those in conventional methods; the heating or reaction time cannot be reduced in conventional methods, and byproducts can form in some reactions. The energy transfer in the case of conventional synthesis is from the glass vessel to the medium followed by the reactants, but in the case of microwave synthesis, dielectric heating occurs, and the glass vessel is the source of heat loss. Therefore, this type of heating via microwave enhances the rate of reaction, as the solvents can be boiled at a temperature higher than their boiling point. The mechanisms involved in microwave irradiation are ionic conduction and dipolar polarization and conduction in the case of conventional heating. There should be direct contact between the glass vessel or reaction vessel and the heating source in the conventional method, whereas in contrast, there should not be any physical contact between the reaction vessel and heating source. Additionally, the selective heating of chemicals in organic reactions can be achieved due to differences in dielectric constants under microwave irradiation. Furthermore, the interaction of microwave energy with different materials provides a better understanding of the progress of the reactions. This depends on the reflection, absorption and transmission of the material under irradiation. In the case of microwave radiation, no energy is introduced into the system, which can be seen in metals as materials; when the dielectric constants of the material are high, heating occurs more quickly, such as in the solvents discussed above; and the transmission of microwaves is possible through quartz, teflon and glass, which have trivial interactions (Lanjekar and Rathod, 2021).

When these interaction theories are clear, it is easier to employ a catalyst which can absorb the radiations well to improve the yield and the rate of reaction in absence of the solvent. In the following sections, the instrumentation, various reactions executed under microwave irradiation, and their advantages and disadvantages are discussed.

2. Named Reactions via Microwave-Assisted Synthesis

2.1. Knoevenagel Condensation

Bhat et al., synthesized benzylidenethiazolidine-2,4-dione derivatives **3** by reacting various aldehydes **1** with thiazolidine-2,4-dione **2**, which is an active methylene compound in equal ratios in water as a solvent and no catalyst. This reaction involved irradiation with microwaves at 120 °C and 250 W for 2-5 minutes (Bhat et al. 2020). The same reaction was carried out by Shaikh and coworkers via conventional methods using piperidine as a catalyst in ethanol and refluxed for 14-18 hours to obtain the desired product (Shaikh et al. 2015). The reaction taking place under microwave irradiation is better than that under conventional methods because it is quicker, yields were improved, and a green solvent was used; additionally, a catalyst was required for conventional method. The reaction is represented in Scheme 1.

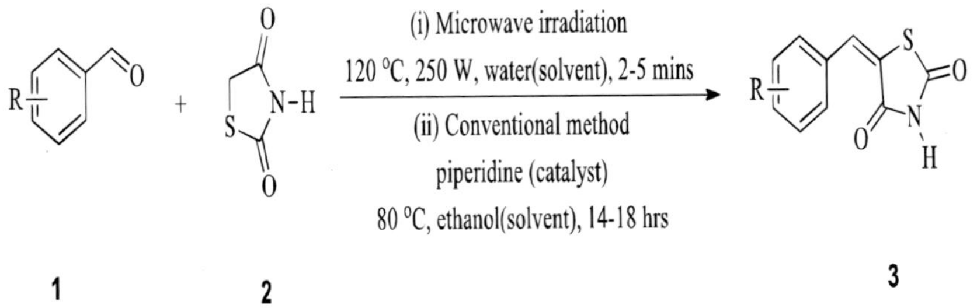

Scheme 1. Knoevenagel condensation reaction of aldehyde and thiazolidine-2,4-dione.

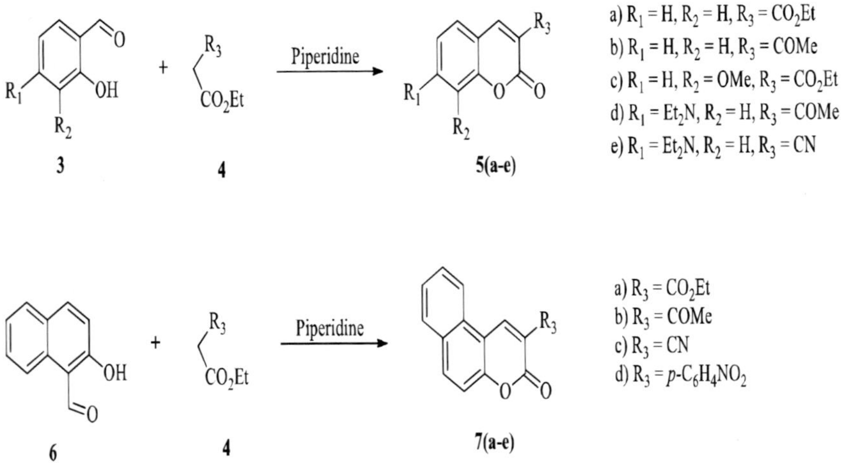

Scheme 2. Coumarin derivatives synthesis by Knoevenagel condensation.

Dariusz Bogdał synthesized coumarin derivatives under microwave irradiation. Substituted hydroxy aldehydes were reacted with carbonyl compound and piperidine was added as a catalyst. The products obtained were recrystallized for purification (Scheme 2). The advantages of carrying out this reaction via microwave irradiation are that the time consumption is reduced, the reaction takes place in the absence of solvent, the yields of the desired products are better, the reaction is safer, and the work up post reaction is also uncomplicated, as it is obtained by recrystallization (Bogdał, 1998).

2.2. Mannich Reactions

These are organic reactions that involve reacting three components, an aldehyde, amino group and acidic hydrogen next to a carbonyl group. This reaction, when developed under microwave irradiation, was more efficient. These reactions are useful because the Mannich bases formed are used in preparing therapeutic drugs; therefore, they are among the most commonly used reactions in drug discovery and development (Cebeci et al. 2022; Mermer et al. 2017; Patel et al. 2020; 2018; Kiss et al. 2023). The synthesis of mannich reactions via microwave irradiation is faster and better. Yildirim and coworkers synthesized a multicomponent reaction by microwave irradiation, yielding thiazolo[3,2-c]pyrimidines. The product was obtained after cooling the reaction mixture, and column chromatography was not required to purify the end product (Scheme 3). The same group of authors synthesized the products by conventional method and found that the reaction time was longer and the yield was slightly lower compared to the microwave method (Yildirim et al., 2014).

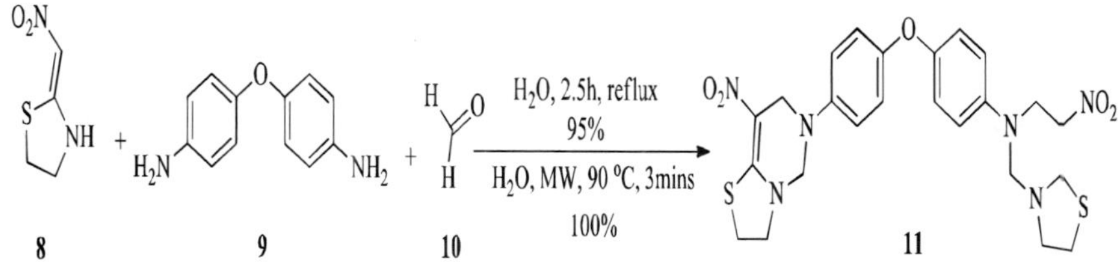

Scheme 3. Synthetic route for nitrothiazolo[3,2-c]pyrimidine.

2.3. Heck Reaction

The Mizoroki–Heck reaction aka Heck reaction, is a palladium-catalyzed reaction that yields substituted alkenes via the cross-coupling of vinyl or aryl halides with alkenes in attendance of a base. The Heck reaction is a significant reaction in organic synthesis in academics and industry because it leads to carbon–carbon double bond formation and high stereoselectivity. This reaction has also been awarded Nobel Prize in the year 2010. Yikai et al. synthesized derivatives by coupling olefins with 3-iodo-benzopyrones with $Pd(OAc)_2$ as a catalyst and triethylamine as a base in DMF and irradiated them at a temperature of 80 °C for 5 minutes, as shown in Scheme 4 (Zhang et al. 2012).

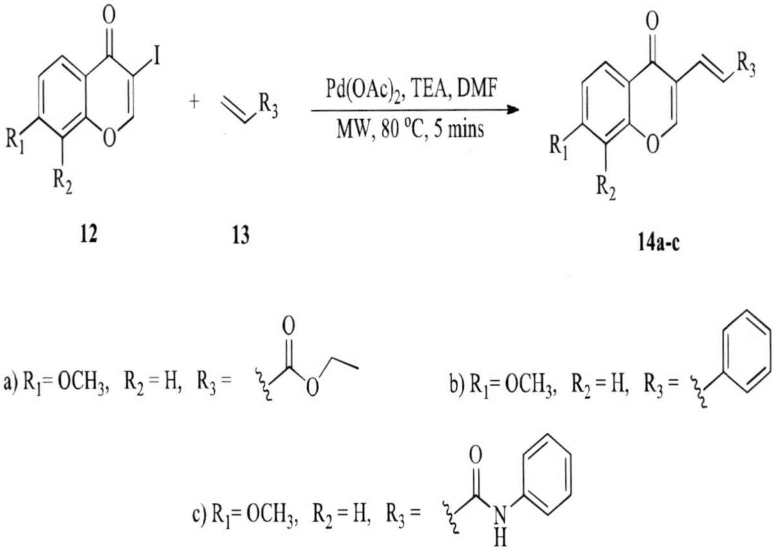

Scheme 4. Synthesis of derivatives via the Heck reaction.

To speed up Heck reaction in ionic liquids, another group of scientists designed a simple approach involving microwave irradiation as represented in Scheme 5. One of the most important benefits of this reaction, apart from its short reaction time and better yields, is that the catalyst can be reused. The reaction was repeatable up to four times with very minimal loss (Xie et al. 2004).

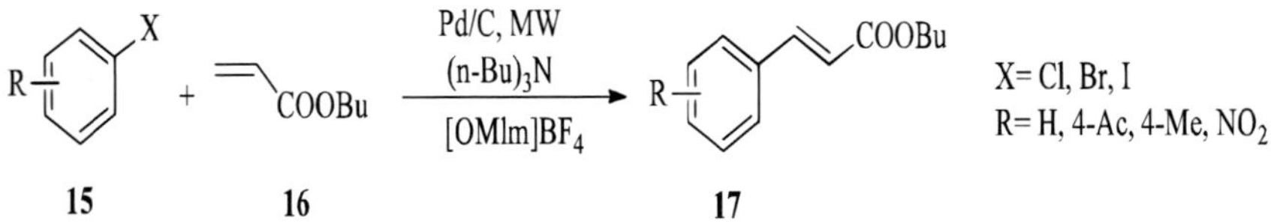

Scheme 5. Microwave assisted Heck reaction in an ionic liquid.

2.4. Sonogashira Coupling

One of the flexible reactions in organic chemistry to yield aromatic alkynes is when terminal alkynes are coupled with vinyl or aryl halides. This coupling reaction was proposed independently by Hagihara, Tohda and Sonogashira in 1975. This reaction uses a copper catalyst and a palladium catalyst simultaneously. The expansion of the Heck, Cassar and Dieck reactions is due to this coupling reaction, which forms the C-C bond. Lei and his group synthesized 1-aryl-2-(trimethylsilyl)acetylene using palladium and copper catalysts via sonogashira coupling with aryl iodides/aryl bromides and trimethylsilylacetylene(TEA) as a base in acetonitrile (Scheme 6). Through microwave irradiation, the yield considerably improved, and the reaction time decreased (Lei et al. 2016).

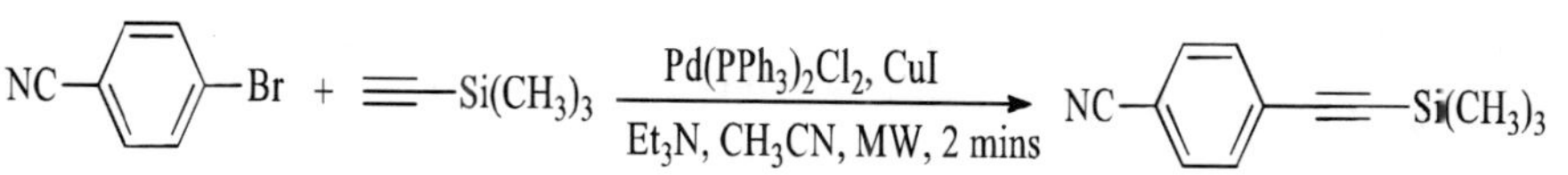

18 **19**

Scheme 6. Synthesis of 4-(trimethylsilyl)ethynyl)benzonitrile.

Although this coupling reaction is more common, one of the drawbacks is that symmetrical compounds are obtained as a result of homocoupling (terminal acetylenes) which is called the Glaser product. A recent paper described the use of a substituent for acetylenes, propiolic acids, in the development of dihydrobenzofurans via microwave synthesis, through a Heck reaction followed by Sonogashira coupling (Lanjekar and Rathod, 2021; Joy et al., 2014). The copper catalyst aids the reaction by generating copper acetylide in situ, which eases the transmetallation part of the reaction in the catalytic cycle; however, this catalyst also promotes the formation of the Glaser product. To avoid these side products, scientists have recently developed several copper-free sonogashira coupling reactions (Ramesh and Satyanarayana, 2019; Hellal and Cuny, 2011; Cívicos et al. 2013). The mechanism for the copper-free sonogashira coupling has been explained by the Gazvoda group via experimental evidence (Gazvoda et al. 2018).

2.5. Suzuki Coupling Reactions

The Suzuki reaction is a transition metal (palladium)-catalyzed reaction between vinyl, alkynyl or aryl organoboronic compounds and pseudo or organic halides. This reaction helps in the synthesis of intermediates, polymers, specialty chemicals and pharmaceuticals. The reaction involves three steps that is oxidative addition, followed by transmetallation, and reductive elimination. Organic compounds bearing boron have less toxicity and can be obtained easily; hence, these reactions are of great importance in the field of organic chemistry (Baviskar et al. 2023). Khormi and coworkers synthesized 2-(4-fluorophenyl)thiophene from **20** and various arylboronic acids in the presence of base and DMF/dioxane as solvents (Scheme 7). The conventional method for the synthesis of this reaction took almost 3-6 hours to complete based on the substituents, whereas under microwave irradiation, the reaction lasted 10-14 minutes. The palladium ligand was initially synthesized by the group, and it exhibited good catalytic activity even at the lowest molar concentrations (Khormi et al. 2022).

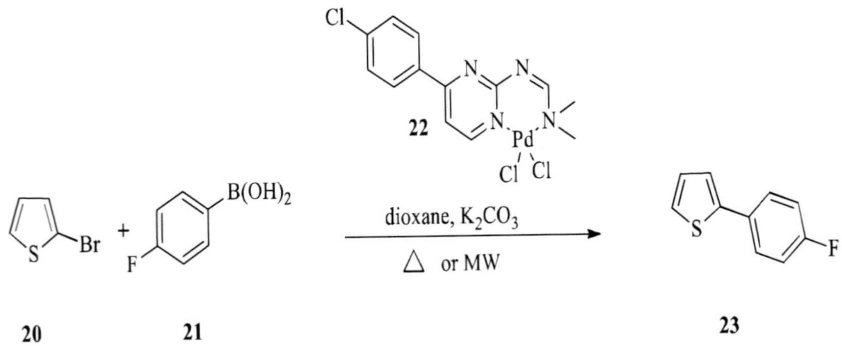

Scheme 7. Synthesis of 2-(4-fluorophenyl)thiophene via the Suzuki coupling reaction.

Another group of scientists designed a novel pyridine-pyrazole based palladium ligand that was used in aqueous media as a catalyst for the Suzuki reaction via microwave irradiation. They optimized the synthesis of various aryl halides with different substituted boronic acids in a water-ethanol mixture to yield the desired products. One such reaction is shown in Scheme 8 (Shen et al. 2013).

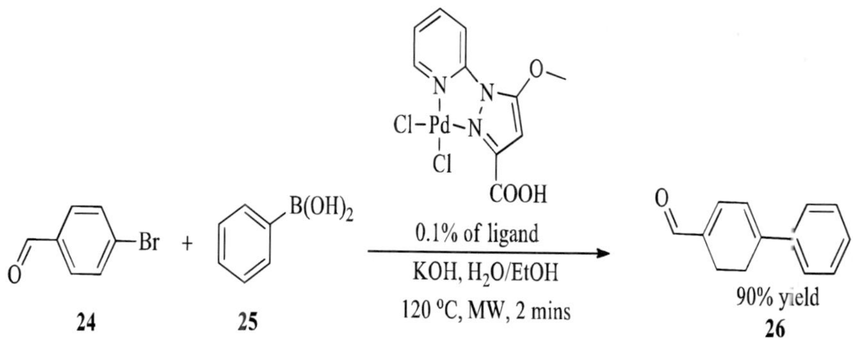

Scheme 8. Suzuki coupling reaction of 4-bromobenzaldehyde (aryl halide) and phenylboronic acid (aryl boronic acid).

Silva et al. reported a phosphine free reaction involving Suzuki coupling to produce biaryls, thienylpyridine, bipyridyls and allylphenols. They optimized the reaction in an ethanol-water mixture, which is a nontoxic solvent, and the yield increased under microwave irradiation. The catalyst used was a polyurea microencapsulated palladium catalyst (Pd EnCat30). The synthetic route is shown in Scheme 9 (Silva et al. 2015).

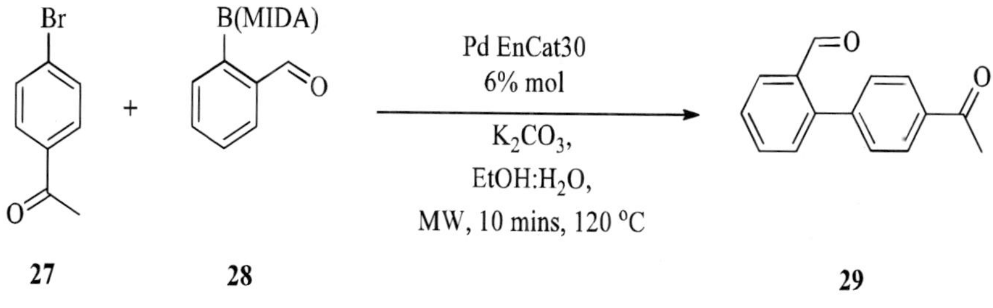

Scheme 9. Suzuki cross-coupling reaction between aryl halides and aryl boronic compounds.

2.6. Strecker Reaction

A Strecker reaction yields α-aminonitriles as its end product after reaction with aldehydes/ketones, a source for cyanide and amine. This is a multicomponent reaction in which compounds bearing α-aminonitriles that are important for therapeutics, agrochemicals and the synthesis of heterocyclic compounds are developed (Madadi, 2021). Sadek and group established the microwave synthesis of new imidazole amine and triazole amine derivatives via the

Strecker reaction, and this pathway follows green chemistry (Sadek et al., 2019). Scheme 10 represents imidazole derivative via Strecker reaction.

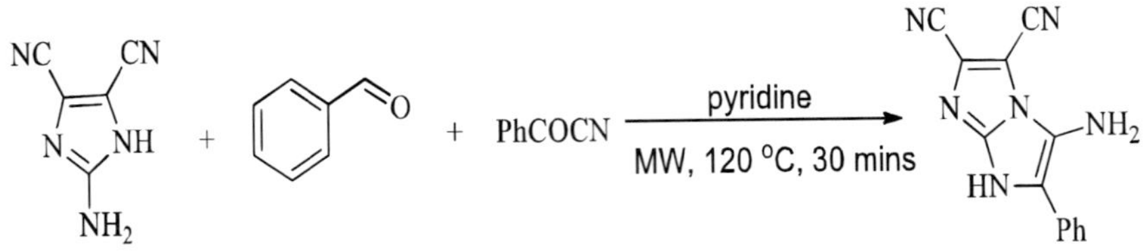

30 **31** **32** **33**

Scheme 10. Synthetic route for 3-amino-2-phenyl-1H-imidazo[1,2-a]imidazole-5,6-dicarbonitrile.

The cyanide source for a Strecker reaction is crucial. When alkaline metal cyanides or gaseous hydrogen cyanide are used, eliminating them during the post reaction phase becomes an issue. On the other hand, when potassium cyanide and sodium cyanide are employed the alkalinity of the reaction increases, which is not preferred by many reactions. Alternatively, some research groups have investigated other sources of cyanide, such as trimethylsilyl cyanide, acetyl cyanide, tributyltin cyanide, and potassium hexacyanoferrate(II) (Harusawa and Shioiri, 2016; Cruz-Acosta et al. 2009; Shah and Singh, 2012; Sipos and Jablonkai, 2009; Mori et al. 1997). However, there are many reports of this synthesis in conventional methods with tedious post reaction work-up, extreme reaction conditions, longer reaction times, etc. An easier way to perform this reaction is by using a microwave synthesizer because it increases the reaction rate, improves the yield and decreases the formation of impurities (Hameed et al. 2017).

2.7. Paal-Knorr Synthesis

Pyrrole, a five-membered heterocycle, is an important scaffold in the pharmaceutical industry. One of the general methods of its synthetic route is Paal-Knorr synthesis. This is one of the oldest methods in organic chemistry and can also produce furans or thiophenes from 1,4-dicarbonyl as a starting material under acidic reaction conditions. Minetto and coworkers synthesized tetrasubstituted pyrroles and trisubstituted furans from commercially accessible β-ketoesters via Paal-Knorr synthesis (Minetto et al. 2004).

Aghapoor et al. synthesized N-substituted pyrroles from hexane-2,5-dione and 4-bromoaniline under microwave irradiation (Scheme 11). They initially optimized the reaction by using various catalyst salts that belong to the alkali and alkaline earth groups; the most potent was found to be $CaCl_2.2H_2O$. The reaction was completed in 10 minutes. The catalyst in this case can also be reused with negligible loss of activity. The reaction proceeds without any solvent and hence is more environmentally friendly (Aghapoor et al., 2018).

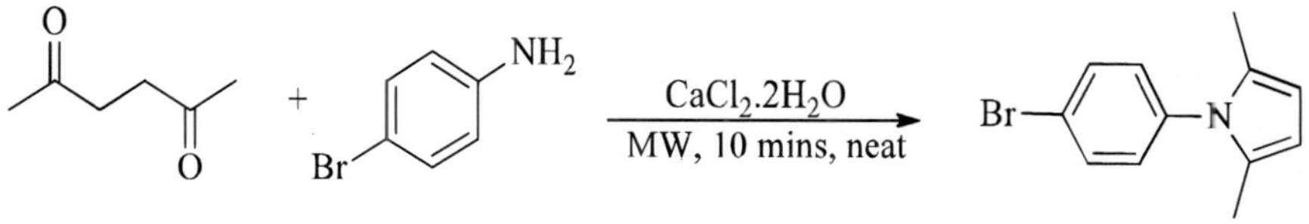

34 **35** **36**

Scheme 11. Synthesis of substituted pyrrole via Paal-Knorr synthesis.

Scientists have synthesized pyrroles by the Paal-Knorr reaction using N-bromosuccinimide (NBS). The reaction was completed in 3-8 minutes in the presence of NBS (Scheme 12), but in the absence of NBS, the reaction took 25 minutes to complete (Andoh-baidoo et al. 2011).

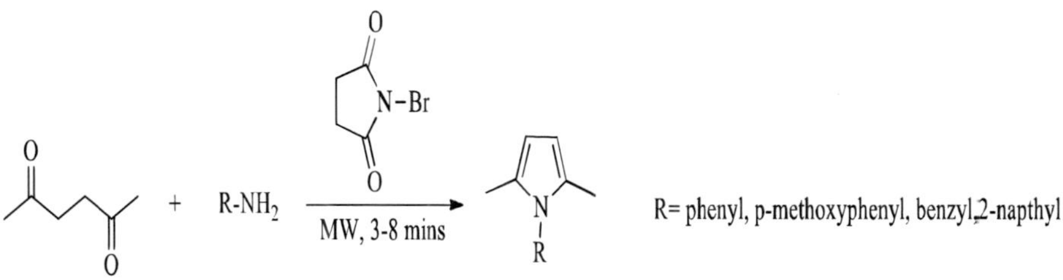

37 **38** **39**

Scheme 12. Paal-Knorr microwave synthesis using NBS as a catalyst.

2.8. Hantzch Synthesis

This is a multicomponent reaction that involves reactions with a β-ketoester, an aldehyde and ammonia. Arthur Hantzsch was a German chemist who explored this reaction and was interested in the field of medicine. The products of this pathway are mainly exploited as active therapeutic agents or as building blocks in drug discovery and development. Therefore, this reaction is vital and is almost a century old. Efforts have since been made to improve or alter the substrates, reaction yield and reaction time and reduce the use of byproducts and energy to increase the environmental friendliness of these materials. The use of ammonia to provide nitrogen in the reaction has been attempted with other sources, such as urea, primary amines and oxahydrazines (Bosica et al. 2020). Zeynizadeh and Gilanizadeh synthesized acridinediones under green conditions and microwave irradiation. Various aldehydes were reacted with dimedone and ammonium acetate in a solvent-free system devoid of catalyst and irradiated at 700 W for 3-8 minutes (Scheme 13). The resulting products were in high yield and achieved quicker without the use of harmful chemicals (Zeynizadeh and Gilanizadeh, 2020).

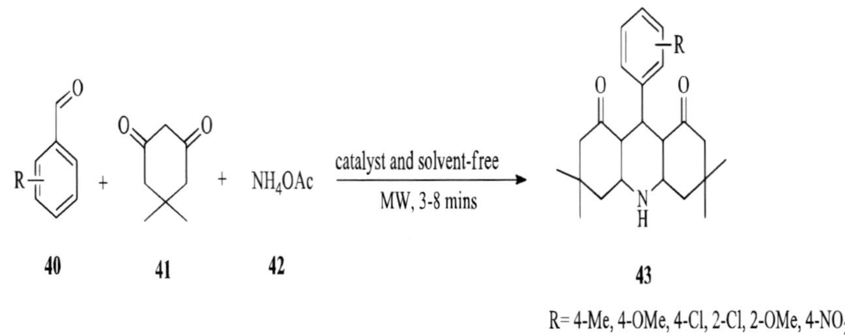

Scheme 13. Synthesis of acridinediones via the Hantzsch reaction.

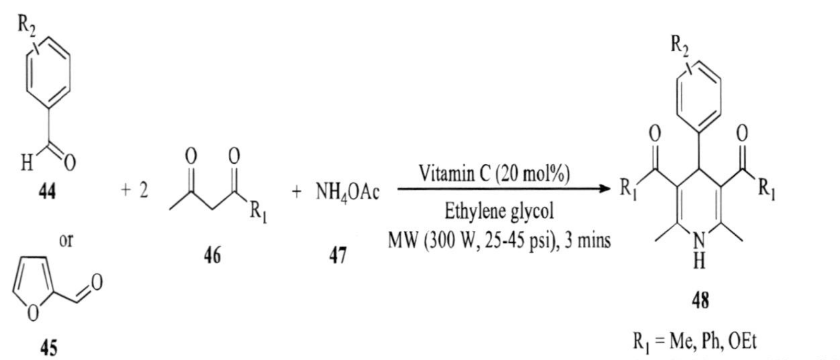

Scheme 14. Synthesis of 1,4-dihydropyridines (DHP) via the Hantzsch reaction.

Ray and coworkers synthesized 1,4-dihydropyridines (DHP) using vitamin C as a catalyst (Scheme 14). Here, there are two advantages: one is that vitamin C imparts its acidic property, which promotes the reaction; the second is the use of microwave irradiation to produce DHP derivatives, which increases the quantity of the products and reduces the reaction time (Ray et al. 2022).

Another group of scientists developed novel 1,4-dihydropyridine derivatives via conventional and microwave irradiation. The reaction was performed under solvent-free conditions, and when the reaction proceeded through microwave radiation, the results were better than those of conventional methods. The single-crystal structure was also determined (Maru et al. 2019).

2.9. Ullmann Reaction

During the 1900s, Fritz Ullmann and Irma Goldberg described and reported aromatic nucleophilic substitution reactions that are mediated by copper. In general, Ullmann reaction is a condensation reaction between an aryl halide and phenol or an amine or thiophenol to give ether or aryl amine or thioether derivatives, respectively, in the presence of copper either in a stoichiometric ratio or in a catalytic amount. The classical reaction of Ullmann is given in Scheme 15 (Lei et al. 2010; Khan et al. 2018).

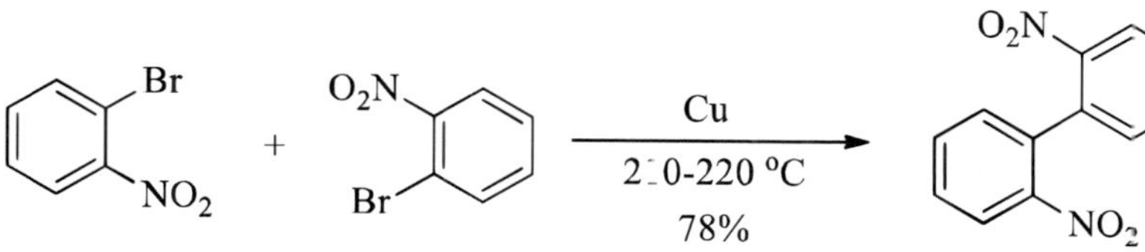

49 **50** **51**

Scheme 15. Original Ullmann reaction.

Shen et al. developed macrocyclic diaryl ethers by intramolecular and/or bimolecular Ullmann coupling. This coupling reaction is one of the first reported coupling reactions under microwave irradiation, as shown in Scheme 16. Another group synthesized copper-catalyzed, ligand-free coupling reactions of halogenated pyridines with different nitrogen-containing nucleophiles under microwave irradiation (Shen et al. 2012; Liu et al. 2011).

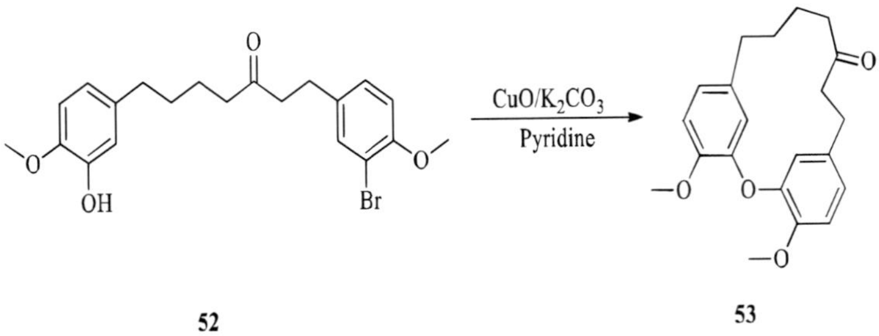

Scheme 16. Synthesis of macrocyclic diaryl ethers via the Ullmann coupling reaction.

2.10. Ugi Reaction

The Ugi reaction is a multicomponent reaction that involves four reactants, carbonyl compounds (aldehyde or ketone), primary amines, isonitrile and carboxylic acid, which give α-aminocayl amides (Scheme 17). The resulting products are generally referred to as peptomers or bis-amides, which possess good therapeutic activity and are structurally diverse. This reaction occurred around approximately 1959, when reactions with isonitrile prevailed. The Ugi reaction can also occur in water (Fouad et al. 2020).

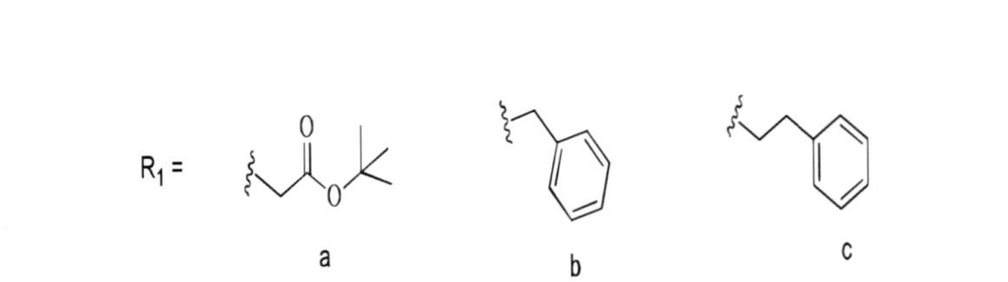

Scheme 17. Ugi reaction via microwave irradiation.

Barreto et al. synthesized a broad range of peptoids by the Ugi reaction under microwave irradiation. They reported a direct method to synthesize different functionalized peptoids in an easy, one-pot Ugi synthesis. The yield of the reaction was excellent (Barreto et al. 2011).

Another set of studies was performed by Lian and coworkers, in which arylboronic acid analogs were synthesized. This four component reaction was synthesized using microwave irradiation under mild conditions. The Ugi reaction proceeded well under microwave irradiation, and the desired products were obtained in good yields. Scheme 18 is shown below (Lian et al. 2014).

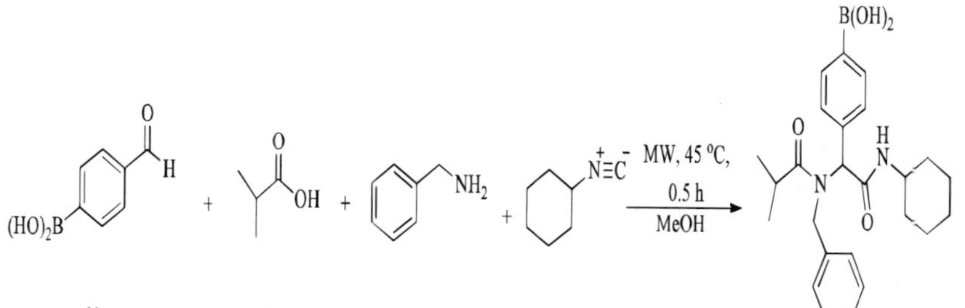

Scheme 18. Synthesis of arylboronic acidvia microwave irradiation.

2.11. Pechmann Reaction

One of the most common ways to prepare coumarins is the Pechmann reaction, in addition to the Knoevenagel, Wittig, Reformatsky and Perkin reactions. Coumarins are important heterocyclic scaffolds in the medicinal, cosmetic, and food industries. It has also been applied as a fluorescent dye in many fields. The general method of synthesis involves the use of phenols with concentrated H_2SO_4 as the catalyst. The traditional protocol of this reaction has several limitations, such as a long reaction time, the formation of byproducts and corrosion issues. Over the years, many attempts have been made to find a more efficient and eco-friendly pathway. Scheme 19 shows the synthesis of coumarin from phenol via both microwave and conventional methods. This particular derivative took 135 minutes to complete the reaction via the conventional method with 55% yield, whereas under microwave irradiation, it took 12 minutes to complete with 65% yield (Rajitha et al. 2006).

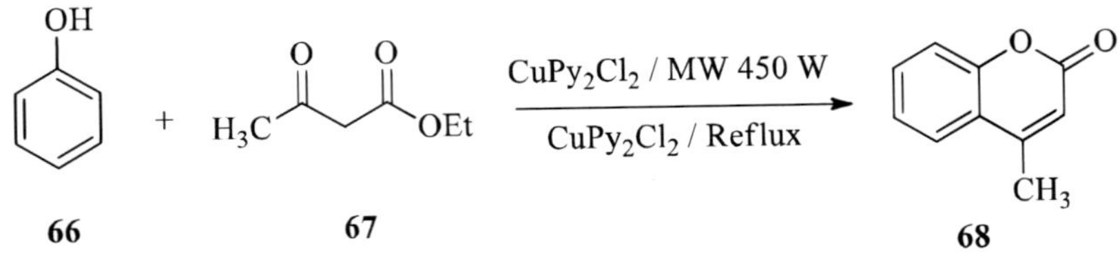

Scheme 19. Coumarin synthesis via the Pechmann reaction.

Similarly, Singh and coworkers synthesized various derivatives of coumarins from phenols and β-ketoesters in the presence of concentrated H_2SO_4. The yields of the reaction increased with increasing speed in the case of microwave heating. The reaction in Scheme 20 reached completion in 2 minutes with 72.1% yield and 12 hours with 62.6% yield upon conventional heating (Singh et al. 1997).

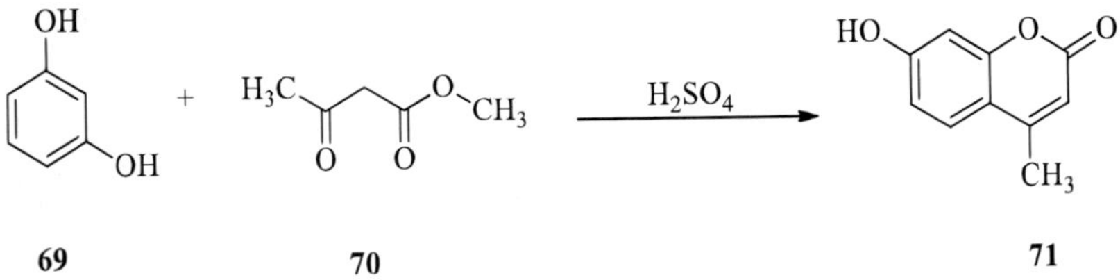

Scheme 20. Coumarin synthesis using sulfuric acid as a catalyst.

Reddy et al. used ceric ammonium nitrate (CAN), which is easily available, inexpensive, and nontoxic, as a catalyst in the Pechmann reaction to yield coumarins. The same reaction described in Scheme 20 was performed by irradiation with microwaves in the presence of CAN; the reaction was complete in 2 minutes with a yield of 96% but took 10 minutes to complete via conventional heating with 94% yield (Reddy et al. 2008).

2.12. Stille Coupling Reactions

The general procedure for Stille cross-coupling is a reaction between an organic halide and an organotin compound. The catalyst used in this reaction was palladium phosphine. Jeon and group reported a microwave method involving the use of 1,4,5-trisubstituted 3-trifluoromethylpyrazoles via the Stille cross-coupling pathway. Initially, bromo-substituted pyrazoles were obtained by reaction with *N*-bromosuccinamide. This mixture was then reacted with allylstannane or arylstannane. The time taken for the reaction to complete was 1-2.5 hours because the different substrates reacted with a very good yield range of 80-97%. The schematic route is given in Scheme 21 (Jeon et al. 2007).

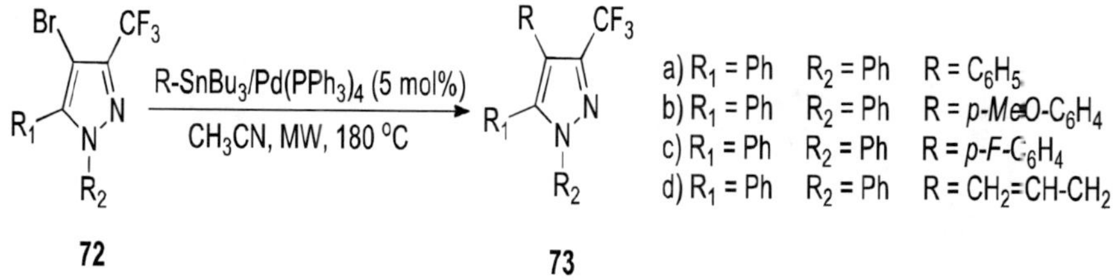

Scheme 21. Stille-coupling reactions with allylstannanes or arylstannes.

Pchalek et al. evaluated the use of microwave synthesis for Stille coupling reactions with homogenous catalysis. The reaction of 3-halo-1,2,4-benzotriazine-1-oxide and Et4Sn was initially optimized. The reaction took place better in acetonitrile, and the use of microwave synthesis aided in overcoming all the disadvantages of conventional heating, such as lower yields, byproduct formation and prolonged reaction times (Pchalek and Hay, 2006).

In another report, polythiophenes were synthesized with the help of microwave irradiation followed by Stille coupling. These polythiophenes were semiconducting, soluble and were prepared in 10 minutes with better yields, high molecular weights, and less polydispersity than the conventional product. It also showed good performance in an organic field-effect transistor (Tierney et al. 2005).

2.13. Biginelli Reaction

This reaction is well known in organic synthesis because it is one of the most essential reactions used in the field of biology for various drug preparations, such as antimalarial, anticancer, and anti-HIV agents. It is a three-component reaction in which aldehyde, β-keto ester and urea react to form 4-aryl-3,4-dihydropyrimidin-2(1H)-one (DHP). To improve the reaction yields, ionic liquids, Lewis acids or Brønsted acids were used as catalysts under different reaction conditions. A simple Biginelli reaction is shown in Scheme 22 (Puripat et al. 2015).

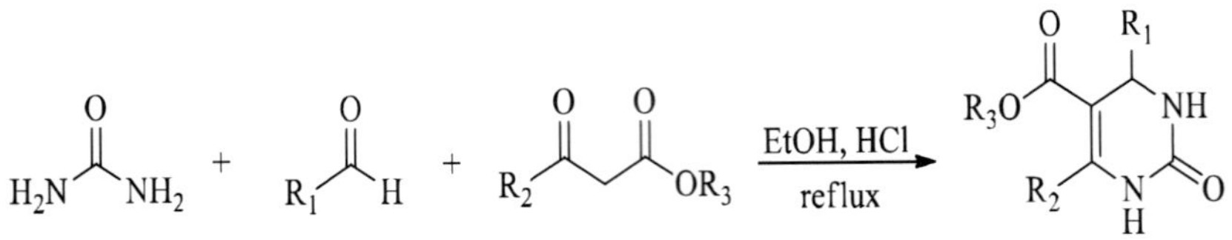

74 75 76 DHP 77

Scheme 22. A simple Biginelli reaction.

Renzhong Fu and coworkers synthesized DHP through microwave synthesis with 3 mol% ionic liquid as the catalyst. The reaction was carried out in a solvent-free environment, and the ionic liquid was reused up to five times. This microwave irradiation route gives better yields in a short reaction span (Fu et al. 2015). Yadav and group synthesized DHP from three common reactants with various acid-based catalysts without solvents. The product was easy to isolate after completion, and the reaction was easy to set and monitor (Yadav et al. 2000).

Choudhary et al. evaluated a Si-MCM-41 supported $FeCl_3$ catalyst for the Biginelli reaction via microwave synthesis. Among the others tested, the above showed better performance via microwave irradiation, and the DHP reaction was completed in 3-5 minutes with excellent yields. One such derivative is shown in Scheme 23 (Choudhary et al. 2003).

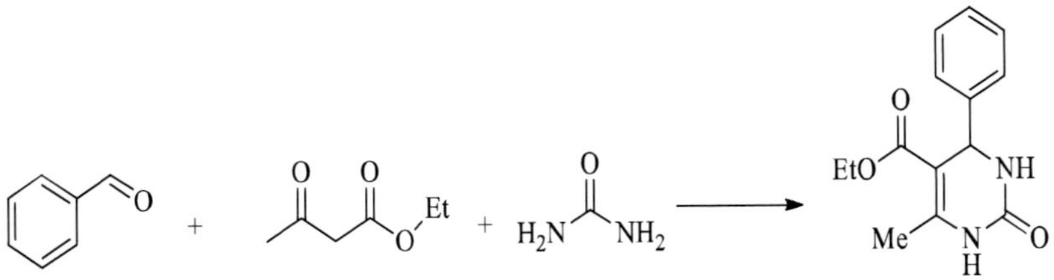

78 79 80 81

Scheme 23. Biginelli reaction with the $FeCl_3$/Si-MCM-41 catalyst.

2.14. Gewald Reaction

Polysubstituted 2-aminothiophenes are derived via the Gewald reaction. It is a multicomponent reaction in which an aldehyde or ketone and an activated nitrile or sulfur react in the presence of a base. Traditionally, in a two step procedure, Knoevenagel-Cope condensation occurs between activated nitrile and carbonyl compounds, followed by a base assisted reaction with sulfur. Later, microwave assisted reactions were employed to improve the reaction time and provide better yields (Puterova et al. 2010).

Sridhar and coworkers synthesized 2-aminothiophenes via a multicomponent reaction under microwave irradiation using KF-Alumina as a catalyst (Scheme 24). The reaction, which took 3.5-7 hours to complete in the conventional heating method, took only 3.5-8 minutes to react under microwave irradiation (Sridhar et al. 2007).

2-Acyl amino thiophenes were synthesized via microwave irradiation of a solid support. This reaction is said to proceed faster with better yields, giving tetrasubstituted thiophenes (Hoener et al. 2002). In another report, imidazole was used as a catalyst to give tetrasubstituted aminothiophenes (Huang et al. 2011).

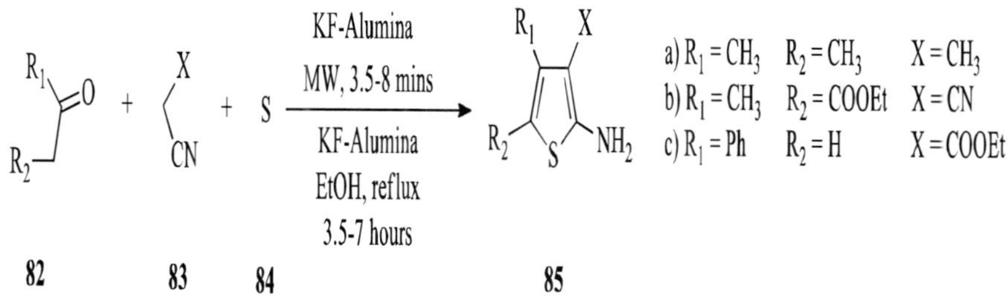

Scheme 24. Synthesis of 2-aminothiophenes via the Gewald reaction.

3. Other Microwave Assisted Reactions

The N-alkylation reaction is one of the most commonly synthesized reactions in organic chemistry (Scheme 25). Alkylated amines, which are aromatic, are used in daily life. The general N-alkylation reaction takes 7-8 hours to complete at room temperature with potassium carbonate as the base in DMF. If the same reaction is performed under microwave irradiation, it occurs faster (Ju and Varma, 2004).

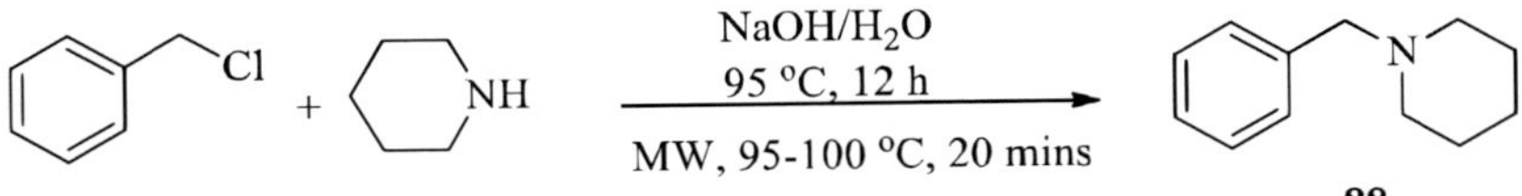

86 87 88

Scheme 25. N-Alkylation of amines.

The microwave-assisted reaction of isatin N-alkylation occurred in 3 minutes under microwave irradiation (Scheme 26). The conventional method requires one hour of heating at 70 °C with 80% yield, whereas under microwave conditions, the yield obtained is 95% (Shmidt et al. 2008).

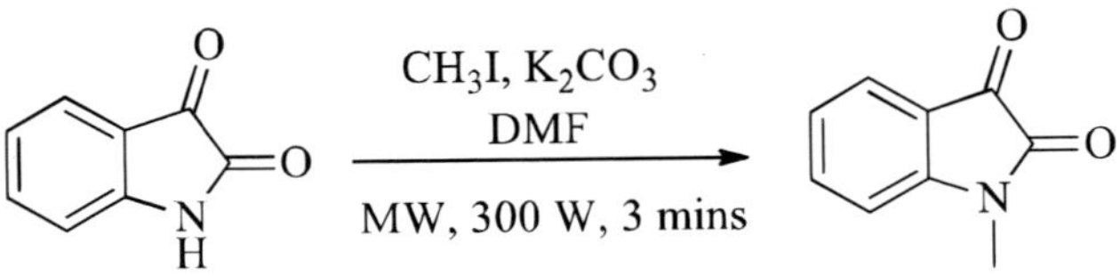

89 90

Scheme 26. N-Alkylation of isatin.

A similar alkylation reaction was performed under microwave irradiation for carbazoles, where the reactions were completed in 4-10 minutes in a dry medium with a yield of 95% (Bogdal et al. 1997).

Benzofuran-2-carboxamide was prepared by an O-alkylation reaction, which was completed in one hour, as there were two steps of 30 minutes each (Scheme 27). The base used was CS_2CO_3 in DMF (Xie et al. 2014).

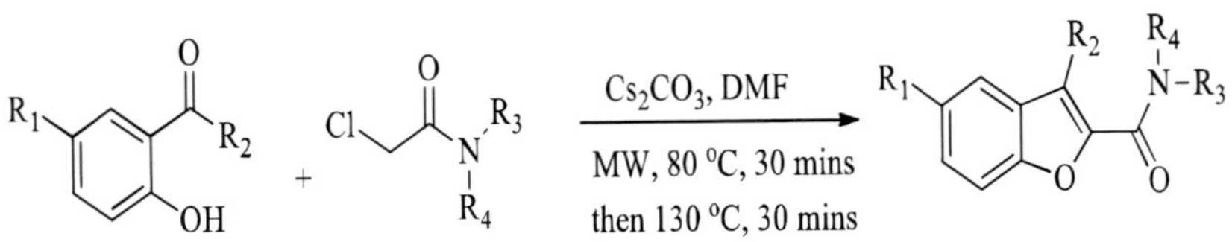

91 92 93

Scheme 27. Synthesis of benzofuran-2-carboxamide via O-alkylation.

Research on nanoparticle synthesis and application has been one of the main areas explored in recent years; hence, their preparation is vital. Microwave irradiation helps in the faster and better synthesis of nanoparticles.

Silver nanoparticles, which can serve as reducing and supporting agents were rapidly synthesized from cellulose nanocrystals via microwave synthesis (Alle et al. 2020). Iron (III) oxide and copper (II) oxide nanoparticles were synthesized in an energy and time efficient manner via microwave irradiation. These nanoparticles displayed good antifungal activity (Devi et al. 2019). The polyol method was employed to synthesize co-ferrite nanoparticles under microwave radiation (Bensebaa et al. 2004). Low et al. synthesized cesium silver bismuth bromide perovskite via microwave heating method in isopropyl alcohol because of its high dielectric heating constant (Low et al. 2023).

Gaonkar et al. synthesized rosiglitazone, an antidiabetic drug, via microwave synthesis. The first step reaction shown in Scheme 28 takes 15 hours to complete via conventional heating to give 96, and the same reaction is completed in 20 minutes under microwave irradiation. Similarly, the next step to obtain the aldehyde was completed in 20 minutes via microwave heating, whereas it took 8 hours to complete under conventional heating (Gaonkar and Shimizu, 2010).

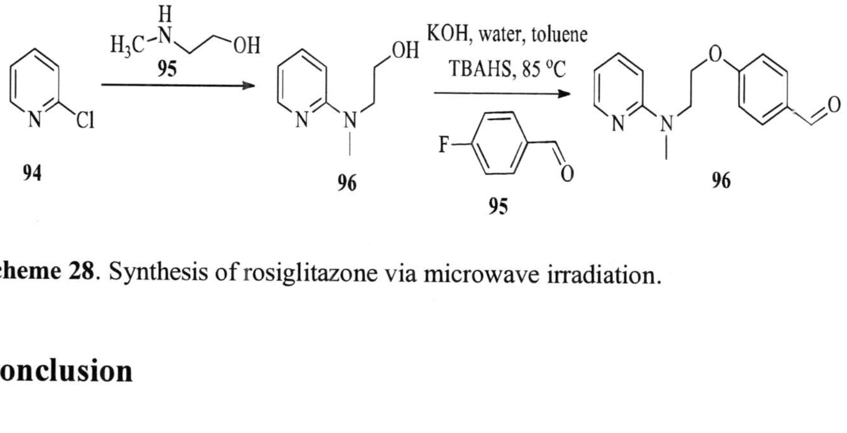

Scheme 28. Synthesis of rosiglitazone via microwave irradiation.

Conclusion

The pros of microwave synthesis in comparison to conventional methods include the following:

1. *Rapid reactions*: The reaction occurs faster, in minutes or seconds, in some cases. In this way, the time consumption can be reduced.
2. *Purity of products*: The products obtained via microwave synthesis are considerably pure most of the time, with very minimal purification techniques - required even if impurities exist.
3. *Improved yields*: The yield of the reaction is significantly increased compared to the conventional heating method.

4. *Fewer byproducts*: Since the reaction time is quicker, the formation of byproducts decreases.
5. *Temperature range:* A wide temperature range can be employed under microwave irradiation, which helps in the dielectric heating of reactants and solvents.
6. *Enhancement of the reaction rate*: The rate of the reaction increases by many-fold as the temperature of the reaction increases rapidly.
7. *Reproducibility*: The catalysts used in microwave synthesis can be recycled several times with very little loss of activity.
8. *Energy saving*: Since microwave irradiation helps in heating the sample alone and not the apparatus as a whole, less energy is consumed.
9. *Uniformity in heating*: The reactants and the solvent are uniformly heated, which is not the case in conventional heating.
10. *Green chemistry:* Microwave irradiation aids in solvent-free reactions, which are environmentally friendly, as harmful solvents can be avoided.

The application of this method of synthesis is very useful in the medicinal and pharmaceutical fields of chemistry and in synthesis involving green chemistry. As there are two sides to a coin, this method also has afew disadvantages, which are discussed below.

The cons of microwave synthesis include the following:

1. *The equipment used is expensive*: The microwave synthesizer is expensive and is not easily available in all places.
2. *Difficult to scale up:* Mass synthesis of compounds cannot be performed in one step. Hence, difficult to use in industries for large-scale products.
3. *Infeasible to monitor the reaction*: The progress of the reaction cannot be monitored as with conventional heating. The instrument should cool down before taking out the reaction mixture to check the TLC.
4. *Evaporation of solvent*: When the reaction proceeds in the presence of solvent, there are chances for the solvent to evaporate; this can affect the reaction content.
5. *Threat to the environment*: With the continuous use of microwave heating to increase the rate of reaction, there is a possibility of radioactive decay. When the pressure inside the sealed tube is higher, it might lead to an explosion.

6. Health hazard: During prolonged exposure to microwave radiation, human organs can be damaged because high-frequency radiation can penetrate through the skin. Radiation can cause harm to cells, tissues and DNA strands.

References

Abdel Hameed, Afaf Mohamed, Moustafa Sherief Moustafa, Saleh Mohammed Al-Mousawi, Reham R. Awed, and Kamal Usef Sadek. 2017. An Efficient and Catalyst-Free Synthesis of N-Arylidene-2-Arylimidazo[1,2-A]Pyridine-3-Ylamine Derivatives via Strecker Reaction under Controlled Microwave Heating. *Green Processing and Synthesis* 6 (4): 371–75.

Adhikari, Aparajita, Sayantika Bhakta, and Tapas Ghosh. 2022. Microwave-Assisted Synthesis of Bioactive Heterocycles: An Overview. *Tetrahedron* 126: 133085.

Aghapoor, Kioumars, Farshid Mohsenzadeh, Hossein Reza Darabi, and Saeed Rastgar. 2018. Microwave-Induced Calcium(II) Chloride-Catalyzed Paal–Knorr Pyrrole Synthesis: A Safe, Expeditious, and Sustainable Protocol. *Research on Chemical Intermediates* 44 (7): 4063–72.

Alle, Madhusudhan, Seung-hwan Lee, and Jin-chul Kim. 2020. Journal of Materials Science & Technology Ultrafast Synthesis of Gold Nanoparticles on Cellulose Nanocrystals via Microwave Irradiation and Their Dyes-Degradation Catalytic Activity. *Journal of Materials Science & Technology* 41: 168–77.

Andoh-baidoo, Rosemarie, Danso, Roberto, Sanghamitra Mukherjee, Debasish Bandyopadhyay, and Bimal Banik K. 2011. "Microwave-Induced N-Bromosuccinimide-Mediated Novel Synthesis of Pyroles via Paal-Knorr Reaction." *Heterocyclic Letters* 1: 107–9.

Barreto, Angélica De Fátima S., Otilie E. Vercillo, Mike A. Birkett, John C. Caulfield, Ludger A. Wessjohann, and Carlos Kleber Z. Andrade. 2011. Fast and Efficient Microwave-Assisted Synthesis of Functionalized Peptoids via Ugi Reactions. *Organic and Biomolecular Chemistry* 9 (14): 5024–27.

Baviskar, Bhushan A., Prashant V. Ajmire, Deshraj S. Chumbhale, Mohammad Sadat Khan, Vitthal G. Kuchake, Madhavi Singupuram, and Purushottam R. Laddha. 2023. Recent Advances in Nickel Catalyzed Suzuki-Miyaura Cross Coupling Reaction via C-O& C-N Bond Activation. *Sustainable Chemistry and Pharmacy* 32 (October 2022): 100953.

Bensebaa, F., F. Zavaliche, P. L. Ecuyer, R. W. Cochrane, and T. Veres. 2004. Microwave Synthesis and Characterization of Co – Ferrite Nanoparticles 277: 104–10.

Bhat, Ajmal R., Mohd H. Najar, Rajendra S. Dongre, and Mohammad S. Akhter. 2020. "Microwave Assisted Synthesis of Knoevenagel Derivatives Using Water as Green Solvent." *Current Research in Green and Sustainable Chemistry* 3 (April): 100008.

Bogdal, D, J. Pielichowski, and K. Jaskot. 1997. New Synthesis Method of N-Alkylation of Carbazole Under Microwave Irradiation in Dry Media. *Synthetic Communications* 27 (9): 1553–60.

Bogdał, Dariusz. 1998. Coumarins: Fast Synthesis by Knoevenagel Condensation under Microwave Irradiation. *Journal of Chemical Research*, no. 8: 468–69.

Bosica, Giovanna, Kaylie Demanuele, José M. Padrón, and Adrián Puerta. 2020. One-Pot Multicomponent Green Hantzsch Synthesis of 1,2-Dihydropyridine Derivatives with Antiproliferative Activity. *Beilstein Journal of Organic Chemistry* 16 (November): 2862–69.

Cebeci, Yıldız Uygun, Şule Ceylan, and Şengül Alpay Karaoğlu. 2022. Conventional and Microwave Irradiated Synthesis, Biological Activity Evaluation of Highly Substituted Indole-Triazole Hybrids. *Journal of Molecular Structure* 1250 (February): 131799.

Choudhary, V. R., V. H. Tillu, V. S. Narkhede, H. B. Borate, and R. D. Wakharkar. 2003. Microwave Assisted Solvent-Free Synthesis of Dihydropyrimidinones by Biginelli Reaction over Si-MCM-41 Supported FeCl3 Catalyst. *Catalysis Communications* 4 (9): 449–53.

Cívicos, José F., Diego A. Alonso, and Carmen Nájera. 2013. Microwave-Promoted Copper-Free Sonogashira-Hagihara Couplings of Aryl Imidazolylsulfonates in Water. *Advanced Synthesis and Catalysis* 355 (1): 203–8.

Cruz-Acosta, Fabio, Alicia Santos-Expósito, Pedro de Armas, and Fernando García-Tellado. 2009. Lewis Base-Catalyzed Three-Component Strecker Reaction on Water. An Efficient Manifold for the Direct Alpha-Cyanoamination of Ketones and Aldehydes. *Chemical Communications (Cambridge, England)*, no. 44 (November): 6839–41.

Devi, Sylvia, Farooq Ahmad, Muhammad Ashraf, Shazia Parveen, and Abdul Hamid. 2019. Spectrochimica Acta Part A : Molecular and Biomolecular Spectroscopy Microwave Synthesis of Nanoparticles and Their Antifungal Activities. *Spectrochimica Acta Part A: Molecular and Biomolecular Spectroscopy* 213: 337–41.

Fouad, Manar Ahmed, Hamida Abdel-Hamid, and Mohammed Salah Ayoup. 2020. Two Decades of Recent Advances of Ugi Reactions: Synthetic and Pharmaceutical Applications. *RSC Advances* 10 (70): 42644–81.

Fu, Renzhong, Yang Yang, Wenchen Lai, Yongfeng Ma, Zhikai Chen, Jian Zhou, Wen Chai, Quan Wang, and Rongxin Yuan. 2015. Efficient and Green Microwave-Assisted Multicomponent Biginelli Reaction for the Synthesis of Dihydropyrimidinones Catalyzed by Heteropolyanion-Based Ionic Liquids Under Solvent-Free Conditions. *Synthetic Communications* 45 (4): 467–77.

Gaonkar, Santosh L., and Hiroki Shimizu. 2010. Microwave-Assisted Synthesis of the Antihyperglycemic Drug Rosiglitazone. *Tetrahedron* 66 (18): 3314–17.

Gazvoda, Martin, Miha Virant, Balazs Pinter, and Janez Košmrlj. 2018. Mechanism of Copper-Free Sonogashira Reaction Operates through Palladium-Palladium Transmetallation. *Nature Communications* 9 (1): 1–9.

Harusawa, Shinya, and Takayuki Shioiri. 2016. Diethyl Phosphorocyanidate (DEPC): A Versatile Reagent for Organic Synthesis. *Tetrahedron* 72 (50): 8125–8200.

Hellal, Malik, and Gregory D. Cuny. 2011. Microwave Assisted Copper-Free Sonogashira Coupling/5-Exo-Dig Cycloisomerization Domino Reaction: Access to 3-(Phenylmethylene)Isoindolin-1- Ones and Related Heterocycles." *Tetrahedron Letters* 52 (42): 5508–11. https://doi.org/10.1016/j.tetlet.2011.08.070.

Hoener, Annabelle P., Bernd Henkel, and Jean-Christophe Gauvin. 2002. "Novel One-Pot Microwave Assisted Gewald Synthesis of 2-Acyl Amino Thiophenes on Solid Support." *Synlett*, no. 1: 0063–0066. https://doi.org/10.1055/s-2003-36229.

Huang, Xian-gui, Jia Liu, Jiangmeng Ren, Tao Wang, Weidong Chen, and Bu-bing Zeng. 2011. A Facile and Practical One-Pot Synthesis of Multisubstituted 2-Aminothiophenes via Imidazole-Catalyzed Gewald Reaction. *Tetrahedron* 67 (34): 6202–5.

Jeon, Sung Lan, Ji Hoon Choi, Bum Tae Kim, and In Howa Jeong. 2007. Synthesis of Novel 1,4,5-Trisubstituted 3-Trifluoromethylpyrazoles via Microwave-Assisted Stille Coupling Reactions. *Journal of Fluorine Chemistry* 128 (10): 1191–97.

Joy, M. Nibin, Yadav D. Bodke, K. K. Abdul Khader, and Ayyiliyath M. Sajith. 2014. A Rapid Approach for the Copper, Amine, and Ligand-Free Sonogashira Coupling of 4-Methyl-7-Nonafluorobutylsulfonyloxy Coumarins under Microwave Irradiation. *Tetrahedron Letters* 55 (15): 2355–61.

Ju, Yuhong, and Rajender S. Varma. 2004. Aqueous N-Alkylation of Amines Using Alkyl Halides: Direct Generation of Tertiary Amines under Microwave Irradiation. *Green Chemistry* 6 (4): 219.

Khan, Faiyaz, Michael Dlugosch, Xin Liu, and Martin G. Banwell. 2018. The Palladium-Catalyzed Ullmann Cross-Coupling Reaction: A Modern Variant on a Time-Honored Process. *Accounts of Chemical Research* 51 (8): 1784–95.

Khormi, Afaf Y., Thoraya A. Farghaly, and Mohamed R. Shaaban. 2022. Microwave-Assisted Synthesis of 2-Aryl and 2,5-Diarylthiophene Derivatives via Suzuki-Miyaura Cross-Coupling Using Novel Palladium Complex as a Catalyst. *Polycyclic Aromatic Compounds* 42 (7): 3893–3907.

Kiss, Márton A., Miroslav Peřina, Laura Bereczki, Ádám Baji, Jakub Bělíček, Radek Jorda, and Éva Frank. 2023. Dihydrotestosterone-Based A-Ring-Fused Pyridines: Microwave-Assisted Synthesis and Biological Evaluation in Prostate Cancer Cells Compared to Structurally Related Quinolines. *The Journal of Steroid Biochemistry and Molecular Biology* 231 (April): 106315.

Lanjekar, Kavita J., and Virendra K. Rathod. 2021. Microwave Catalysis in Organic Synthesis. In *Green Sustainable Process for Chemical and Environmental Engineering and Science*, 1–50. Elsevier.

Lei, Aiwen, Wei Liu, Chao Liu, and Mao Chen. 2010. Arylation of Unactivated Arenes. *Dalton Transactions* 39 (43): 10352.

Lei, Yonghua, Tianhan Hu, Xingsen Wu, Yue Wu, Hua Xiang, Haopeng Sun, Qidong You, and Xiaojin Zhang. 2016. Microwave-Assisted Copper- and Palladium-Catalyzed Sonogashira-Type Coupling of Aryl Bromides and Iodides with Trimethylsilylacetylene. *Tetrahedron Letters* 57 (10): 1100–1103.

Lian, Ru Ching, Meng Hsuan Lin, Pin Hsuan Liao, Jia Jie Fu, Meng Ju Wu, Yang Chang Wu, Fang Rong Chang, Chin Chung Wu, and Po Shen Pan. 2014. Direct Synthesis of the Arylboronic Acid Analogues of Phenylglycine via Microwave-Assisted Four-Component Ugi Reaction. *Tetrahedron* 70 (9): 1800–1804.

Liu, Zhen-Jiang, Jean-Pierre Vors, Ernst R. F. Gesing, and Carsten Bolm. 2011. Microwave-Assisted Solvent- and Ligand-Free Copper-Catalysed Cross-Coupling between Halopyridines and Nitrogen Nucleophiles. *Green Chem.* 13 (1): 42–45.

Low, Y. J, J. Y. C. Liew, M. A. Kamarudin, H. N. Lim, F. D. Muhammad, K. P. Lim, M. H. M. Zaid, et al. 2023. Synthesis of Cesium Silver Bismuth Bromide Double Perovskite Nanoparticles via a Microwave-Assisted Solvothermal Method. *Materials Today Chemistry* 29: 101477.

Madadi, Elaheh. 2021. Strecker Synthesis of α-Aminonitriles Facilitated by N-Methyl Imidazolium Acetate. *Alinteri Journal of Agriculture Sciences* 36 (1): 323–27.

Maru, Minaxi S., P. K. Sudhadevi Antharjanam, and Noor-ul H. Khan. 2019. Catalyst-Free Solid Phase Microwave-Assisted Synthesis of 1,4-Dihydropyridine Derivatives and Their Single Crystal Structure Determination. *ChemistrySelect* 4 (3): 774–82.

Mermer, Arif, Serpil Demirci, Serap Basoglu Ozdemir, Ahmet Demirbas, Serdar Ulker, Faik Ahmet Ayaz, Fatma Aksakal, and Neslihan Demirbas. 2017. Conventional and Microwave Irradiated Synthesis, Biological Activity Evaluation and Molecular Docking Studies of Highly Substituted Piperazine-Azole Hybrids. *Chinese Chemical Letters* 28 (5): 995–1005.

Minetto, Giacomo, Luca F. Raveglia, and Maurizio Taddei. 2004. Microwave-Assisted Paal-Knorr Reaction. A Rapid Approach to Substituted Pyrroles and Furans. *Organic Letters* 6 (3): 389–92.

Mori, Maki, Hironori Imma, and Takeshi Nakai. 1997. Asymmetric Catalytic Cyanosilylation of Aldehydes Using a Chiral Binaphthol-Titanium Complex. *Tetrahedron Letters* 38 (35): 6229–32.

Patel, Vatsal M., Navin B. Patel, Manuel J. Chan-Bacab, and Gildardo Rivera. 2018. Synthesis, Biological Evaluation and Molecular Dynamics Studies of 1,2,4-Triazole Clubbed Mannich Bases. *Computational Biology and Chemistry* 76 (June): 264–74.

Patel, Vatsal M., Navin B. Patel, Manuel J. Chan-Bacab, and Gildardo Rivera. 2020. N-Mannich Bases of Benzimidazole as a Potent Antitubercular and Antiprotozoal Agents: Their Synthesis and Computational Studies. *Synthetic Communications* 50 (6): 858–78.

Pchalek, Karin, and Michael P. Hay. 2006. Stille Coupling Reactions in the Synthesis of Hypoxia-Selective 3-Alkyl-1,2,4-Benzotriazine 1,4-Dioxide Anticancer Agents. *Journal of Organic Chemistry* 71 (17): 6530–35.

Puripat, Maneeporn, Romain Ramozzi, Miho Hatanaka, Waraporn Parasuk, Vudhichai Parasuk, and Keiji Morokuma. 2015. The Biginelli Reaction Is a Urea-Catalyzed Organocatalytic Multicomponent Reaction. *The Journal of Organic Chemistry* 80 (14): 6959–67.

Puterova, Zita, Alžbeta Krutošíková, and Daniel Végh. 2010. Gewald Reaction: Synthesis, Properties and Applications of Substituted 2-Aminothiophenes. Edited by Viktor V. Zhdankin. *Arkivoc* 2010 (1): 209–46.

Rajitha, B., V. Naveen Kumar, P. Someshwar, J. Venu Madhav, P. Narsimha Reddy, and Y. Thirupathi Reddy. 2006. Dipyridine Copper Chloride Catalyzed Coumarin Synthesis via Pechmann Condensation under Conventional Heating and Microwave Irradiation. *Arkivoc* 2006 (12): 23–27.

Ramesh, Karu, and Gedu Satyanarayana. 2019. Domino [Pd]-Catalysis: Heck Followed by Decarboxylative Sonogashira Couplings under Microwave Irradiation in Aqueous Medium. *Journal of Organometallic Chemistry* 890: 58–71.

Ray, Devalina, Ram Naresh, and Bimal Krishna. 2022. Results in Chemistry Vitamin C-Catalyzed Hantzsch Reaction under Microwave Condition : A Greener Access to 1, 4-Dihydropyridines. *Results in Chemistry* 4 (March): 100330.

Reddy, Y. Thirupathi, Vijayakumar N. Sonar, Peter A. Crooks, Pavan K. Dasari, P. Narsimha Reddy, and B. Rajitha. 2008. Ceric Ammonium Nitrate (CAN): An Efficient Catalyst for the Coumarin Synthesis via Pechmann Condensation Using Conventional Heating and Microwave Irradiation. *Synthetic Communications* 38 (13): 2082–88.

Sadek, Kamal Usef, Afaf Mohamed Abdel-Hameed, Hisham A Abdelnabi, and Yasser Meleigy. 2019. An Efficient Green Synthesis of Novel 1H-Imidazo[1,2-a]Imidazole-3-Amine and Imidazo[2,1-c][1,2,4]Triazole-5-Amine Derivatives via Strecker Reaction under Controlled Microwave Heating. *Green Processing and Synthesis* 8 (1): 297–301.

Sanghi, Rashmi. 2000. Microwave Irradiation. *Resonance* 5 (3): 77–81.

Shah, Sakshi, and Baldev Singh. 2012. Catalyst-Free, Facile, and an Efficient Synthesis of α-Aminonitriles Employing Zn(CN) 2 as an Ecofriendly Cyanating Agent. *Tetrahedron Letters* 53 (2): 151–56.

Shaikh, Faiyazalam M., Navin B. Patel, Giuseppina Sanna, Bernardetta Busonera, Paolo La Colla, and Dhanji P. Rajani. 2015. Synthesis of Some New 2-Amino-6-Thiocyanato Benzothiazole Derivatives Bearing 2,4-Thiazolidinediones and Screening of Their in Vitro Antimicrobial, Antitubercular and Antiviral Activities. *Medicinal Chemistry Research* 24 (8): 3129–42.

Shen, Li, Charles J Simmons, and Dianqing Sun. 2012. Microwave-Assisted Synthesis of Macrocycles via Intramolecular and / or Bimolecular Ullmann Coupling. *Tetrahedron Letters* 53 (32): 4173–78.

Shen, Liqun, Suyu Huang, Yuanmei Nie, and Fuhou Lei. 2013. An Efficient Microwave-Assisted Suzuki Reaction Using a New Pyridine-Pyrazole/Pd(II) Species as Catalyst in Aqueous Media. *Molecules* 18 (2): 1602–12.

Shmidt, María Sol, Ana María Reverdito, Lautaro Kremenchuzky, Isabel Amalia Perillo, and María Mercedes Blanco. 2008. Simple and Efficient Microwave Assisted N-Alkylation of Isatin. *Molecules (Basel, Switzerland)* 13 (4): 831–40.

Silva, Joaquim F. M. Da, Andres F. Yepes Perez, and Natália P. De Almeida. 2015. Phosphine-Free Suzuki Cross-Coupling Reaction Using an Efficient and Reusable Pd Catalyst in an Aqueous Medium under Microwave Irradiation. *Synthetic Communications* 45 (17): 1995–2004.

Singh, Vasundhara, Jasvinder Singh, Kanwal Preet Kaur, and Goverdhan L. Kad. 1997. Acceleration of the Pechmann Reaction by Microwave Irradiation: Application to the Preparation of Coumarins. *Journal of Chemical Research - Part S*, no. 2: 58–59.

Sipos, Szabolcs, and István Jablonkai. 2009. One-Pot Synthesis of α-Aminonitriles from Alkyl and Aryl Cyanides: A Strecker Reaction via Aldimine Alanes. *Tetrahedron Letters* 50 (16): 1844–46.

Sridhar, Madabhushi, Rayankula Mallikarjuna Rao, Nanduri H. K. Baba, and Ravindra M Kumbhare. 2007. Microwave Accelerated Gewald Reaction: Synthesis of 2-Aminothiophenes. *Tetrahedron Letters* 48 (18): 3171–72.

Tierney, Steve, Martin Heeney, and Iain McCulloch. 2005. Microwave-Assisted Synthesis of Polythiophenes via the Stille Coupling. *Synthetic Metals* 148 (2): 195–98.

Xie, Xingang, Jiangping Lu, Bo Chen, Junjie Han, Xuegong She, and Xinfu Pan. 2004. Pd/C-Catalyzed Heck Reaction in Ionic Liquid Accelerated by Microwave Heating. *Tetrahedron Letters* 45 (4): 809–11.

Xie, Yong Sheng, Deepak Kumar, V. D. Vijaykumar Bodduri, P. Srivastava Tarani, Bao Xiang Zhao, Jun Ying Miao, Kiwan Jang, and Dong Soo Shin. 2014. Microwave-Assisted Parallel Synthesis of Benzofuran-2-Carboxamide Derivatives Bearing Anti-Inflammatory, Analgesic and Antipyretic Agents. *Tetrahedron Letters* 55 (17): 2796–2800.

Yadav, J. S., B. V. Subba Reddy, E. Jagan Reddy, and T Ramalingam. 2000. Microwave-Assisted Efficient Synthesis of Dihydro Pyrimidines: Improved High Yielding Protocol for the Biginelli Reaction. *Journal of Chemical Research* 2000 (7): 354–55.

Yildirim, Muhammet, Derya Çelikel, Yaşar Dürüst, David W. Knight, and Benson M. Kariuki. 2014. A Rapid and Efficient Protocol for the Synthesis of Novel Nitrothiazolo[3,2-c]Pyrimidines via Microwave-Mediated Mannich Cyclisation. *Tetrahedron* 70 (12): 2122–28.

Zeynizadeh, Behzad, and Masumeh Gilanizadeh. 2020. Microwave-Promoted Three-Component Hantzsch Synthesis of Acridinediones under Green Conditions. *Current Chemistry Letters* 9: 71–78.

Zhang, Yikai, Zhiliang Lv, Hanyu Zhong, Mingfeng Zhang, Tao Zhang, Wannian Zhang, and Ke Li. 2012. Efficient Heck Cross-Coupling of 3-Iodo-Benzopyrones with Olefins under Microwave Irradiation without Phosphine. *Tetrahedron* 68 (47): 9777–87.

Index

A

ablation, v, 29, 30, 31, 38, 39, 40, 41, 43, 49, 50, 51, 52, 53, 54, 55, 56, 57, 58, 59
absorbent, 15, 89, 95, 101, 105, 111, 123, 124, 135, 142, 143, 144, 145, 152, 154
acid-basicity, 166
activation, v, vi, 2, 7, 8, 9, 19, 142, 148, 163, 164, 176, 177, 187, 213, 214, 215, 216, 217, 218, 220, 223, 225, 226, 227, 229, 231, 234, 235, 258, 268, 289
air drying, 16, 26, 63, 67, 68, 70, 71, 72, 73, 75, 77, 78, 79, 81, 84, 85
antenna, 11, 23, 30, 33, 36, 37, 38, 41, 47, 52, 53, 54, 57, 59

B

Biginelli reaction, 283, 284, 290, 292, 294
biochar, vi, 90, 91, 100, 101, 115, 116, 117, 124, 155, 163, 164, 165, 166, 167, 168, 169, 170, 171, 172, 173, 174, 175, 176, 177, 178, 180, 181, 182, 183, 184, 185, 186, 187, 189, 190, 192, 193, 194, 195, 196, 197, 198, 199, 201, 202, 203, 204, 205, 206, 207, 208, 210, 211, 212, 213, 216, 217, 218, 219, 220, 221, 222, 223, 224, 225, 226, 227, 228, 229, 230, 231, 232, 233, 234, 235, 236, 265
biodiesel, 12, 21, 24, 25
biomaterials, 17
blood, 29, 30, 31, 41, 42, 47, 48, 49, 51, 54, 55, 58, 59
blood glucose testing, 47
building materials, 115, 182, 226, 232

C

cancer, 30, 31, 32, 38, 50, 51, 52, 53, 54, 55, 57, 58, 59, 291
cancer cells, 38, 53
catalyst, 91, 119, 120, 122, 123, 142, 146, 147, 148, 149, 150, 151, 152, 153, 154, 156, 158, 159, 161, 175, 180, 181, 194, 212, 213, 224, 238, 242, 255, 256, 261, 263, 264, 269, 270, 272, 273, 274, 275, 277, 278, 281, 282, 284, 285, 289, 290, 291, 292, 293
catalytic, 24, 25, 26, 90, 115, 117, 119, 123, 144, 147, 148, 149, 150, 152, 153, 154, 155, 156, 157, 158, 159, 160, 161, 187, 194, 220, 221, 222, 224, 228, 229, 233, 234, 239, 262, 264, 265, 273, 274, 279, 289, 292
cereals, 73
char production, vi, 163, 218
chemical(s), 1, 4, 7, 8, 9, 10, 11, 21, 22, 24, 25, 44, 45, 51, 61, 68, 84, 114, 115, 120, 121, 123, 147, 156, 157, 159, 160, 161, 163, 164, 173, 174, 176, 184, 213, 215, 216, 217, 218, 219, 220, 221, 222, 223, 224, 225, 227, 228, 229, 231, 233, 234, 235, 237, 238, 239, 254, 257, 258, 262, 263, 264, 268, 269, 274, 278, 289, 290, 291, 292, 293, 294
coagulation, 29, 42, 49
coaxial cable, 33, 38
component(s), 1, 11, 22, 38, 57, 95, 114, 119, 121, 123, 124, 125, 128, 129, 131, 132, 133, 135, 137, 139, 140, 144, 145, 149, 150, 151, 153, 154, 160, 165, 170, 171, 172, 173, 175, 237, 238, 239, 240,

243, 244, 245, 246, 247, 248, 250, 251, 253, 254, 255, 257, 261, 271, 281, 283, 290, 291, 294
composition(s), 24, 44, 61, 120, 121, 123, 125, 129, 130, 131, 132, 133, 134, 136, 137, 138, 139, 140, 141, 142, 150, 151, 157, 170, 172, 180, 184, 207, 219, 222, 235, 238, 240, 254, 255, 257, 259, 260, 261
conductivity, 15, 175, 184, 186
consumption, 17, 21, 69, 70, 72, 79, 113, 123, 148, 174, 184, 220, 240, 246, 267, 270, 287
conventional, v, 2, 3, 4, 5, 8, 9, 10, 11, 16, 21, 23, 24, 34, 46, 68, 69, 74, 75, 77, 78, 86, 112, 122, 155, 156, 157, 158, 159, 184, 188, 191, 194, 195, 206, 207, 224, 225, 227, 229, 262, 267, 268, 269, 271, 274, 276, 279, 281, 282, 283, 285, 286, 287, 288, 290, 292, 293
cooking, 62, 81, 82, 87, 112, 268
cooling, 10, 11, 14, 15, 22, 27, 36, 38, 82, 232, 241, 242, 249, 250, 271

D

dehydration, 20, 68, 69, 71, 72, 80, 84, 85, 86, 87, 122, 166, 216
detection, 45, 48, 50, 52, 54, 55, 56, 59, 60, 189, 224, 236
diagnostic(s), v, 29, 30, 45, 48, 49, 50, 57, 59
dielectric heating, 2, 24, 27, 39, 63, 116, 269, 287, 288
disinfection, 44, 53, 54
drying, v, 15, 16, 17, 24, 25, 26, 61, 62, 63, 64, 65, 66, 67, 68, 69, 70, 71, 72, 73, 74, 75, 76, 77, 78, 79, 80, 81, 83, 84, 85, 86, 87, 90, 113, 114, 188, 238

E

efficiency, 16, 21, 25, 26, 30, 34, 42, 45, 65, 68, 69, 70, 71, 72, 73, 75, 77, 78, 80, 81, 82, 83, 114, 116, 123, 142, 144, 148, 160, 174, 176, 178, 179, 181, 188, 191, 192, 193, 198, 202, 204, 205, 217, 222, 225, 237, 239, 256
electric, 4, 7, 8, 10, 12, 19, 22, 27, 39, 61, 62, 63, 66, 67, 184, 191, 223, 242
electrochemical, 175, 186, 189, 213, 227, 232, 236
electromagnetic, v, 1, 2, 8, 22, 30, 38, 39, 40, 46, 47, 48, 49, 52, 53, 54, 57, 63, 65, 67, 84, 86, 114, 184, 185, 188, 191, 194, 230, 268
electron(s), 11, 23, 187, 264
element(s), 4, 56, 124, 134, 160, 170, 172, 192, 205, 232
environment(s), 2, 15, 17, 53, 55, 64, 66, 68, 71, 89, 90, 112, 155, 164, 166, 192, 213, 222, 223, 225, 226, 227, 229, 230, 231, 232, 238, 249, 263, 264, 265, 284, 288
enzymatic, 16, 34
equipment, 11, 36, 38, 43, 66, 120, 217, 238, 288
equivalence ratio, 237, 238, 239, 240, 242, 244, 245, 246, 249, 250, 252, 253, 254, 261, 263, 265

F

food application, 15
food drying, v, 61, 62, 63, 70, 81, 83
formation, v, 2, 3, 15, 19, 21, 30, 122, 128, 129, 133, 146, 152, 167, 171, 183, 207, 216, 228, 229, 237, 238, 242, 251, 265, 271, 273, 276, 281, 283, 288
freeze drying, 62, 63, 64, 68, 69, 71, 74, 76, 78, 79, 81, 85, 87
frequency, v, 1, 2, 3, 4, 6, 7, 8, 11, 12, 15, 30, 33, 39, 48, 52, 58, 59, 60, 63, 64, 71, 115, 184, 188, 289
fruit, 26, 74, 75, 84, 101, 114, 124, 157, 189, 195, 199, 226, 229

G

gasification, vi, 7, 22, 90, 116, 159, 160, 188, 220, 222, 226, 229, 233, 235, 237, 238, 239, 240, 241, 242, 243, 244, 245, 246, 247, 248, 249, 250, 251, 252, 253, 254, 255, 256, 257, 258, 259, 260, 261, 262, 263, 264, 265
generators, 2, 11, 27
geopolymer(s), 18, 19, 20, 24
Gewald reaction, 285, 291, 292, 293

H

Hantzch synthesis, 277
heavy metals, 176, 177, 178, 180, 181, 227, 229, 234
Heck reaction, 271, 272, 273, 294
high heating value, 163, 164, 174, 197, 206, 218
high voltage, 11, 14
high-conductivity, 20
homogeneous, 11, 15, 23, 51, 114, 257, 258, 262
hyperthermia, v, 29, 31, 32, 36, 50, 53, 54, 55, 57, 58, 59

I

imaging, 25, 29, 30, 32, 45, 46, 47, 49, 50, 51, 52, 53, 54, 55, 56, 57, 59, 60
improvement, 33, 34, 75, 76, 116, 123, 129, 151, 164, 183, 184, 198, 206, 208, 210, 218, 226, 244
infrared, v, 2, 22, 54, 55, 57, 65, 174, 188, 223, 231

K

keratinized cells, 40
Knoevenagel condensation, 269, 270, 290

L

leakage, 22, 184

M

magnetron, 11, 12, 13, 14, 22, 23, 25, 26, 27, 65, 66, 67
Mannich reactions, 271
meat, 72, 73, 86
mechanism, 3, 7, 31, 42, 49, 55, 63, 119, 148, 152, 154, 157, 180, 181, 219, 222, 227, 229, 231, 251, 273, 290
medical treatment(s), v, 29, 30
medical waste, 44, 45, 50, 52, 53, 54, 59
microorganism(s), 45, 170, 180
microwave absorbent, v, 89, 90, 96, 100, 101, 105, 108, 111, 113, 119, 120, 123, 124, 128, 135, 142, 143, 144, 145, 146, 152, 154, 157, 191, 192, 194, 201, 239
microwave applications, 1, 11, 22
microwave heating, 1, 2, 3, 4, 10, 12, 15, 16, 17, 18, 21, 22, 25, 26, 34, 35, 36, 53, 63, 65, 67, 68, 72, 75, 81, 82, 90, 91, 93, 94, 95, 97, 99, 100, 101, 102, 103, 104, 105, 107, 109, 110, 111, 113, 114, 115, 142, 156, 158, 159, 188, 191, 194, 224, 228, 234, 250, 257, 263, 268, 282, 287, 288, 289, 293, 294
microwave heating mechanism, 1, 3, 4
microwave-assisted pyrolysis, 24, 89, 91, 95, 100, 102, 111, 112, 113, 115, 116, 123, 131, 155, 156, 157, 158, 159, 161, 163, 164, 192, 194, 197, 198, 201, 202, 206, 208, 210, 212, 218, 223, 224, 225, 228, 229, 231, 233, 235, 236, 242, 262, 263
molecule(s), 3, 6, 7, 9, 11, 19, 22, 24, 26, 30, 35, 39, 45, 63, 72, 81, 82, 90, 141, 149, 166, 191, 201, 212, 217, 251, 252, 267, 293
monitoring, v, 18, 24, 29, 30, 45, 47, 50, 51, 52, 54, 55, 57, 58, 59, 71, 83, 174, 231, 250

N

neuroplasticity, 42
neurostimulation, 42

non-thermal effect(s), v, 1, 7, 9, 10, 22, 23, 26, 27, 29, 49

O

obstacle(s), 22, 62
oil production, vi, 90, 113, 116, 119, 120, 123, 138, 146, 149, 155, 156, 157, 160, 161, 192, 264
organic synthesis, vi, 52, 267, 268, 271, 283, 290, 291
organic wastes, v, vi, 89, 90, 111, 119, 123, 154, 163, 194, 199, 218, 237, 238
overheating, 15, 20, 36, 68, 71, 79, 83

P

Paal-Knorr synthesis, 276, 277
particle(s), v, 2, 3, 4, 22, 44, 46, 65, 91, 93, 95, 96, 97, 98, 99, 100, 101, 102, 107, 108, 112, 113, 114, 121, 142, 160, 172, 194, 212, 220, 222, 228, 250, 251, 252, 257, 262
pasteurization, 62, 82, 84, 85, 86, 87
Pechmann reaction, 281, 282, 293
peptide, 29, 34, 35, 49
physical, 22, 40, 48, 61, 68, 84, 110, 111, 163, 164, 173, 188, 192, 194, 201, 213, 214, 216, 217, 218, 229, 231, 257, 269
plastic(s), v, vi, 20, 33, 89, 90, 95, 96, 97, 98, 99, 100, 105, 108, 109, 110, 111, 112, 113, 114, 115, 116, 119, 120, 121, 122, 123, 133, 134, 135, 136, 137, 138, 140, 141, 142, 143, 144, 145, 146, 147, 148, 149, 150, 151, 152, 153, 154, 155, 156, 157, 158, 159, 160, 161, 182, 237, 238, 239, 240, 248, 250, 255, 256, 261, 262, 263, 264, 265
polar, v, 2, 3, 4, 6, 7, 9, 11, 19, 22, 26, 30, 39, 45, 90, 173, 191
polarization, 3, 22, 184, 268, 269
porous, vi, 68, 80, 168, 173, 176, 180, 183, 206, 213, 221, 222, 224, 225, 226, 227, 228, 230, 231, 234
properties, vi, 15, 18, 20, 22, 25, 26, 45, 46, 52, 54, 56, 61, 76, 77, 81, 84, 114, 115, 116, 122, 140, 155, 158, 159, 160, 161, 163, 164, 165, 166, 167, 172, 173, 175, 182, 183, 184, 185, 186, 194, 197, 206, 208, 216, 217, 218, 219, 220, 221, 222, 225, 226, 227, 228, 229, 230, 231, 232, 233, 234, 235, 238, 292

R

radiation, v, 2, 3, 7, 15, 16, 21, 22, 27, 48, 51, 54, 57, 58, 62, 63, 69, 72, 119, 134, 148, 154, 183, 191, 194, 269, 279, 287, 289
ratio, 5, 20, 69, 74, 79, 89, 96, 108, 109, 111, 119, 123, 145, 150, 151, 152, 154, 189, 190, 194, 205, 218, 239, 245, 246, 247, 250, 252, 253, 256, 261, 279
remediation, 175, 180, 184, 222, 226, 228, 232
retention time, 237, 239, 240, 247, 248, 261
rheumatism, 48

S

seafood, 62, 69, 81
silicon carbide, 10, 25, 113, 114, 144, 237, 240, 244, 245, 246, 261
soil, 164, 172, 175, 178, 180, 181, 184, 194, 206, 213, 216, 219, 221, 222, 223, 224, 225, 226, 228, 229, 232, 234, 236
Sonogashira coupling, 52, 273, 290, 291, 292
specific surface area, 163, 164, 168, 169, 175, 176, 177, 194, 206, 207, 208, 209, 210, 211, 212, 213, 216, 218
spectrum, 2, 22, 45, 52, 63, 69, 174
spouted bed, 63, 65, 66, 69, 73, 74, 76, 78, 79, 85, 86, 87
sterilization, 29, 44, 45, 49, 55, 83
Stille coupling reactions, 282, 283, 291, 292
Strecker reaction, 275, 276, 289, 290, 293

structural, 27, 33, 55, 58, 61, 75, 78, 147, 194, 205, 207, 222, 224
surface, 3, 7, 11, 15, 17, 36, 53, 65, 66, 77, 81, 122, 148, 158, 166, 168, 169, 172, 173, 176, 177, 180, 181, 184, 187, 191, 193, 206, 208, 209, 210, 211, 212, 213, 216, 217, 218, 225, 227, 230, 232, 233, 236, 263
Suzuki coupling reactions, 274
syngas, vi, 91, 155, 226, 227, 235, 236, 237, 238, 239, 240, 242, 243, 244, 245, 246, 247, 248, 249, 250, 251, 252, 253, 254, 255, 260, 261, 263, 264, 265
synthesis, vi, 23, 29, 34, 49, 122, 157, 194, 227, 228, 229, 230, 233, 237, 238, 264, 267, 268, 269, 270, 271, 272, 273, 274, 275, 276, 277, 278, 280, 281, 282, 283, 284, 285, 286, 287, 288, 289, 290, 291, 292, 293, 294

T

thermal effect, v, 1, 7, 9, 10, 22, 26, 29, 30, 45
tomography, 46, 47, 51, 52, 53, 57, 59
transformer(s), 11, 14, 22
treatment(s), v, 15, 17, 20, 21, 23, 30, 31, 32, 33, 36, 40, 41, 44, 45, 47, 50, 52, 53, 54, 55, 56, 57, 58, 59, 70, 77, 78, 79, 81, 82, 86, 115, 160, 174, 175, 176, 177, 180, 181, 194, 216, 217, 223, 228, 231, 233, 236, 263, 268
tumor tissue, 31, 38, 40

U

Ugi reaction, 280, 281, 289, 290, 291
Ullmann reaction, 279
ultrasound, 32, 48, 79, 81, 86
urethroplasty, 33

V

vacuum, 6, 11, 62, 63, 64, 65, 66, 67, 68, 69, 70, 74, 75, 76, 77, 78, 79, 80, 81, 84, 85, 86, 87, 180, 191, 192, 224, 228, 230
vacuum frying dehydration, 63, 66, 70
vegetable, 23, 76, 77, 86, 87
vehicle fuel, 21

W

wastewater, 44, 175, 176, 177, 178, 180, 226, 227, 228, 229, 231, 232, 233, 234
waveguide(s), 11, 12, 14, 22, 25, 26, 27, 67, 223
wavelength(s), v, 1, 2, 22, 30, 56, 188